AF333742

# Maize Kernel Development

# Maize Kernel Development

Edited by Brian A. Larkins

**CABI is a trading name of CAB International**

| | |
|---|---|
| CABI | CABI |
| Nosworthy Way | 745 Atlantic Avenue |
| Wallingford | 8th Floor |
| Oxfordshire OX10 8DE | Boston, MA 02111 |
| UK | USA |
| | |
| Tel: +44 (0)1491 832111 | Tel: +1 (617) 682-9015 |
| Fax: +44 (0)1491 833508 | E-mail: cabi-nao@cabi.org |
| E-mail: info@cabi.org | |
| Website: www.cabi.org | |




**British Library Cataloguing-in-Publication Data**

A catalogue record for this book is available from the British Library.

**Library of Congress Cataloging-in-Publication Data**

Names: Larkins, B.A. (Brian A.), editor.
Title: Maize kernel development / edited by Brian Larkins.
Description: Boston, MA : CABI, [2017] | Includes bibliographical
  references and index.
Identifiers: LCCN 2017023565 (print) | LCCN 2017024621 (ebook) |
  ISBN 9781786391223 (ePDF) | ISBN 9781786391230 (ePub) |
  ISBN 9781786391216 (hbk : alk. paper)
Subjects: LCSH: Corn--Seeds. | Seeds--Development.
Classification: LCC SB191.M2 (ebook) | LCC SB191.M2 M1223 2017
  (print) | DDC 633.1/5--dc23
LC record available at https://lccn.loc.gov/2017023565


ISBN: 978 1 78639 121 6 (hardback)
       978 1 78639 122 3 (e-book)
       978 1 78639 123 0 (e-pub)


Commissioning editor: Rachael Russell
Editorial assistant: Emma McCann
Production editor: Alan Worth


Typeset by SPi, Pondicherry, India
Printed and bound in the UK by CPI Group (UK) Ltd, Croydon, CR0 4YY

# Contents

# Preface


Brian A. Larkins

*Associate Vice Chancellor for Life Sciences,
University of Nebraska, Lincoln, Nebraska*


This book is intended to be a guide to understanding maize kernel development. It was not my vision to provide a thorough review of past research that informs our current understanding of kernel development; rather, I asked the authors to provide a synopsis of what is known and to explain what we do not know. What are the important gaps in our knowledge and how might one go about filling them? Some features of the maize seed (caryopsis), such as its size, differ from those of other cereal grains (e.g. rice, wheat, and barley), but many aspects of its structure, cell biology, physiology, and biochemistry are shared. I am confident that anyone interested in understanding the amazing biology associated with seed development, particularly in cereal grains, will value reading this book.

Humans domesticated cereals approximately 7000–10,000 years ago. Seeds of these plants are nutritious, with a high content of starch, protein, and oil, as well as some vitamins and minerals. By growing these crops, people were able to live in the same location for extended periods, as cereal seeds can be stored and later converted to food. Flour, made by grinding them, can be moistened to produce dough that when cooked or baked creates nutritious food with a relatively long shelf life. These grains could also have been fermented to create alcoholic beverages for entertainment and religious experiences. Today, cereals are the most widely grown crops, accounting for more than 50% of human caloric intake. But they have many other uses, including animal feed, raw materials for manufacturing, alcoholic beverages, and biofuels. They are an important part of the multi-billion-dollar seed industry. Traditionally, rice was the predominant cereal crop, but in recent years land devoted to maize production has begun to exceed that of rice, even in China (Ranum *et al.*, 2014). This is a consequence of the higher yield potential of maize and its value for feeding livestock.

The knowledge and insight in this book are the culmination of centuries of genetic and biochemical work directed at increasing the yield and nutritional value of maize. Arguably, this work began around 10,000 years ago, when pioneering farmers/plant breeders (presumed to be women) began selecting teosinte for larger seeds and improved functional characteristics of kernels for making food. Initially, these changes would have been subtle, but ultimately the transition from teosinte to maize was dramatic (Chapter 1). Over the course of time, perhaps centuries, Native Americans began to soak the dry kernels in a solution of lime (calcium oxide/calcium hydroxide). They discovered this procedure was useful for softening and removing the seed coat (pericarp), which facilitated grinding the grain to create meal suitable for food applications. This process, called nixtamalization, also had the non-obvious benefit of releasing vitamin B3 (niacin) from a form that cannot be used by the human body. Niacin deficiency, which can occur from a predominantly maize-based diet, causes pellagra, a skin disease first described in 1735 by a Spanish physician, Gaspar Casal. It was not until 1937 that Conrad Elvehjema, a biochemistry professor at the University of Wisconsin, demonstrated that niacin could cure pellagra in dogs, and later it was shown to do the same in humans.

Another limitation of a maize-based diet is lysine deficiency. In the early 1900s, nutritionists found that rats could not survive on a diet with zein, the major storage protein in maize kernels, as the principal source of protein. However, the rats recovered when fed milk casein protein. This observation led William C. Rose, a professor at the University of Illinois, to the realization that humans require several so-called essential amino acids in their diet, one of which is lysine. Zein accounts for 60–70% of the protein in maize kernels and is essentially devoid of lysine. In the mid-1960s, Oliver Nelson and Edwin Mertz, professors at Purdue University, demonstrated that certain mutants of maize, notably *opaque2*, contain reduced amounts of zein and larger amounts of lysine-containing proteins (Mertz *et al.*, 1964). Their discovery ultimately led to the development of Quality Protein Maize (Nelson, 2001; Chapter 14, this volume).

Genetic research on maize can be dated to at least the 1860s, with studies by Gregor Mendel, who, while more famous for his work with peas, demonstrated that the same genetic principles apply to maize (see review by Rhoades (1984) for this and the following details). Prior to 1900, genetic experiments by Hugo de Vries and Carl Correns led to the rediscovery of Mendel's laws of genetic inheritance. De Vries and Correns also investigated the phenomenon of "xenia" in maize (Chapters 16 and 17) and are considered among the first pioneers of maize genetics. However, it is Rollins A. Emerson, once an Assistant Professor in the Horticulture Department at the University of Nebraska, who is generally considered the founder of modern maize genetics. Emerson was recruited to Cornell University in 1914, where he and his students began to systematically identify maize mutations and place them on chromosome maps. Over the next 30 years, Emerson's group, in collaboration with many others, created a genetic linkage map of the 10 maize chromosomes showing the positions of mutations affecting various parts of the plant, including the kernel. During this period (1920s–1930s), research on maize cytogenetics led by Emerson's associates, Barbara McClintock, Marcus Rhoades, and Charles Burnham, established the physical correspondence of linkage groups with the genetic map of maize chromosomes.

Some of the mutations mapped by Emerson and his colleagues correspond to mutants with obvious changes in kernel structure and composition, notably starch and protein content. However, most of these researchers were not biochemists, and at that time methods for isolating and characterizing enzymes were cumbersome. Consequently, the biochemical basis of mutant phenotypes was largely unknown. But this began to change in the 1950s with the development of starch gel electrophoresis (Smithies, 1955). This technique permitted the separation and assay of isozymes, which provided insight into the genetic complexity of phenotypic traits as well as how enzymes work. Starch gel electrophoresis was followed shortly thereafter (1966–1970) by electrophoretic separation in agarose and acrylamide gels (notably SDS-polyacrylamide gel electrophoresis), which accelerated the characterization of nucleic acids and proteins, respectively. These techniques ushered in the era (1960–1980) of maize "biochemical genetics" (Schwartz, 1960; Nelson, 1967), which focused on the chemical basis of mutant phenotypes.

The development of DNA cloning and sequencing technologies in the 1970s would eventually impact maize genetics in the 1980s. Genes encoding zein storage proteins were among the first to be isolated and sequenced (Chapter 14), and they were quickly followed by many others, among them genes associated with kernel and embryo development (Chapters 2, 3, 4, 7, 9 and 10); the function of specialized endosperm cell types (Chapters 5 and 6);

processes associated with storage metabolite accumulation (Chapters 11, 12, 13 and 14); and plant and seed interactions (Chapters 8, 15, 16 and 17). Within a few years, most every type of mutant and plant tissue had been investigated by gene cloning. Methods of gene tagging by transposon mutagenesis greatly facilitated this process, ultimately allowing the identification of genes responsible for specific traits and biochemical pathways.

Not long after the period of gene cloning, methods for creating transgenic maize were developed, making it possible not only to test the function of genes, but also to create novel phenotypes. Beginning in 2005, improvements in the efficiency and throughput of DNA sequencing dramatically reduced its cost, making it possible to sequence complete genomes, including those of maize and other cereals. This led to new strategies for identifying genes and characterizing their mutations. Collectively, these technical achievements, along with significant financial support from federal governments and agricultural biotechnology companies, led to our current understanding of maize kernel development. But while we have a relatively deep comprehension of many aspects of the biology and biochemical processes underlying this process, many questions remain to be answered.

Although this book provides a comprehensive treatise of many areas of maize kernel research, several topics are not covered because they either have not been studied or are at early stages of investigation. Among them are the following:

**1.** Development of the nucellus, its differentiation of a megaspore-mother-cell, and the meiotic process leading to formation of the female gametophyte.
**2.** Programmed cell death/degradation of the nucellus and acquisition of its metabolites by the embryo/endosperm.
**3.** The biochemical genetics of mineral acquisition and storage. There has been research on phosphate accumulation as phytic acid (Raboy, 2009a,b), and more recent efforts to understand accumulation of other important minerals, e.g. iron and zinc (Baxter *et al.*, 2013; Asaro *et al.*, 2016).
**4.** The mechanisms that control vitamin synthesis and accumulation. Maize, like most cereals, is deficient in a number of vitamins, including A, E, K, riboflavin, niacin, pantothenate, pyridoxine and folate (Fitzpatrick *et al.*, 2012). There has been research to increase the level of provitamin A ($\beta$-carotene) (Vallabhaneni and Wurtzel, 2009; Owens *et al.*, 2014), but the biochemical genetics underlying the accumulation of most other vitamins are open to discovery.
**5.** The process of seed desiccation and dormancy. Mutations causing vivipary, which could be associated with these processes, have been studied for many years, but a systematic analysis of how the seed loses water, becomes metabolically arrested and remains viable (dormant), is open to investigation.

Molecular biologists are generally reductionists, believing that by understanding the biochemistry and regulation of genetic and cellular processes it is possible to modify them in ways that better suit our needs. With current methods of plant transformation, recombinant DNA techniques, and genome editing, rationally designed new and improved alleles can be inserted into breeding germplasm to manipulate developmental and biochemical pathways. The potential impacts on yield and nutritional properties of the kernel could exceed what Native Americans likely never imagined! The authors of chapters in this book have done an excellent job of illuminating the way forward.

## References

Asaro, A. Ziegler, G., Ziyomo, C. Hoekenga, O.A., Dilkes, B.P. and Baxter, I. (2016) The interaction of genotype and environment determines variation in the maize kernel ionome. *G3: Genes, Genomes, Genetics* 6, 4175–4183. DOI:10.1534/g3.116.034827

Baxter, I.R, Gustin, J.L., Settles, A.M. and Hoekenga, O.A. (2013) Ionomic characterization of maize kernels in the intermated B73×Mo17 population. *Crop Science* 53, 208–220.

Fitzpatrick, T.B., Basset, G.J.C., Borel, P., Carrari, F., DellaPenna, D., *et al.* (2012) Vitamin deficiencies in humans: Can plant science help? *Plant Cell* 24, 395–414.

Mertz, E.T., Bates, L.S. and Nelson, O.E. (1964) Mutant gene that changes protein composition and increases lysine content of maize endosperm. *Science* 145, 279–280.

Nelson, O.E. (1967) Biochemical genetics of higher plants. *Annual Review of Genetics* 1, 245–268.

Nelson, O.E. (2001) Maize: the long trail to QPM. In: Reeve, E.C.R. and Black, I. (eds.) *Encyclopedia of Genetics*. Fitzroy Dearborn, New York and London, pp. 657–660.

Owens, B.F., Lipka, A.E., Magallanes-Lundback, M., Tjede, T., Diepenbrock, C.H., *et al.* (2014) A foundation for provitamin A biofortification of maize: genome-wide association and genomic prediction models of carotenoid levels. *Genetics* 198, 1699–1716. Available at: https://doi.org/10.1534/genetics.114.169979 (accessed June 7, 2017).

Raboy, V. (2009a) Approaches and challenges to engineering seed phytate and total phosphorus. *Plant Science* 177, 281–296.

Raboy, V. (2009b) Seed total phosphate and phytic acid. In: Kriz, A.L. and Larkins, B.A. (eds.) *Molecular Genetic Approaches to Maize Improvement*. Springer, Berlin, Heidelberg, pp. 41–53.

Ranum, P., Peña-Rosas, J.P. and Garcia-Casal, M.N. (2014) Global maize production, utilization, and consumption. *Annals of the New York Academy of Science* 1312, 105–112.

Rhoades, M.M. (1984) The early years of maize genetics. *Annual Review of Genetics* 18, 1–29.

Schwartz, D. (1960) Genetic studies of mutant enzymes in maize: synthesis of hybrid enzymes by heterozygotes. *Proceedings of the National Academy of Sciences of the United States of America* 46, 1210–1215.

Smithies, O. (1955) Zone electrophoresis in starch gels: group variations in serum proteins of normal human adults. *Biochemical Journal* 61, 629–641.

Vallabhaneni, R. and Wurtzel, E.T. (2009) Timing and biosynthetic potential for carotenoid accumulation in genetically diverse germplasm of maize. *Plant Physiology* 150, 562–572.

# Contributors

Philip Becraft
Genetics, Development and Cell Biology Department/Agronomy Department, Iowa State University, Ames, IA 50011, USA
Tel: (001) 515-294-2903; email: becraft@iastate.edu

James A. Birchler
Division of Biological Sciences, University of Missouri, Columbia, MO 65211, USA
Tel: (001) 573-882-4905; email: birchlerj@missouri.edu

Susan Boehlein
University of Florida, Gainesville, FL 32611
Tel: (001) 352-392-3991; email: sboehlei@ufl.edu

Ljudmilla Borisjuk
Department of Molecular Genetics, IPK-Gatersleben, Gatersleben, Germany
Tel: (0039) 482 5687; email: borisjuk@ipk-gatersleben.de

Prem S. Chourey
U.S. Department of Agriculture, Agricultural Research Service, and University of Florida, Gainesville, FL 32608, USA
Email: psch@ufl.edu

Janice K. Clark
Biology Department, University of North Dakota, Grand Forks, ND 58202–9019
Tel: (001) 701-777-2621; email: janclark@gra.midco.net

Joanne M. Dannenhoffer
Department of Biology, Central Michigan University, Mount Pleasant, MI 48859, USA
Tel: (001) 989-774-2509; email: danne1jm@cmich.edu

Matthew Evans
Department of Plant Biology, Carnegie Institution for Science, Stanford, CA 94305, USA
Email: mevans@carnegiescience.edu; web: https://dpb.carnegiescience.edu/labs/evans-lab

Sherry Flint-Garcia
U.S. Department of Agriculture, Agricultural Research Service, Columbia, MO 65211, USA
Tel: (001) 573-884-0116; email: Sherry.Flint-Garcia@ARS.USDA.GOV

Bryan C. Gontarek
Genetics, Development & Cell Biology Department, Iowa State University, Ames, Iowa 50011, USA
Email: bryan.gontarek@pioneer.com

José F. Gutiérrez-Marcos
School of Life Sciences, University of Warwick, Coventry, UK
Email: J.F.Gutierrez-Marcos@warwick.ac.uk

Jeffrey E. Habben
DuPont Pioneer, Johnston, IA 50131, USA
Email: jeffrey.habben@pioneer.com

L. Curtis Hannah
University of Florida, Gainesville, FL 32611, USA
Tel: (001) 352-392-6957; email: lchannah@ufl.edu

Tracie A. Hennen-Bierwagen
Roy J. Carver Department of Biochemistry, Biophysics and Molecular Biology, Iowa State University, Ames, IA 50011, USA
Tel: (001) 515-291-1067; email: tabier@iastate.edu

Gregorio Hueros
Departamento de Biomedicina y Biotecnologia, Universidad de Alcalá, Madrid, Spain
Tel: (0034) 91 883 9219; email: gregorio.hueros@uah.es

Gwyneth C. Ingram
Laboratoire Reproduction et Développement des Plantes, Université de Lyon, ENS de Lyon, UCB Lyon 1, CNRS, INRA, F-69342, Lyon, France
Tel: (0033) 04 72 72 8982; email: gwyneth.ingram@ens-lyon.fr

Adam F. Johnson
Division of Biological Sciences, University of Missouri, Columbia, MO 65211, USA
Email: afj8c8@mail.missouri.edu

Shawn K. Kaeppler
Campbell-Bascom Professor of Agronomy, Department of Agronomy and Great Lakes Bioenergy Research Center, University of Wisconsin, Madison, WI 53706, USA
Tel: (001) 608-262-9571; email: smkaeppl@wisc.edu

Karen E. Koch
Department of Horticultural Sciences, University of Florida, Gainesville, FL 32611, USA
Tel: (001) 352-273-4833; email: kekoch@ufl.edu

Brian A. Larkins
Department of Agronomy and Horticulture, University of Nebraska-Lincoln, Lincoln, NE 68588-0355, USA
Tel: (001) 520-603-5166; email: larkins@email.arizona.edu

Fangfang Ma
Department of Horticultural Sciences, University of Florida, Gainesville, FL 32611, USA
Tel: (001) 607-220-7382; email: mafang@ufl.edu

Donald R. McCarty
Department of Horticultural Sciences, University of Florida, Gainesville, FL 32611, USA
(001) 352-273-4846; email: drm@ufl.edu

Joachim Messing
University Professor of Molecular Biology, Waksman Institute of Microbiology, Rutgers, The State University of New Jersey, Piscataway, NJ 08854-8020, USA
Tel: (001) 848-445-4257; email: messing@waksman.rutgers.edu

Alan Myers
Roy J. Carver Department of Biochemistry, Biophysics and Molecular Biology, Iowa State University, Ames, IA 50011, USA
Tel: (001) 515-294-9548; email: albergo@mac.com

Keith Roesler
DuPont-Pioneer, 7300 Northwest 62nd Avenue, Johnston, IA 50131, USA
Email: Keith.roesler@pioneer.com

Hardy Rolletschek
Department of Molecular Genetics, IPK-Gatersleben, Gatersleben, Germany
Tel: (0039) 482-5686; email: rollet@ipk-gatersleben.de

Paolo Sabelli
School of Plant Sciences, University of Arizona, Tucson, AZ 85721, USA
Email: psabelli1@gmail.com

Jeffrey Schussler
DuPont-Pioneer, 7300 Northwest 62nd Avenue, Johnston, IA 50131, USA
Email: jeff.schussler@pioneer.com

Bo Shen
DuPont-Pioneer, 7300 Northwest 62nd Avenue, Johnston, IA 50131, USA
Email: bo.shen@pioneer.com

William F. Sheridan
Department of Biology, University of North Dakota, Grand Forks, ND 58202, USA
Tel: (001) 701-777-4479; email: william.sheridan@email.und.edu

Rentao Song
National Maize Improvement Center of China, China Agricultural University, Beijing, China 100193
Email: rentaosong@cau.edu.cn

Erik Vollbrecht
Department of Genetics, Development and Cell Biology, Iowa State University, Ames, IA 50011-3260, USA
Tel: (001) 515-294-9009; email: vollbrec@iastate.edu

Thomas Widiez
Laboratoire Reproduction et Développement des Plantes, Université de Lyon, ENS de Lyon, UCB Lyon 1, CNRS, INRA, F-69342, Lyon, France
Tel: (0033) 04 72 72 8608; email: thomas.widiez@ens-lyon.fr

Yongrui Wu
National Key Laboratory of Plant Molecular Genetics, Institute of Plant Physiology & Ecology, Shanghai Institutes for Biological Sciences, Chinese Academy of Sciences, Shanghai 200032, China
Email: yrwu@sibs.ac.cn

Ramin Yadegari
School of Plant Sciences, University of Arizona, Tucson, AZ 85721, USA
Tel: (001) 520-621-1616; email: yadegari@email.arizona.edu

Junpeng Zhan
School of Plant Sciences, University of Arizona, Tucson, AZ 85721, USA
Email: zhan@email.arizona.edu

Xia Zhang
Biotechnology Research Institute, Chinese Academy of Agricultural Sciences, Beijing 100081, China
Email: zhangxia@caas.cn

# 1 Kernel Evolution: From Teosinte to Maize


Sherry A. Flint-Garcia*

*U.S. Department of Agriculture, Agricultural Research Service, Columbia, Missouri, USA*


## 1.1 Introduction

Maize is the most productive and highest value commodity crop in the U.S. and around the world: over 1 billion tons were produced each year in 2013 and 2014 (FAO, 2016). Together, maize, rice, and wheat comprise over 60% of the world's caloric intake (http://www.fao.org). The importance of maize in terms of production and caloric intake is not a recent development. In fact, Native Americans have relied on maize and its ancestor for more than 9000 years. The "Columbian exchange" allowed maize to spread around the world, to adapt to new environments and become a major crop that feeds large portions of the human population. Maize, and the kernel in particular, has undergone dramatic changes over the past 9000 years. The biology of maize seed size and its starch, protein, oil content, and food characteristics, are described in other chapters of this book. Here I review the evolution of maize from teosinte (the wild ancestor) to landraces (locally adapted, open-pollinated farmer varieties) to modern maize (inbreds and hybrids), and discuss changes in kernel composition and size during this process.

## 1.2 Domestication

Maize, like all the world's major agricultural crop plant and animal species, underwent domestication from a wild relative. The suite of phenotypic traits that were modified during domestication is referred to as the "domestication syndrome" (Hammer, 1984) and usually includes traits related to productivity (e.g. increased seed number and size), harvestability (e.g. non-shattering and fewer seed-bearing structures), and consumption (reduced toxicity and improved palatability) among other species-specific traits (Olsen and Wendel, 2013). Evolution of the seed was central to domestication, as were traits facilitating harvest.

Genetic and archeological evidence suggest maize was domesticated from teosinte (*Zea mays* ssp. *parviglumis*) approximately 9000 years ago in the Central Balsas River Valley in southwestern Mexico in the states of Guerrero and Michoacán (Matsuoka *et al.*, 2002; Piperno *et al.*, 2009). *Zea mays* ssp. *parviglumis* (hereafter parviglumis) is an annual diploid species endemic to southwestern Mexico (Doebley and Iltis, 1980). There are several other species of teosinte with different ploidy levels, perenniality,

---


*Corresponding author e-mail: Sherry.Flint-Garcia@ARS.USDA.GOV

and/or special regional adaptation to higher elevations or lower latitudes (Fukunaga *et al.*, 2005), but these will not be discussed in any detail. Hereafter, whenever teosinte is mentioned, the reader may assume parviglumis unless otherwise noted.

There are dramatic differences in plant, ear, and kernel morphology between maize and teosinte (reviewed in Doebley, 2004). Parviglumis plants, when grown under the short-day conditions typical of central Mexico, are bushy and composed of many stalks (tillers) with long lateral branches ending in male inflorescences (Fig. 1.1A). In contrast, most modern maize plants are unbranched, with a single stalk and short lateral branches (ear shanks) ending in female inflorescences (Fig. 1.1B). Teosinte plants are capable of producing over 100 ear structures, each of which contains 5 to 12 seeds stacked and without a cob (Fig. 1.1C). Modern maize plants usually produce one or two ears with cobs that bear several hundred kernels in eight or more rows around the ear (Fig. 1.1D). Teosinte kernels are very small (approximately one-tenth the weight of maize kernels) and are enclosed in a hardened fruitcase (Fig. 1.1E) absent in modern maize (Fig. 1.1F). Teosinte ears shatter and disperse their seeds upon maturation, a characteristic absent in maize.

**Fig. 1.1.** Teosinte (A) and maize (B) differ greatly in terms of number of stalks and male and female inflorescences. Teosinte ears (C) contain 5–12 kernels without the familiar cob structure characteristic of maize (D). The small teosinte seeds (E) are enclosed in a hard fruitcase, while maize kernels (F) are naked and weigh approximately ten times more than those of teosinte.

It is something of a mystery how native peoples of Mexico used teosinte prior to domestication. There were no large domesticated animals in North America at the time, so it is unlikely teosinte was a forage crop. Modern maize is used primarily for grain, and a natural assumption is that teosinte was used similarly. However, its hard fruitcase would be a formidable deterrent, along with the limited amount of food obtained from the small seeds. George Beadle devised a method to create "teo-tortillas" using a primitive metate (grinding stone) and a water-based method to float off the broken fruitcases. Beadle also proposed that natives could have popped teosinte, similar to modern popcorn (Beadle, 1939). Others have proposed Native Americans chewed or sucked out sugars stored in the pithy teosinte stalks (Iltis, 2000) or created fermented beverages (Smalley and Blake, 2003).

### 1.2.1 Archeological evidence

The oldest archeological ear/cob samples are from 6200 years ago, originating in Guilá Naquitz Cave in Oaxaca (Benz, 2001), and 5500-year-old samples from the San Marcos Cave in the Tehuacán Valley in Puebla (Long *et al.*, 1989). Unfortunately, these samples are too old to bear kernels, but they do show non-shattering cobs with two to four rows of naked (no fruitcase) kernels. The oldest kernel samples, though not intact, include microfossils dated to 8700 years old and found on grinding stones from the Xihuatoxtla Shelter in Guerrero (Piperno *et al.*, 2009). Analysis of starch grains found on these stones revealed maize was the primary species processed and included popcorn and other hard/flinty kernel types. Sequence analysis of ancient DNA obtained from 660–4405-year-old ear samples from New Mexico and Mexico indicated that alleles representative of modern maize were present 4400 years ago (Jaenicke-Després *et al.*, 2003). So, it is clear primitive maize with morphologically distinct ears and kernels, though perhaps not quite resembling modern maize, was grown within a few

thousand years of domestication and was an important part of the Native American diet.

### 1.2.2 The master regulators of domestication

Beginning in the 1800s, there were various hypotheses concerning the origin of corn that involved an extinct progenitor species, teosinte, tripsacum, pod corn, corngrass, and combinations thereof. During the 1930s, debates revolved around the extreme phenotypic differences between maize and teosinte. In an effort to understand inheritance of these differences, Beadle examined the phenotypes of over 50,000 $F_2$ plants derived from a cross between maize and teosinte (Beadle, 1972). He determined that approximately 1 in 500 plants looked like very teosinte-like, or very maize-like, with a ratio that suggested four or five genes control the main morphological differences between maize and teosinte.

Indeed, Beadle's calculation of a handful of genes has been largely supported by quantitative trait locus (QTL) mapping studies of morphological differences between maize and teosinte. In an $F_2$ population derived from a cross of a maize landrace with a more distantly related teosinte subspecies (*Zea mays* ssp. *mexicana*, hereafter mexicana), six major QTLs (chromosomes 1–5) were found to underlie key traits that differentiate maize and teosinte: lateral branch length and inflorescence architecture, and secondary sex traits such as the hard fruitcase and paired floral spikelets (Doebley *et al.*, 1990). The QTL analysis of a second $F_2$ population derived from a primitive landrace crossed with parviglumis revealed the same genomic regions, suggesting domestication from teosinte to a primitive maize landrace could be accomplished by modifying a few key genes or gene regions (Doebley and Stec, 1993).

Since then, several QTL have been fine mapped and cloned, revealing the importance of transcription factors controlling key steps in domestication. The important regulator of apical dominance, *teosinte branched 1* (*tb1*), is located on the long arm of chromosome 1 (Doebley *et al.*, 1995). The domesticated allele of this transcription factor contains a *Hopscotch* transposable element 63 kb upstream of the start codon (Studer *et al.*, 2011) that results in higher expression of a lateral branch repressor (Doebley *et al.*, 1997). Thus, maize represses growth of lateral branches, resulting in fewer tillers. Also on chromosome 1 (short arm) is a QTL controlling prolificacy: in teosinte, the long lateral branches bear many ears, while the maize lateral branch bears a single terminal ear. The QTL controlling prolificacy was fine mapped to *grassy tillers 1*, a homeodomain leucine zipper transcription factor (Wills *et al.*, 2013) that was previously demonstrated to control tillering (Whipple *et al.*, 2011). The QTL on chromosome 5 originally thought to be a master controller of a number of ear-related traits (kernel row number, ear diameter, pedicellate spikelet length, and shattering) fractionated into multiple independent factors (Lemmon and Doebley, 2014). More recently, fine mapping and cloning of a shattering QTL in sorghum identified a YABBY-like transcription factor as a candidate gene for the QTL on chromosome 5 (Lin *et al.*, 2012). The genes responsible for the QTLs on chromosomes 2 and 3 have yet to be cloned.

The QTL on chromosome 4 is of particular interest to kernel evolution, since it controls development of the hardened fruitcase enclosing the teosinte seed and is absent or severely reduced in maize. The QTL underlying this trait, *teosinte glume architecture 1*, was mapped to chromosome 4 (Dorweiler *et al.*, 1993) and encodes a transcription factor in the squamosa promoter binding-protein family (Wang *et al.*, 2005); the causative lesion was later determined to be a single amino acid change affecting dimerization (Wang *et al.*, 2015). In teosinte, the fruitcase is composed of (i) a cup-shaped segment of the stem, the "cupule," in which the seed is seated, and (ii) a hardened bract or glume that is hinged onto the cupule that completely encloses the seed. The maize allele represses formation of these structures, such that the cupule and glume no longer surround the seed; these structures were evolutionarily repurposed to form the hard sections of the maize cob.

### 1.2.3 A thousand small effect genes underlie domestication

While QTL studies are useful as a forward genetics approach to determine genomic regions underlying a phenotype, reverse genetics approaches can be used to scan the genome for signatures of selection that could result in a phenotype related to the domestication syndrome. Selection during domestication results in a reduction of nucleotide diversity relative to the progenitor and an excess of rare variants as populations recover from selection, and can be measured using a variety of population genetic statistics. For example, an analysis of sequence diversity of 21 genes on chromosome 1 revealed only *tb1* as a target of selection (Tenaillon *et al.*, 2001).

A large-scale selection scan suggested approximately 2–4% of maize genes could have been targets of selection during domestication and/or modern breeding (Wright *et al.*, 2005). Assuming 35,000 genes in maize, this translates to 700–1400 genes that could be responsible for the transformation of teosinte into modern maize. Using the HapMap2 dataset of 55 million single nucleotide polymorphisms (SNPs) (Chia *et al.*, 2012), Hufford *et al.* (2012) found approximately 1000 genes experienced selection, with the strongest selection occurring during domestication rather than during modern breeding. The finding that so many genes were involved in domestication obviously conflicts with the five-gene hypothesis of Beadle (1939) and the early QTL mapping studies by the Doebley lab. But this paradox can be resolved by invoking the theory that a handful of master regulators can orchestrate a cascade involving intermediate and small effect genes that control a wide range of traits targeted by domestication.

## 1.3 Modern Breeding

As primitive corn was carried from central Mexico, north and south across the Americas, the outbreeding nature of maize and large population sizes allowed maize to adapt to new environments, e.g. day-length, climate, soil types, and human uses (dietary preferences and religious purposes). For example, gene flow from mexicana, a highland teosinte, allowed maize to adapt to higher elevations within Mexico (van Heerwaarden *et al.*, 2011). Maize moved into the Southwestern USA by 4000 years ago, initially via a highland route through Mexico, followed approximately 2000 years later by gene flow from lowland races from the Pacific coast (Fonseca *et al.*, 2015). From the Southwestern USA, maize spread north to Canada (Vigouroux *et al.*, 2008) and became the dominant crop species of North America by 800 AD (Smith, 1989). For the southward expansion, highland maize spread to the lowland tropics of southern Mexico and Guatemala, through the Isthmus of Panama, and into Colombia. From Colombia, maize spread to the Caribbean via the Lesser Antilles and also into the rest of South America, including an independent adaptation to highlands of the Andes (Takuno *et al.*, 2015). Maize was carried to Europe, Asia, and Africa by Columbus and the early explorers, and continued to adapt (Mir *et al.*, 2013). Each landrace has distinct plant, ear, and kernel characteristics that have been used to identify and classify them (Goodman and Brown, 1988) and define their uses around the world.

Maize inbreeding began at the end the 1800s and subsequent hybridization of the early cycle inbreds (Shull, 1909) led to the hybrid seed industry and evolution of heterotic groups. Today, in the U.S. Corn Belt, there are three main heterotic groups: stiff stalks, non-stiff stalks, and iodents (Troyer, 1999). Breeding programs usually focus on specific traits relevant to the target environment: cold tolerance for northern climates, drought tolerance for the high plains, disease and insect resistance in the south, etc.

### 1.3.1 Dent corn

The vast majority of corn grown in the U.S. is a commodity referred to as "Number 2 Yellow Dent." In general, yield is the primary driver of dent corn, and seed quality is of secondary importance. There are regions of the USA that cater to specialty food-grade dent corn markets, such as white food corn, where producers contract their crop directly to processors and for which white food corn varieties were tested until 2002 (Darrah *et al.*, 2002). While all teosintes have white endosperm, there is wide variability in landraces and inbred lines for endosperm color, including orange and yellow (from carotenoids) and red and purple (from anthocyanins). Yellow predominates in commodity corn due to the higher nutritional value of carotenoids for animal feed, while white is preferred for human consumption in many regions around the world (Poneleit, 2001). A survey of the *y1* (*phytoene synthase*) locus revealed classic signatures of selection, in particular much lower diversity in yellow relative to white lines (Palaisa *et al.*, 2003). Anthocyanin kernel pigments appear to have been targeted by post-domestication selection for the ability to produce red and purple pigments via the *colored aleurone 1* locus (Hanson *et al.*, 1996). Together, these results suggest kernel color traits were targets of selection.

The most recognizable types of food corn are sweet corn and popcorn, where flavor and kernel quality are of highest importance. Another example, baby corn, is simply an immature ear harvested as silks begin developing; it is primarily produced in Thailand (Aekatasanawan, 2001). Each of these specialty corns has a different set of ear-kernel phenotypes and underlying genetics, some of which is discussed in detail in other chapters of this book. There has been continued evolution, breeding, and refinement of the genetics underlying these kernel phenotypes, and breeding efforts have kept the associated germplasm separate. Phylogenetic analysis of the NC7 (Ames, IA) Plant Introduction Station collection of 2800 maize inbred lines showed clear germplasm separation (Romay *et al.*, 2013): the popcorn and sweet corn accessions form very distinct germplasm groups; the stiff stalk and non-stiff stalk inbreds within the temperate germplasm have intermediate separation from each other; the tropical germplasm also forms a very distinct group. Analysis of marker data for inbred lines divided by era showed continued separation of the major heterotic groups of corn belt maize and decreased diversity in the ancestry of the heterotic pools (van Heerwaarden *et al.*, 2012).

### 1.3.2 Sweet corn

Cultures across the Americas have eaten "green corn" for millennia, enjoying standard starchy corn that is picked at the "milk stage" of kernel development. Green corn is not a result of sweet corn mutations, but rather owes its low-level sweetness to sugars not yet converted to starch. Modern sweet corn is the result of precise breeding, utilizing mutations in the starch biosynthetic pathway (Chapter 12) to produce specific market classes of sweet corn ranging from the original sugary varieties to the newer synergistic, augmented, and supersweet varieties. There are only eight genes used in commercial sweet corn production, with three predominating the market at present (reviewed in Tracy, 1994): *sugary 1* (*su1*) mutations affect a starch debranching enzyme, resulting in phytoglycogen accumulation; *sugary enhancer 1* (*se1*) has an unknown function, but causes the sweet phenotype when used in conjunction with *su1* (Schultz and Juvik, 2004); *shrunken 2* (*sh2*) mutations block all complex carbohydrates (starch and phytoglycogen), causing an accumulation of sugars. While not widely grown as compared to non-sugary varieties, sweet corn (primarily *su1* types) has been grown and consumed in confections and alcoholic beverages since before the arrival of Columbus (Wellhausen *et al.*, 1952).

Among commercially important sweet corn mutations, *su1* has an interesting evolutionary history related to the diffusion of landraces across the Americas. Sequence analysis of 57 accessions of *su1* germplasm from six geographic regions of the Americas revealed five independent origins of *su1* sweet corn (Tracy *et al.*, 2006). Of these, three different alleles are caused by single amino acid changes in conserved residues of what is considered the active site of the isoamylase enzyme, and are spatially clustered in Northwestern Mexico and throughout the U.S. A fourth allele was caused by a transposon insertion in the first exon, and was found in two Mexican Maiz Dulce accessions. The causative lesions could not be determined for the fifth allele, which was identified in two Peruvian highland accessions of Chullpi. Selection for and maintenance of the first *sugary 1* mutations by Native Americans led to the success of modern breeding for additional mutations and secondary flavor and texture traits. The starch mutants were found in limited genetic resources, originating from the ancestral group of "Northern Flints" and resulting in the tight population structure of the U.S. maize germplasm collection, as discussed earlier (Romay *et al.*, 2013).

### 1.3.3 Popcorn

Popcorn is another favorite food corn around the world. The primary traits that make popcorn unique are the explosion of the kernel upon exposure to heat and the subsequent expansion of starch to form large "flakes" (reviewed in Ziegler, 1994). During popping, the moisture contained in the kernel expands until the pericarp can no longer withstand the pressure and bursts. Starch of the hard endosperm gelatinizes with the released steam, expands due to heat, and dries and hardens into flakes. Flake production is related to a higher ratio of hard to soft starch and a thicker pericarp that can withstand building pressure from steam, traits absent from dent corn. While popcorn kernel colors range from yellow and white (the most commercially important) to red, blue, purple and nearly black, there are only two kernel shapes: rice types with long, slender kernels and a pointed tip; and pearl types with round kernels and a smooth top. Once popped, there are two main flake shapes (with intermediate variation) that appear to be under genetic control: butterfly flakes are irregularly shaped but with many wings; mushroom flakes are round with only a few wings.

As discussed earlier, Native Americans probably enjoyed pop-teosinte prior to domestication. It is likely many primitive landraces were popcorns selected from earlier flint types for larger popping expansion. By the time of Columbus, popcorn was prevalent in both North and South America. As popcorn became a distinct industry in the 1880s (Erwin, 1949), modern breeding methods were employed to improve agronomic traits and popcorn-specific traits: pericarp strength, popping volume, and flavor. Interestingly, a single gene has played a key role in maintaining distinct popcorn germplasm—the gametophyte factor known as *ga1*. The dominant strong allele, *Ga1-s*, which confers nearly perfect cross-incompatibility with non *Ga1-s* pollen, is present in nearly all modern popcorn germplasm (Nelson, 1952). While this gene does not affect kernel phenotypes per se, it does maintain the already distinct popcorn kernel phenotypes by preventing pollen contamination by dent maize, which typically carries the *ga1* allele.

## 1.4 Seed Size and Kernel Composition

It is clear that the kernel was a central focus during domestication and breeding—humans selected large seeds that are easy to harvest and consume. In the course of evolution, there have been drastic changes in seed composition. The typical chemical composition of teosinte, landraces, and inbred lines is shown in Table 1.1. Of note is the large increase in starch (34%) and large decrease in protein (–58%) during domestication (Flint-Garcia *et al.*, 2009a). Since these values are expressed as a percentage of total kernel weight, it is no surprise that various traits are correlated, regardless of the underlying biochemistry. The biology, genetics, and biochemistry of kernel composition traits and seed size are described in other chapters of this book. The objective here is to discuss evolution of these traits, which are intertwined with other traits.

### 1.4.1 Seed size

Increasing seed size/weight was undoubtedly valuable to the survival and prosperity of early Native Americans. Indeed, maize kernels (either landraces or modern inbred lines, excepting popcorns) weigh almost ten times more than teosinte seeds (Flint-Garcia *et al.*, 2009a), and this increase occurred during domestication. After selection to reduce and open up the fruitcase, primarily acting through *tga1*, seed volume was no longer limited by space inside the fruitcase. Enlarged seed size was probably the most important domestication trait to Native Americans, but very little is known about the genetics underlying the evolution of the process. In a QTL analysis of the same landrace × teosinte $F_2$ populations described earlier (Doebley *et al.*, 1990; Doebley and Stec, 1993), six and four QTL were found to control seed weight during the transition from teosinte to landraces, where all the teosinte alleles decreased seed weight (Doebley *et al.*, 1994). In a backcross 1-derived mapping population of parviglumis in the W22 background, six QTLs were identified for kernel weight (Briggs *et al.*, 2007). A similar result of a handful of QTLs controlling seed weight was also seen in a population of near isogenic lines (NILs) derived from ten parviglumis donors in the B73 background (Liu *et al.*, 2016); there was a total of eight QTLs across the entire population, with a range of

Table 1.1. Kernel composition and seed traits for a panel of teosinte (parviglumis) accessions, landraces, and inbred lines. Data summarized from Flint-Garcia *et al.* (2009a).

| Germplasm | $N$ | Protein % | Fat % | Fiber % | Ash % | Carbohydrate % | Seed Wt. (g) | Percent endosperm |
|---|---|---|---|---|---|---|---|---|
| Teosinte | 11 | 28.71 | 5.61 | 0.91 | 2.24 | 52.92 | 0.03 | 90.18 |
| Landraces | 17 | 12.13 | 4.40 | 1.75 | 1.55 | 71.16 | 0.28 | 90.13 |
| Inbred lines | 27 | 11.11 | 4.12 | 1.80 | 1.40 | 72.37 | 0.26 | 91.85 |

two to six QTLs per donor. Many of the QTLs identified in these studies overlapped, and, as expected, the majority of the teosinte alleles caused a decrease in seed weight; however, one of the teosinte alleles for the QTL on chromosome 2 appears to increase seed weight (Liu *et al.* 2016). While this allelic effect remains to be validated, its potential use in breeding is attractive.

There has been limited progress identifying genes underlying teosinte kernel weight QTLs and establishing that they are related to domestication. Interestingly, *prolamin-box binding factor 1* (*pbf1*) is a strong candidate for a QTL on chromosome 2, and it will be discussed below in Section 1.4.3 on kernel proteins. For a QTL on chromosome 1, a gene with homology to *GS3* from rice was proposed as a selection candidate in maize, as *OsGS3* was found to be a domestication gene controlling grain size in rice (Takano-Kai *et al.*, 2009). Although the maize ortholog of *GS3* has lower sequence diversity in maize than teosinte, selection tests revealed it is a neutrally-evolving gene (Li *et al.*, 2010) and did not play a role in kernel evolution from teosinte, despite being a potential candidate gene underlying kernel weight.

### 1.4.2  Starch

Starch synthesis and accumulation in the seed involves a complex biochemical system with an array of sugars and starches, a number of plant organs and structures, and temporal regulation (Chapter 12). To explain the system briefly, and in a highly oversimplified way, a series of enzymes including sucrose synthases (e.g. *shrunken 1*) and invertases (e.g. *mn1*) break down the sucrose entering the endosperm via the basal endosperm transfer layer (BETL) into glucose and fructose; a series of enzymes including ADP-glucose pyrophosphorylase (e.g. *brittle 2=bt2* and *shrunken 2*) convert the glucose to ADP-glucose; and finally starch synthases (e.g. *waxy 1*), starch branching enzymes (e.g. *amylose extender 1=ae1*) and debranching enzymes (e.g. *su1*) act on the ADP-glucose to

convert it into the two primary forms of starch (Chapters 5 and 12).

Population genetic analysis of six genes in the starch pathway revealed that three genes — *bt2*, *su1*, and *ae1* — show a signature of selection. This suggests that the starch pathway was targeted by selection (Whitt *et al.*, 2002). However, because DNA sequence data were collected from inbred lines and teosinte accessions, but no landraces, it was difficult to determine whether selection occurred during domestication or during breeding. Recently, an analysis of 348 genes in archeological landrace samples from the Southwestern USA dating back to 750–4000 years ago and Mexican samples dating back to 1400–5900 years ago showed selection for several composition genes, including *ae1* and particularly *su1* (Fonseca *et al.*, 2015). The results of this study suggest selection on *su1* was more recent, approximately 1000–1200 years ago, which coincided with the appearance of larger cobs and floury endosperm texture. Both of these genes (*ae1* and *su1*) affect the structure of amylopectin and are involved in pasting properties important for making porridge and tortillas (Whitt *et al.*, 2002; Wilson *et al.*, 2004). Again, it is not a surprise that starch synthesis was affected by domestication, because as seed size increased, starch content also increased.

### 1.4.3  Protein

The nature of proteins in the maize kernel is described in Chapter 14. Briefly, approximately 10–20% of the proteins are globulins found in the embryo; the remaining 80–90% occur in the endosperm. Prolamins, or zeins ($\alpha$, $\beta$, $\gamma$, and $\delta$), are the principal endosperm storage proteins and are found in protein bodies (Boston and Larkins, 2009). Native Americans developed a process called "nixtamalization," in which corn kernels were soaked in an alkaline solution (lime; calcium hydroxide) prior to cooking. This process allows easy removal of the pericarp and improves texture by gelatinizing the starch; most importantly, it improves the nutritional value of the resulting masa by degrading the protein bodies and releasing niacin (vitamin B3) (Gomez *et al.*, 1989). Without this treatment, diets based largely on maize lead to a skin disease known as Pellagra.

Swarup *et al.* (1995) found that exotic maize and wild members of the genus *Zea* exhibit higher levels of methionine-rich δ-zeins than maize inbreds, leading the authors to hypothesize that the high methionine trait was lost in the course of domestication. Indeed, an HPLC-based survey of the zein profiles in a panel of teosinte, landrace, and inbred accessions showed higher levels of δ-zeins as well as β-zeins in landraces and teosinte (Flint-Garcia *et al.*, 2009a). A number of classical kernel mutants affect zein synthesis and/or formation of protein bodies. For example, *opaque 2* encodes a bZIP transcription factor that, when mutated, results in a severe reduction of the lysine-poor zeins and a concomitant increase in other storage proteins and free amino acids, including lysine (Schmidt *et al.*, 1990). *Opaque 1*, *floury 1*, and *floury 2* are all involved in aspects of zein trafficking in the endoplasmic reticulum. There is no evidence these genes or any of the zein genes were selected during domestication or breeding (Hufford *et al.*, 2012).

Several of the zeins (27 kDa γ-zein and 22 kDa α-zein) are regulated by *pbf1*, an endosperm-specific transcription factor (Vicente-Carbajosa *et al.*, 1997). DNA sequence analysis of *pbf1* in 660–4405-year-old ear samples from New Mexico and Mexico showed the modern maize haplotype was nearly fixed in these landrace samples (Jaenicke-Després *et al.*, 2003). This evidence of a selective sweep strongly suggests protein quality could have been under selection. The absence of a knockout mutant in *pbf1* suggests this gene is critical. Lang *et al.* (2014) used heterozygosity in a NIL carrying a teosinte *pbf1* allele to determine the target trait. They found twofold higher expression of the teosinte *pbf1* allele and a slight increase in seed weight, but no change in zein composition. This positive allelic effect on seed weight was not seen in the original maize × teosinte QTL study (Doebley

*et al.*, 1994), but is consistent with the effect we observed for one of our ten donors (Liu *et al.*, 2016). The authors of the former study hypothesized that the reduction in seed weight from the maize allele was a negative pleiotropic effect of selection at *pbf1* for some unknown aspect of kernel composition.

Because zeins are so abundant, they impact the amino acid composition of the kernel, limiting the content of the essential amino acids lysine, tryptophan, and methionine (Prasanna *et al.*, 2001). However, there is variability in free amino acids (Moro *et al.*, 1996). In two large-scale selection scans, three genes involved in amino acid metabolism were identified as being selected (Wright *et al.*, 2005; Yamasaki *et al.*, 2005): *chorismate mutase, cysteine synthase*, and *dihydrodipicolinate synthase*. These results prompted an in-depth analysis of amino acid pathways (Flint-Garcia *et al.*, 2009b). Of the 15 additional amino acid metabolism genes tested, only four showed weak evidence of selection: *aspartate kinase – homoserine dehydrogenase 1 – AK domain, glutamate dehydrogenase, proline dehydrogenase*, and *sam synthetase II*. However, none of the selected genes cluster in pathways that make a convincing argument for evolutionary selection.

### 1.4.4 Oil

The typical maize kernel contains 4.3–4.5% oil, a high energy component of the grain. Generally, the mature embryo is 10% of the total kernel mass and contains about 85% of the kernel lipids, primarily as triacylglycerols (Chapter 13). In a survey of kernel traits across *Zea mays* germplasm, there was a significant decrease (−26%) in kernel oil content between teosinte and maize landraces/inbred lines (Flint-Garcia *et al.*, 2009a). Although the reduction in oil content during domestication (−21%) was small compared to the starch increase and protein decrease, it represents a major change in kernel composition. Interestingly, no change was found in the endosperm-to-embryo ratio between teosinte and landraces, suggesting it may be possible to increase oil content by using teosinte alleles without a negative pleiotropic effect of increased embryo size.

One of the best characterized QTLs for kernel oil content is on chromosome 6 (Laurie *et al.*, 2004). It was mapped to a BAC with five genes, one of which is *DGAT1-2* (Zheng *et al.*, 2008). In the 2008 study, an association analysis identified a 3-bp insertion at position 469, resulting in an extra phenylalanine (F469) as the causative factor conferring high oil. The F469 allele was found in all teosinte accessions analyzed, and thus is considered ancestral (Zheng *et al.*, 2008). A follow-up study showed the high-oil allele is present in most of the Southwestern USA, Northern Flint, and Southern Dent landraces, at a moderate frequency in Corn Belt Dent, and nearly absent in the early inbred lines. Two hypotheses were offered to explain diversity at *DGAT1-2*: (i) the high oil F469 allele was lost due to genetic drift when a small number of Corn Belt Dent populations were chosen to develop inbred lines; or (ii) the F469 allele was selected against because of pleiotropy with other favorable agronomic traits, such as high starch content (Chai *et al.*, 2012). Indeed, *DGAT1-2* was associated with both oil and starch content in the Nested Association Mapping population (Cook *et al.*, 2012).

One unappealing aspect of using genome-wide selection scans as a reverse-genetic approach is that there may not be an immediate connection with the target trait. Among the 48 genes identified as selection candidates by Wright *et al.* (2005) and Yamasaki *et al.* (2005), most did not have obvious target traits associated with the gene. In an effort to identify the phenotypic effects associated with these selected genes, 32 genes were tested in an association analysis of two teosinte populations scored for a panel of phenotypic traits (Weber *et al.*, 2009). Interestingly, a gene with homology to an ankyrin-repeat-like protein, *AY106616*, associated most strongly with kernel oil content, but also with starch content. The ankyrin-repeat-like protein is involved in carbohydrate metabolism and allocation in tobacco and Arabidopsis (Weber *et al.*, 2009); thus, a plausible target trait for carbon cycling within the kernel has been established.

## 1.5 Lingering Questions and Prospects for Maize Improvement

The evolutionary history of the maize kernel presents geneticists and breeders with a series of questions from how domestication occurred to prospects for maize improvement.

### 1.5.1 Relationships between composition and seed size traits

As noted, there are correlations among many of the size and kernel composition traits, especially between germplasm groups: teosinte, landraces, and inbred lines (Flint-Garcia *et al.*, 2009a). For example, there is positive correlation of seed weight with kernel starch content, which begs the question from an evolutionary perspective: which came first, the chicken or the egg? Did liberation of the seed from the fruitcase allow the kernel to expand in size due to a subsequent increase in starch accumulation? Or, did selection for high starch alleles occur first and help drive expansion of the seed out of the fruitcase? Would reintroduction of all the fruitcase alleles (*tga1* and other minor QTLs, if any) limit the size of the kernel and change kernel composition, e.g. decreased starch and increased protein and oil?

The question of pleiotropy versus linkage of QTLs is not an evolution-specific one, but it is still very relevant. Because composition and seed size traits are so highly correlated, are there specific genes that mechanistically contribute to variation for multiple traits? Or are there multiple genes linked (tightly or not) in a single QTL that control different traits independently? Can these traits be manipulated independently?

### 1.5.2 How many of the 1000 selected genes are involved in kernel traits?

Seed size was obviously an important trait during domestication, and one would expect a large number of the 1000 selected genes could influence seed size genes (Chapter 16). Alternatively, because of the strong correlations between seed size and composition traits, one could also expect a large number of the selected genes to be kernel composition genes. The genome-wide selection scan of Hufford *et al.* (2012) provided an excellent starting point to answer this question; however, in my opinion, poor genome annotation has been the primary impediment of progress. Of the 1000 selected genes, the vast majority are not annotated. Nevertheless, a simple query of the selection candidates in Hufford *et al.* (2012) using the 464 genes from the classical gene list (Schnable and Freeling, 2011) identified eight interesting new selection candidates that could be involved in kernel traits (Table 1.2). These genes can be tested rigorously for signatures of selection (e.g. HKA tests, coalescent simulations, etc.) and their phenotypic effects determined in both maize and teosinte germplasm.

### 1.5.3 Do teosinte alleles have value for improving corn?

Long ago—9000 years—humans began modifying teosinte to improve harvestability. Selection resulted in reduced genetic variation in genes underlying these traits; consequently, modern maize shows little variation. Additionally, every gene across the genome has lost some diversity because of demographic events (bottlenecks, random sampling, etc.), even if these are neutrally-evolving genes.

Today, we are growing corn in very different environments using different agronomic practices than those practiced 9000 years ago during domestication, or 1000 years ago as corn became the predominant crop in the USA or even 100 years ago when modern breeding began. Traits that were relevant 9000, 1000, or 100 years ago may not be useful today; therefore, alleles selected 9000, 1000, or 100 years ago that persist in modern germplasm may not be optimal today. This reduction in genetic variation is irreversible—especially if the current practice of recycling germplasm in breeding programs is continued—unless of course variation is reintroduced from teosinte and/or landraces.

A straightforward goal would be to try to modify our current corn for specific traits. Novel sources of genetic resistance to the foliar diseases grey leaf spot (Lennon *et al.*, 2016) and southern leaf blight (Lennon *et al.*, 2017) were identified in parviglumis. Introgression of mexicana into maize resulted in lines with significantly higher protein content, as well as higher lysine, methionine, and/or phenylalanine content (Wang *et al.*, 2008). Thus, teosinte has potential to improve many traits in maize.

If we strive for the more extreme goal of introducing large portions of the teosinte genome into modern maize germplasm, what genes/alleles should we target? Genes showing signatures of selection would provide the greatest return on investment, as they harbor allelic diversity in teosinte not present in maize. Clearly, we do not want the hard fruitcase trait back, so we will avoid *tga1*! However, perhaps a plant with a single ear is not the best ideotype in today's agronomic system where we no longer harvest

**Table 1.2.** Potential new selection candidates with effects on kernel traits. Results were obtained by merging the candidate gene lists from Hufford *et al.* (2012) with the Classical Gene List (Schnable and Freeling, 2011).

| Gene ID | Gene name | Possible target trait |
| --- | --- | --- |
| GRMZM2G348551 | *su2; sugary 2* | Starch |
| GRMZM2G394450 | *ivr1; invertase 1* | Starch |
| GRMZM2G089836 | *ivr2; invertase 2* | Starch |
| GRMZM2G110175 | *bm1; brown midrib 1* | Starch |
| AC196475.3_FG004 | *bm3; brown midrib 3* | Starch |
| GRMZM2G098298 | *ccp1; cysteine protease 1* | Protein |
| GRMZM2G138727 | *zp27; 27-kDa zein protein* | Protein & Amino acids |
| GRMZM2G087612 | *SDP1; sugar dependent1* | Oil |

corn manually and where combines are capable of harvesting many ears per plant. Reintroducing the branching and prolificacy alleles at *tb1* and *gt1* from teosinte would be first steps to increase prolificacy. However, reintroduction of the teosinte alleles will likely disrupt the source–sink balance (see Chapter 16) that has been established in modern germplasm. Incorporating teosinte alleles of the various starch biosynthetic genes could also be useful in reprogramming corn.

One interesting question to ask: if we had a thousand years to rerun a domestication experiment, using our knowledge of plant biology, genetics, and breeding/statistics and specifically the genes that have been selected to create the crop we currently call corn, would we be able to re-domesticate a "new corn" from teosinte with the optimal alleles for our environmental conditions and agronomic practices?

## References

Aekatasanawan, C. (2001) Baby corn. In: Hallauer, A.R. (ed.) *Specialty Corns* (2nd edn.). CRC Press, Boca Raton, Florida, pp. 275–292.

Beadle, G.W. (1939) Teosinte and the origin of maize. *Journal of Heredity* 30, 245–247.

Beadle, G.W. (1972) The mystery of maize. *Field Museum of Natural History Bulletin* 43, 2–11.

Benz, B.F. (2001) Archaeological evidence of teosinte domestication from Guilá Naquitz, Oaxaca. *Proceedings of the National Academy of Sciences of the United States of America* 98, 2104–2106.

Boston, R.S. and Larkins, B.A. (2009) The genetics and biochemistry of maize zein storage proteins. In: Bennetzen, J.L. and Hake, S. (eds.) *Handbook of Maize: Genetics and Genomics*. Springer, New York, pp. 715–730.

Briggs, W.H., McMullen, M.D., Gaut, B.S. and Doebley, J. (2007) Linkage mapping of domestication loci in a large maize teosinte backcross resource. *Genetics* 177, 1915–1928.

Chai, Y., Hao, X., Yang, X., Allen, W.B., Li, J., *et al.* (2012) Validation of DGAT1-2 polymorphisms associated with oil content and development of functional markers for molecular breeding of high-oil maize. *Molecular Breeding* 29, 939–949.

Chia, J.-M., Song, C., Bradbury, P.J., Costich, D., de Leon, N., *et al.* (2012) Maize HapMap2 identifies extant variation from a genome in flux. *Nature Genetics* 40, 803–807.

Cook, J.P., McMullen, M.D., Holland, J.B., Tian, F., Bradbury, P., *et al.* (2012) Genetic architecture of maize kernel composition in the nested association mapping and inbred association panels. *Plant Physiology* 158, 824–834.

Darrah, L.L., Maddux, L.D., Hibbard, B.E., Wilmont, D.B., Lee, E.A., *et al.* (2002) White food corn 2002 performance tests. *Special Report 547*, USDA-ARS and Agricultural Experiment Station, University of Missouri-Columbia.

Doebley, J. (2004) The genetics of maize evolution. *Annual Review of Genetics* 38, 37–59.

Doebley, J. and Iltis, H.H. (1980) Taxonomy of *Zea* (Gramineae). I. A subgeneric classification with key to taxa. *American Journal of Botany* 67, 982–993.

Doebley, J. and Stec, A. (1993) Inheritance of the morphological differences between maize and teosinte: comparison of results for two $F_2$ populations. *Genetics* 134, 559–570.

Doebley, J., Stec, A., Wendel, J. and Edwards, M. (1990) Genetic and morphological analysis of a maize-teosinte $F_2$ population: implications for the origin of maize. *Proceedings of the National Academy of Sciences of the United States of America* 87, 9888–9892.

Doebley, J., Bacigalupo, A. and Stec, A. (1994) Inheritance of kernel weight in two maize-teosinte hybrid populations: implications for crop evolution. *Journal of Heredity* 85, 191–195.

Doebley, J., Stec, A. and Gustus, C. (1995) *Teosinte branched1* and the origin of maize: evidence for epistasis and the evolution of dominance. *Genetics* 141, 333–346.

Doebley, J., Stec, A. and Hubbard, L. (1997) The evolution of apical dominance in maize. *Nature* 386, 485–488.

Dorweiler, J., Stec, A., Kermicle, J. and Doebley, J. (1993) Teosinte glume architecture 1: a genetic locus controlling a key step in maize evolution. *Science* 262, 233–235.

Erwin, A.T. (1949) The origin and history of popcorn, *Zea mays* L. var. *indurata* (Sturt.) Bailey mut. *everta* (Sturt.) Erwin. *Agronomy Journal* 41, 53–56.

FAO (2016) Statistics at FAO. Available at: http://faostat.fao.org (accessed December 14, 2016).

Flint-Garcia, S.A., Bodnar, A.L. and Scott, M.P. (2009a) Wide variability in kernel composition, seed characteristics, and zein profiles among diverse maize inbreds, landraces, and teosinte. *Theoretical and Applied Genetics* 119, 1129–1142.

Flint-Garcia, S.A., Guill, K.E., Sanchez-Villeda, H., Schroeder, S.G. and McMullen, M.D. (2009b) Maize amino acid pathways maintain high levels of genetic diversity. *Maydica* 54, 375–386.

Fonseca, R.R.D., Smith, B.D., Wales, N., Cappellini, E., Skoglund, P., *et al.* (2015) The origin and evolution of maize in the southwestern United States. *Nature Plants* 1, 14003.

Fukunaga, K., Hill, J., Vigouroux, Y., Matsuoka, Y., Sanchez, G., *et al.* (2005) Genetic diversity and population structure of teosinte. *Genetics* 169, 2241–2254.

Gomez, M.H., McDonough, C.M., Rooney, L.W. and Waniska, R.D. (1989) Changes in corn and sorghum during nixtamalization and tortilla baking. *Journal of Food Science* 54, 330–336.

Goodman, M.M. and Brown, W.L. (1988) Races of corn. In: Sprague, G.F. and Dudley, J.W. (eds.) *Corn and Corn Improvement Agronomy No. 18, Third Edition*. ASA-CSSA-SSSA, Madison, Wisconsin, pp. 33–79.

Hammer, K. (1984) The domestication syndrome. *Die Kulturpflanze* 32, 11–34.

Hanson, M.A., Gaut, B.S., Stec, A.O., Fuerstenberg, S.I., Goodman, M.M., *et al.* (1996) Evolution of anthocyanin biosynthesis in maize kernels: the role of regulatory and enzymatic loci. *Genetics* 143, 1395–1407.

Hufford, M.B., Xu, X., van Heerwaarden, J., Pyhajarvi, T., Chia, J.-M., *et al.* (2012) Comparative population genomics of maize domestication and improvement. *Nature Genetics* 44, 808–811.

Iltis, H.H. (2000) Homeotic sexual translocations and the origin of maize (*Zea mays*, Poaceae): a new look at an old problem. *Economic Botany* 54, 7–42.

Jaenicke-Després, V., Buckler, E.S., Smith, B.D., Gilbert, M.T.P., Cooper, A., *et al.* (2003) Early allelic selection in maize as revealed by ancient DNA. *Science* 302, 1206–1208.

Lang, Z., Wills, D.M., Lemmon, Z.H., Shannon, L.M., Bukowski, R., *et al.* (2014) Defining the role of prolamin-box binding factor1 gene during maize domestication. *Journal of Heredity* 105, 576–582.

Laurie, C.C., Chasalow, S.D., LeDeaux, J.R., McCarroll, R., Bush, D., *et al.* (2004) The genetic architecture of response to long-term artificial selection for oil concentration in the maize kernel. *Genetics* 168, 2141–2155.

Lemmon, Z.H. and Doebley, J.F. (2014) Genetic dissection of a genomic region with pleiotropic effects on domestication traits in maize reveals multiple linked QTL. *Genetics* 198, 345–353.

Lennon, J.R., Krakowsky, M., Goodman, M., Flint-Garcia, S. and Balint-Kurti, P.J. (2016) Identification of alleles conferring resistance to gray leaf spot in maize derived from its wild progenitor species teosinte. *Crop Science* 56, 209–218.

Lennon, J.R., Krakowsky, M.D., Goodman, M., Flint-Garcia, S. and Balint-Kurti, P.J. (2017) Identification of teosinte (*Zea mays* ssp. parviglumis) alleles for resistance to southern leaf blight in near isogenic maize lines. *Crop Science* 57, 1973–1983. DOI:10.2135/cropsci2016.12.0979

Li, Q., Yang, X., Bai, G., Warburton, M.L., Mahuku, G., *et al.* (2010) Cloning and characterization of a putative *GS3* ortholog involved in maize kernel development. *Theoretical and Applied Genetics* 120, 753–763.

Lin, Z., Li, X., Shannon, L.M., Yeh, C.-T., Wang, M.L., *et al.* (2012) Parallel domestication of the Shattering1 genes in cereals. *Nature Genetics* 44, 720–724.

Liu, Z., Cook, J., Melia-Hancock, S., Guill, K., Bottoms, C., *et al.* (2016) Expanding maize genetic resources with predomestication alleles: maize–teosinte introgression populations. *The Plant Genome* 9. DOI: 10.3855/plantgenome2015.07.0053

Long, A.B., Benz, B.F., Donahue, D.J., Jull, A.J.T. and Toolin, L.J. (1989) First direct AMS dates on early maize from Tehuacán, Mexico. *Radiocarbon* 31, 1035–1040.

Matsuoka, Y., Vigouroux, Y., Goodman, M.M., Sanchez, G.J., Buckler, E., *et al.* (2002) A single domestication for maize shown by multilocus microsatellite genotyping. *Proceedings of the National Academy of Sciences of the United States of America* 99, 6080–6084.

Mir, C., Zerjal, T., Combes, V., Dumas, F., Madur, D., *et al.* (2013) Out of America: tracing the genetic footprints of the global diffusion of maize. *Theoretical and Applied Genetics* 126, 2671–2682.

Moro, G.L., Habben, J.E., Hamaker, B.R. and Larkins, B.A. (1996) Characterization of the variability in lysine content for normal and opaque2 maize endosperm. *Crop Science* 36, 1651–1659.

Nelson, O.E. (1952) Non-reciprocal cross-sterility in maize. *Genetics* 37, 101–124.

Olsen, K.M. and Wendel, J.F. (2013) A bountiful harvest: genomic insights into crop domestication phenotypes. *Annual Review of Plant Biology* 64, 47–70.

Palaisa, K.A., Morgante, M., Williams, M. and Rafalski, A. (2003) Contrasting effects of selection on sequence diversity and linkage disequilibrium at two phytoene synthase loci. *Plant Cell* 15, 1795–1806.

Piperno, D.R., Ranere, A.J., Holst, I., Iriarte, J. and Dickau, R. (2009) Starch grain and phytolith evidence for early ninth millennium B.P. maize from the Central Balsas River Valley, Mexico. *Proceedings of the National Academy of Sciences of the United States of America* 106, 5019–5024.

Poneleit C.G. (2001) Breeding white endosperm corn. In: Hallauer, A.R. (ed.) *Specialty Corns* (2nd edn.). CRC Press, Boca Raton, Florida, pp. 235–273.

Prasanna, B.M., Vasal, S.K., Kassahun, B. and Singh, N.N. (2001) Quality protein maize. *Current Science* 81, 1308–1319.

Romay, M.C., Millard, M.J., Glaubitz, J.C., Peiffer, J.A., Swarts, K.L., *et al.* (2013) Comprehensive genotyping of the USA national maize inbred seed bank. *Genome Biology* 14, R55.

Schmidt, R.J., Burr, F.A., Aukerman, M.J. and Burr, B. (1990) Maize regulatory gene *opaque-2* encodes a protein with a "leucine-zipper" motif that binds to zein DNA. *Proceedings of the National Academy of Sciences of the United States of America* 87, 46–50.

Schnable, J.C. and Freeling, M. (2011) Genes identified by visible mutant phenotypes show increased bias toward one of two subgenomes of maize. *PLOS ONE* 6, e17855.

Schultz, J.A. and Juvik, J.A. (2004) Current models for starch synthesis and the *sugary enhancer1* (*se1*) mutation in *Zea mays*. *Plant Physiology and Biochemistry* 42, 457–464.

Shull, G.H. (1909) A pure line method of corn breeding. *American Breeders Association Report* 5, 51–59.

Smalley, J. and Blake, M. (2003) Sweet beginnings: stalk sugar and the domestication of maize. *Current Anthropology* 44, 675–703.

Smith, B.D. (1989) Origins of agriculture in eastern North America. *Science* 246, 1566–1571.

Studer, A., Zhao, Q., Ross-Ibarra, J. and Doebley, J. (2011) Identification of a functional transposon insertion in the maize domestication gene *tb1*. *Nature Genetics* 43, 1160–1163.

Swarup, S., Timmermans, M.C., Chaudhuri, S. and Messing, J. (1995) Determinants of the high-methionine trait in wild and exotic germplasm may have escaped selection during early cultivation of maize. *Plant Journal* 8, 359–368.

Takano-Kai, N., Jiang, H., Kubo, T., Sweeney, M., Matsumoto, T., *et al.* (2009) Evolutionary history of *GS3*, a gene conferring grain length in rice. *Genetics* 182, 1323–1334.

Takuno, S., Ralph, P., Swarts, K., Elshire, R.J., Glaubitz, J.C., *et al.* (2015) Independent molecular basis of convergent highland adaptation in maize. *Genetics* 200, 1297–1312.

Tenaillon, M.I., Sawkins, M.C., Long, A.D., Gaut, R.L., Doebley, J.F., *et al.* (2001) Patterns of DNA sequence polymorphism along chromosome 1 of maize (*Zea mays* ssp. *mays* L.). *Proceedings of the National Academy of Sciences of the United States of America* 98, 9161–9166.

Tracy, W.F. (1994) Sweet corn. In: Hallauer, A.R. (ed.) *Specialty Corns*. CRC Press, Boca Raton, Florida, pp. 147–187.

Tracy, W.F., Whitt, S.R. and Buckler, E.S. (2006) Recurrent mutation and genome evolution: example of *Sugary1* and the origin of sweet maize. *Crop Science* 46, S49–S54.

Troyer, A.F. (1999) Background of U.S. hybrid corn. *Crop Science* 39, 601–626.

van Heerwaarden, J., Doebley, J., Briggs, W.H., Glaubitz, J.C., Goodman, M.M., *et al.* (2011) Genetic signals of origin, spread, and introgression in a large sample of maize landraces. *Proceedings of the National Academy of Sciences of the United States of America* 108, 1088–1092.

van Heerwaarden, J., Hufford, M.B. and Ross-Ibarra, J. (2012) Historical genomics of North American maize. *Proceedings of the National Academy of Sciences of the United States of America* 109, 12420–12425.

Vicente-Carbajosa, J., Moose, S.P., Parsons, R.L. and Schmidt, R.J. (1997) A maize zinc-finger protein binds the prolamin box in zein gene promoters and interacts with the basic leucine zipper transcriptional activator Opaque2. *Proceedings of the National Academy of Sciences of the United States of America* 94, 7685–7690.

Vigouroux, V., Glaubitz, J.C., Matsuoka, Y., Goodman, M.M., Sánchez, G., *et al.* (2008) Population structure and genetic diversity of New World maize races assessed by DNA microsatellites. *American Journal of Botany* 95, 1240–1253.

Wang, H., Nussbaum-Wagler, T., Li, B., Zhao, Q., Vigouroux, Y., *et al.* (2005) The origin of the naked grains of maize. *Nature* 436, 714–719.

Wang, H., Studer, A.J., Zhao, Q., Meeley, R. and Doebley, J.F. (2015) Evidence that the origin of naked kernels during maize domestication was caused by a single amino acid substitution in *tga1*. *Genetics* 200, 965–974.

Wang, L., Xu, C., Qu, M. and Zhang, J. (2008) Kernel amino acid composition and protein content of introgression lines from *Zea mays* ssp. *mexicana* into cultivated maize. *Journal of Cereal Science* 48, 387–393.

Weber, A.L., Zhao, Q., McMullen, M.D. and Doebley, J.F. (2009) Using association mapping in teosinte to investigate the function of maize selection-candidate genes. *PLOS ONE* 4, e8227.

Wellhausen, E.J., Roberts, L.M. and Hernandez, X.E. (1952) *Races of Maize in Mexico.* Bussey Institute, Harvard University, Cambridge, Massachusetts.

Whipple, C.J., Kebrom, T.H., Weber, A.L., Yang, F., Hall, D., *et al.* (2011) *grassy tillers1* promotes apical dominance in maize and responds to shade signals in the grasses. *Proceedings of the National Academy of Sciences of the United States of America* 108, E506–E512.

Whitt, S.R., Wilson, L.M., Tenaillon, M.I., Gaut, B.S. and Buckler, E.S. (2002) Genetic diversity and selection in the maize starch pathway. *Proceedings of the National Academy of Sciences of the United States of America* 99, 12959–12962.

Wills, D.M., Whipple, C.J., Takuno, S., Kursel, L.E., Shannon, L.M., *et al.* (2013) From many, one: genetic control of prolificacy during maize domestication. *PLOS Genetics* 9, e1003604.

Wilson, L.M., Whitt, S.R., Ibáñez, A.M., Rocheford, T.R., Goodman, M.M., *et al.* (2004) Dissection of maize kernel composition and starch production by candidate gene association. *Plant Cell* 16, 2719–2733.

Wright, S.I., Vroh Bi, I., Schroeder, S.G., Yamasaki, M., Doebley, J.F., *et al.* (2005) The effects of artificial selection on the maize genome. *Science* 308, 1310–1314.

Yamasaki, M., Tenaillon, M.I., Vroh Bi, I., Schroeder, S.G., Sanchez-Villeda, H., *et al.* (2005) A large-scale screen for artificial selection in maize identifies candidate agronomic loci for domestication and crop improvement. *Plant Cell* 17, 2859–2872.

Zheng, P., Allen, W.B., Roesler, K., Williams, M.E., Zhang, S., *et al.* (2008) A phenylalanine in DGAT is a key determinant of oil content and composition in maize. *Nature Genetics* 40, 367–372.

Ziegler, K.E. (1994) Popcorn. In: Hallauer, A.R. (ed.) *Specialty Corns.* CRC Press, Boca Raton, Florida, pp. 189–223.

# 2 Gametophyte Interactions Establishing Maize Kernel Development


Erik Vollbrecht[1] and Matthew M.S. Evans[2,*]

[1]*Department of Genetics, Development and Cell Biology, Iowa State University, USA;*
[2]*Department of Plant Biology, Carnegie Institution for Science, Stanford, California, USA*


## 2.1 Introduction

This chapter focuses on tissue- and cell-level interactions required to set in motion foundational processes that lead to and promote maize kernel development. After pollination, key cell biological, genetic, and epigenetic interactions occur, including those between the male gametophyte and the pistil, between the male and female gametophytes, and between the female gametophyte and the other seed tissues, ultimately leading to successful fertilization and initiation of kernel development (Fig. 2.1). The unicellular pollen tube germinates and grows through the transmitting tract of the silk until it reaches the ovule. It is guided by chemical cues to the ovule's micropyle and the female gametophyte's synergid cell. Upon interaction with the synergid, the pollen tube penetrates it and ruptures, releasing two sperm cells. Double fertilization occurs subsequently via a short series of fusion events. First, one sperm cell fuses with the egg cell and the other fuses with the central cell. Then, the sperm nuclei fuse with the corresponding egg and central cell nuclei to initiate development of the embryo and endosperm, respectively. Some gametophytic cells persist during seed development after fertilization, providing an opportunity for interactions between gametophytic cells and the developing embryo and endosperm.

## 2.2 Interactions between the Pollen Tube and the Silk

### 2.2.1 Pollen tube growth and guidance in the silk

Pollination occurs when a pollen grain lands on a silk and germinates a pollen tube that must navigate tissues of the pistil to reach the embryo sac and achieve fertilization. In maize, the pollen tube first grows along a silk hair, penetrates the silk, and then grows within one of the two transmitting tracts of the silk (Fig. 2.2). The maize silk can be quite long, requiring a pollen tube path of up to 40 cm; multiple pollen grains typically land on each silk, yet ultimately only one pollen grain fertilizes the embryo sac. The competition between pollen grains imposes strong selection for growth. Growth rates as high as 1cm/hr have been reported (Barnabas and Fridvalszky, 1984), and this must occur through the existing


*Corresponding author e-mail: mevans@carnegiescience.edu

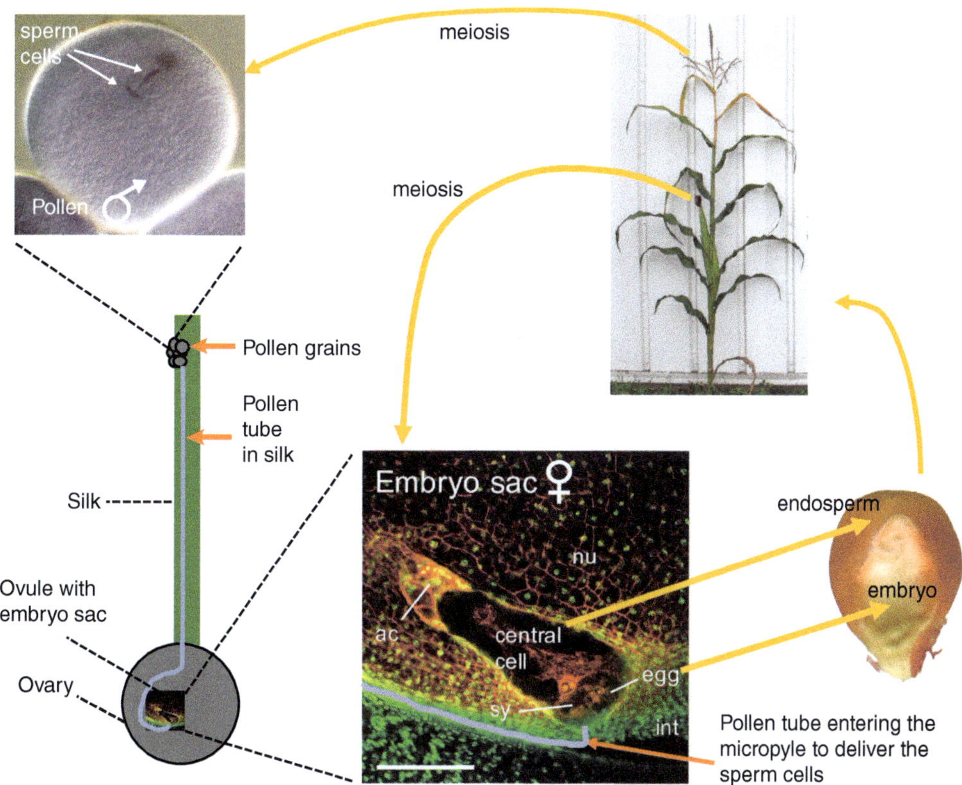



**Fig. 2.1.** Maize male (pollen grain) and female (embryo sac) gametophytes in the context of the maize life cycle. Pollen was stained with hematoxylin according to Kindiger (1994) and embryo sacs were stained with Acriflavine and Periodic Acid according to Chettoor and Evans (2015). A potential pollen tube entry path is drawn in the embryo sac image. ac, antipodal cells; int, integument; nu, nucellus; sy, synergid; Scale bar in pollen = 40µm. Scale bar in embryo sac = 100µm.

cell walls of the transmitting tract. Consequently, in addition to cellular activities required for pollen tube tip growth, such as actin dynamics, vesicle trafficking, and calcium signaling (Estruch *et al.*, 1994; Gibbon *et al.*, 1999; Kovar *et al.*, 2000; Safadi *et al.*, 2000; Arthur *et al.*, 2003; Cole and Fowler, 2006; Xu and Dooner, 2006; Iwano *et al.*, 2009), growing pollen tubes need additional functions to interact with and penetrate the female tissues *in vivo*. The pollen grain produces, or carries in its coat from secretions of the anther's tapetal cells, proteins that are necessary for this process, some of which have been identified by biochemical and genetic means. These proteins have been shown to have potent cell wall loosening activity (Cosgrove *et al.*, 1997). A pollen coat xylanase, for example, is required for penetration of the silk but not for *in vitro* pollen germination (Suen and Huang, 2007). Mutations in the *Expb1* gene, encoding the most abundant isoform of the major pollen allergen Zea m1, reduce pollen tube growth *in vivo* but not *in vitro* (Valdivia *et al.*, 2007).

In delivering sperm cells for fertilization, pollen tube elongation follows directional chemical cues leading to the ovule. A gradient of a chemoattractant(s) is proposed to be responsible for pollen tube guidance through the transmitting tract, and some attractants functioning in the style have been identified in both monocots and dicots. These include chemocyanin, a small basic cell wall protein in lily, a related plantacyanin in Arabidopsis, a small cysteine-rich

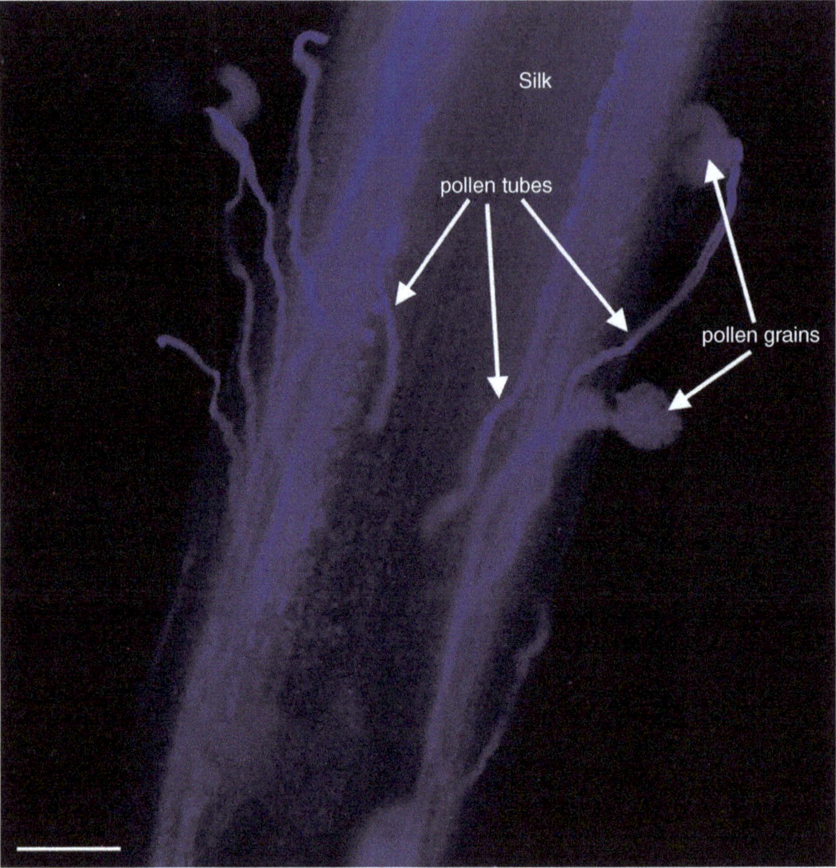


**Fig. 2.2.** Pollinated maize silk showing pollen tubes that have entered and are growing through the transmitting tracts of the silk. Pollen tubes were stained for callose using Aniline Blue and visualized according to Lu *et al.* (2014). Scale bar = 100μm.

peptide (CRP) in lily, and TTS, an arabino-galactan protein in *Nicotiana* (Wu *et al.*, 2000; Kim *et al.*, 2003; Dong *et al.*, 2005). Compounds that direct pollen tube guidance in the style of maize and other grass species have not been identified. Since plantacyanin proteins appear to be involved in pollen tube guidance in lily and Arabidopsis, it raises the possibility that a related protein performs the same function in maize. These directional cues are not limited to the transmitting tract of the maize silk, since incompatible pollen tubes that leave the transmitting tract still grow generally downward toward the base of the silk and the ovule (Lu *et al.*, 2014). Endocytosis is critical for pollen penetration of the pistil in Arabidopsis (Hao *et al.*,

2016), and the S-RNase expressed in the style of *Solanum* is taken up by pollen tubes and causes pollen rejection (Luu *et al.*, 2000). This raises the possibility that pollen uptake of stylar proteins may be a more general mechanism of pollen–pistil interactions.

Additionally, in some species, such as *Arabidopsis thaliana* and *Torenia fournieri*, interaction of the pollen tube with the style elicits changes in the pollen tube that confer the ability to later target the ovule micropyle and discharge the two sperm cells (Higashiyama *et al.*, 1998; Palanivelu and Preuss, 2006; Qin *et al.*, 2009). In Arabidopsis, these changes include increased mRNA levels of the MYB transcription factors,

*MYB97*, *MYB101*, and *MYB120*, which are required for sperm cell release (Leydon *et al.*, 2013). Whether or not pollen–pistil interactions are required to activate later pollen tube functions in maize has not been shown.

### 2.2.2 Pollen tube acceptance or rejection by the silk

In *Zea mays*, genetic cross-incompatibility, which also influences mate selection, occurs during interactions between the male gametophyte and the sporophyte style. Cross-incompatibility (CI) at any of three loci, when present, prevents hybrid formation and promotes reproductive isolation between different *Zea mays* populations. By contrast, reproductive isolation by hybrid sterility/seed abortion as occurs in many species, including rice (Ouyang and Zhang, 2013), has not been reported between *Zea mays* lines of matched ploidy. The three CI systems are the so-called *Gametophyte factors* (*Ga*): *Ga1* and *Ga2*, and *Teosinte crossing barrier1* (*Tcb1*) (Mangelsdorf and Jones, 1926; Nelson, 1994; Evans and Kermicle, 2001; Kermicle and Evans, 2010; Dresselhaus *et al.*, 2011) (see Chapter 1, this volume). These CI systems have two functions: a female function that produces a barrier to non-self-type pollen in the silk, and a male function that enables pollen of the appropriate genotype to overcome that barrier. These two functions, a priori, could have been encoded by a single gene or two separate genes; in each system, a single locus confers both functions, but the locus structures have not been elucidated. *Ga1* on chromosome 4 was the first characterized (Mangelsdorf and Jones, 1926) and has been intensively studied (Mangelsdorf and Jones, 1926; Lausser *et al.*, 2010; Bloom and Holland, 2011; Zhang *et al.*, 2012). Plants carrying the *Ga1-strong* (*Ga1-s*) haplotype have male and female functions. *Ga1-s* females reject *ga1* pollen but accept *Ga1-s* pollen. The same nomenclature system is used for the *ga2* and the *tcb1* systems— plants carrying male and female functions

of *ga2* and *tcb1* are termed *Ga2-s* and *Tcb1-s*, respectively. All three crossing barrier systems have naturally occurring haplotypes that carry only the male function and lack the female function. These haplotypes are termed *Ga1-male (Ga1-m)*, *Ga2-m* and *Tcb1-m* (e.g. *Tcb1-m* can fertilize *Tcb1-s* females but accepts *tcb1* pollen) (Kermicle, 2006; Kermicle *et al.*, 2006; Kermicle and Evans, 2010). Both *Ga1-s* and *Ga2-s* were originally identified in domesticated maize lines, while *Tcb1-s* (and the male-only *Tcb1-m*) has only been found in teosinte (or partially teosinte) lines. Pollen tube growth arrest is the mechanism for the pollen rejection conferred by the *Gametophyte factor1-s* (*Ga1-s*), *Ga2-s*, and *Tcb1-s* systems (House and Nelson, 1958; Lausser *et al.*, 2010; Dresselhaus *et al.*, 2011; Zhang *et al.*, 2012; Lu *et al.*, 2014). Thus, *tcb1* pollen tubes germinate successfully on *Tcb1*-s (or *Ga1*-s or *Ga2*-s) silks but grow more slowly than compatible pollen tubes and arrest before reaching the ovule. The molecular identity of these factors is unknown, despite their practical utility for maize breeding. Most temperate maize lines lack the CI haplotypes and so have no restriction to interbreeding. An exception to this is the presence of *Ga1-s* in popcorn varieties, which prevents pollination by non-popcorn varieties, thus maintaining popcorn lines as distinct and separate from other maize lines.

## 2.3 Interactions between the Pollen Tube and the Ovule/Embryo Sac

### 2.3.1 Pollen tube attraction to the micropyle and the embryo sac

When the pollen tube reaches the bottom of the silk and nears the ovule, it is guided to the micropyle by cues from diploid ovule tissue and the haploid female gametophyte (Hulskamp *et al.*, 1995; Baker *et al.*, 1997; Ray *et al.*, 1997). The mature, unfertilized maize female gametophyte, or embryo sac, consists of two synergids near the micropyle, one egg cell, a large central cell with two partially fused polar nuclei, and a cluster of antipodal cells at the chalazal end (Fig. 2.1). It is surrounded by the diploid ovule tissues of the nucellus and the two integuments. In Arabidopsis, GABA (gamma-aminobutyric acid) is involved in signaling from the diploid tissues to the pollen tube, while nitric oxide produced by ovary tissues acts as a pollen tube repellent to reorient pollen tube growth (Palanivelu *et al.*, 2003; Prado *et al.*, 2006). Laser ablation studies in *Torenia* showed the source of the embryo sac signals to be the synergid cell (Higashiyama *et al.*, 2001). Disruption of synergid development in the Arabidopsis *myb98* mutant demonstrates that the synergids are required for pollen tube attraction in Arabidopsis as well (Kasahara *et al.*, 2005). The Arabidopsis *central cell guidance* mutant demonstrates the central cell can also influence pollen tube attraction (Chen *et al.*, 2007), although this could be via indirect effects on the synergid. The first embryo-sac-expressed pollen tube attractants discovered were the DEFENSIN-related cysteine-rich proteins, LURE1 and 2 of *Torenia* (Okuda *et al.*, 2009). Many DEFENSIN/LURE (DEFL) proteins are produced by the female gametophyte of maize (Chettoor *et al.*, 2014). However, none of the maize proteins have been shown to act as pollen tube attractants. Maize and Arabidopsis GEX3 is expressed in sperm and egg cells, and downregulation of Arabidopsis GEX3 in the female disrupts pollen tube guidance to the micropyle (Alandete-Saez *et al.*, 2008). In maize, the only pollen tube attractant identified to date, ZMEGGAPPARATUS1 (EA1), is unrelated to the LURE family and is produced by the synergids and the egg cell, instead of just the synergid (Marton *et al.*, 2005). EA1 protein interacts directly with the pollen tube tip in a species-specific manner to attract the pollen tube (Marton *et al.*, 2012; Uebler *et al.*, 2013). Little is known about the receptors on the pollen tube for these attractants. In Arabidopsis, the Receptor-Like Cytoplasmic Kinases, LOST IN POLLEN TUBE GUIDANCE1 (LIP1) and LIP2, are required for response to AtLURE1 (Liu *et al.*, 2013).

### 2.3.2 Pollen tube reception

Upon encountering the synergid, the pollen tube ceases growing and ruptures to release the two sperm cells inside the synergid. Factors expressed in the female and male gametophyte are required for this interactive process. Little is currently known about the proteins involved in this process in maize, and this presents exciting opportunities given that several important molecular players have been identified in Arabidopsis. The synergid-expressed FERONIA/SIRENE receptor-like kinase (RLK), GPI-anchored LORELEI, and MLO-like NORTIA proteins are required for pollen tube growth arrest and sperm release in Arabidopsis (Huck *et al.*, 2003; Rotman *et al.*, 2003; Escobar-Restrepo *et al.*, 2007; Capron *et al.*, 2008; Kessler *et al.*, 2010; Liu *et al.*, 2016). Interestingly, pollen tubes also fail to arrest in interspecific pollinations in some taxa (Muller *et al.*, 2016), indicating a role for these factors in compatibility. On the male side the FER/SIR related proteins, ANXUR1 and ANXUR2, prevent rupture of the pollen tube until it reaches the egg apparatus (Boisson-Dernier *et al.*, 2009; Miyazaki *et al.*, 2009). Three maize RLKs similar to FERONIA/SIRENE and ANXUR1 and 2, increase expression after entry of a pollen tube into the embryo sac, while one is downregulated (Wang *et al.*, 2014), providing hints of potentially conserved functions. Pollen tube overgrowth also occurs when both the male and female gametophytes of Arabidopsis lack the peroxin, ABSTINENCE BY MUTUAL CONSENT (Boisson-Dernier *et al.*, 2008). Interestingly, many of the components involved in reception of the pollen tube by the female gametophyte have shared functions or homologs with functions in disease responses.

In maize, the DEFENSIN-like protein ZmES4, expressed in the female gametophyte, is required for pollen tube growth arrest and burst, reminiscent of activities of FER/SIR and LRE in Arabidopsis (Amien *et al.*, 2010). ZmES4 protein is found in the synergids, egg, and central cell, with the highest concentration in the filiform apparatus that is located at the micropylar end of the synergids. Furthermore, ZmES4 acts in a species-dependent manner to open the K⁺ channel KZM1 in the pollen tube. ZmES4 peptides can be released after unknown pollen–embryo sac interactions and activate KZM1, although direct interaction of KZM1 and ZmES4 has not been demonstrated. Maize Pectin Methylesterase Inhibitor 1 (ZmPMEI1), which is expressed in male and female gametophytes, also promotes pollen tube rupture (Woriedh *et al.*, 2013). After penetration of the synergid by the pollen tube, the synergid fluoresces much more intensely following fixation with formaldehyde, indicating broad chemical changes in the degenerating synergid (Fig. 2.3A,B). Interestingly, the penetrated synergid also persists next to the developing embryo for several days. RNA-seq analysis of maize ovules before and after fertilization identified 221 transcripts expressed exclusively after pollen tube entry, with signaling, RNA binding and transcription, and lipid metabolism among the most common GO terms (Wang *et al.*, 2014).

## 2.4 Gamete Interactions, Fusion, and Initiation of Development

The processes of gamete interaction and the initiation of seed development present a research frontier in maize, in that currently we know little about the mechanisms involved. First, the gamete cells must fuse (egg cell with sperm cell, central cell with sperm cell) in a process termed plasmogamy, then each set of colocalized gamete nuclei must fuse in the process of karyogamy. In Arabidopsis, release of the sperm cells from the pollen tube causes exocytosis of EGG CELL1 (EC1) protein from the egg cell into the region of the degenerated synergid where the sperm cells have been released (Sprunck *et al.*, 2012). EC1 then induces redistribution of the gamete fusogen, HAPLESS2/GENERATIVE CELL SPECIFIC1 (HAP2/GCS1). Gamete attachment and fusion (plasmogamy) requires the transmembrane protein GEX2 as well as HAP2/GCS1 (Mori *et al.*, 2006, 2014; Mori *et al.*, 2014). HAP2 function is

conserved in *Chlamydomonas* and so is likely required for gamete fusion in all higher plants (Liu *et al.*, 2008), although it has not been studied in maize. After plasmogamy, the F-actin network is required to bring the sperm and egg cell nuclei together for karyogamy (Kawashima *et al.*, 2014). In maize, karyogamy does not depend upon the position of the polar nuclei within the central cell, as *baseless1* mutants with abnormal polar nuclei placement are fertilized and produce seed as frequently as wild-type (Gutiérrez-Marcos *et al.*, 2006). Gametophyte-encoded functions may also occur in the initial stages of development after fertilization is complete. In Arabidopsis, delivery of *SHORT SUSPENSOR* (*SSP*) transcripts by the sperm cell is proposed to activate the YODA MAP kinase cascade to regulate the first asymmetric division of the zygote (Bayer *et al.*, 2009). Whether or not a similar mechanism is required for setting up polarity in the maize embryo has yet to be determined.

## 2.5 Post-fertilization Gametophyte Interactions

In maize, cells of the female gametophyte persist after fertilization, presenting the possibility of direct interaction between the products of fertilization and the remnants of the haploid phase. Notably, the antipodal cells persist at the crown of the endosperm long after fertilization (Fig. 2.3C) (Weatherwax, 1926; Randolph, 1936). Given that the antipodal cells are a region of auxin signaling before fertilization (Chettoor and Evans, 2015), they could act as a signaling center after fertilization as well, and influence endosperm development. For example, a continued source of signaling molecules could provide positional information for endosperm growth and development. Ablation of the antipodal cells, either genetically or by manipulation, would help to determine what, if any, role they have after fertilization. This hypothesis has not yet been tested in maize or Arabidopsis, which also has persistent antipodal cells (Song *et al.*, 2014). Similarly, the

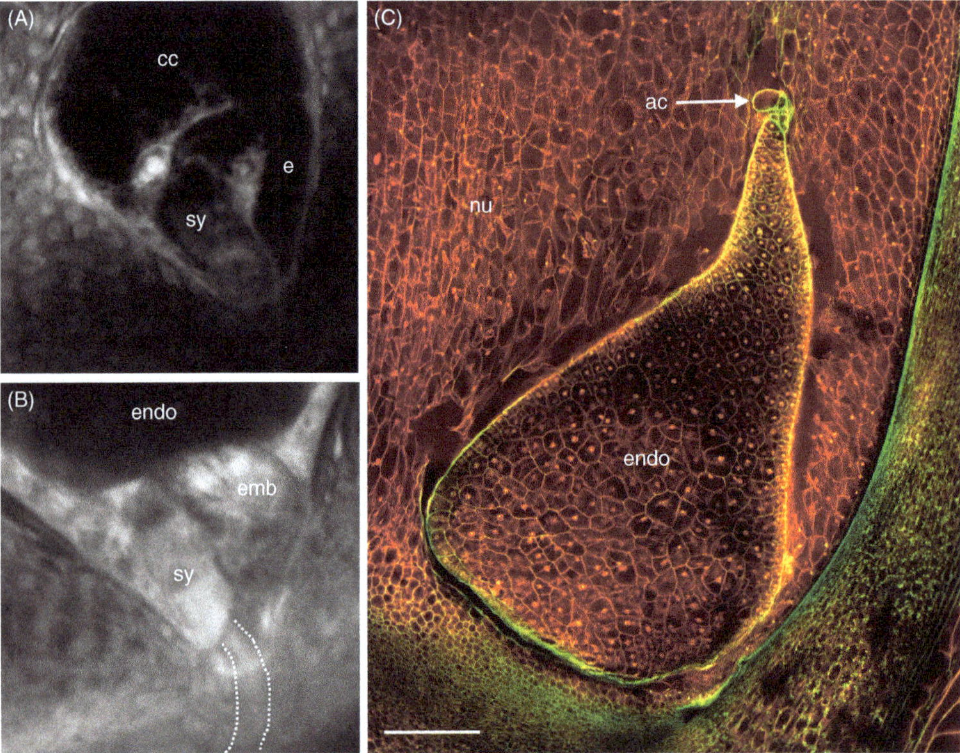



**Fig. 2.3.** Gametophyte changes and persistence after fertilization. (A) Micropylar end of a mature, unfertilized embryo sac. cc, central cell; e, egg cell; sy, synergid. (B) Micropylar end of a developing seed 4 days after pollination. emb, embryo; endo, endosperm; pt, pollen tube that has entered the micropyle marked with dashed lines. (C) Developing maize endosperm 5 days after pollination. The antipodal cell cluster (ac) is still intact at the crown of the developing endosperm (endo). The embryo is out of the plane of focus for this image. Most of the nucellus (nu) is still intact at this stage. Scale bar = 50µm in (A) and (B); 200µm in (C).

remnants of the synergid that was penetrated by the pollen tube persist next to the developing embryo for at least 5 days after pollination, at which stage it is as large as the embryo itself. Thus, signals from this synergid remnant can exist, which could influence development of the adjacent embryo.

Gametophytes also play key roles in programming the epigenetic states of egg, central and sperm cell nuclei, and thus in establishing the epigenetic states present in the endosperm and embryo after karyogamy. Work from other species, including Arabidopsis, indicates that on a global scale, epigenetic reprogramming plays out broadly in regulating the status of transposable element (TE) activity. In Arabidopsis, within the gametophyte cells that have "nurse cell" functions, i.e. the pollen vegetative cell and the central cell, TEs become hypomethylated and their expression activated (Slotkin *et al.*, 2009; Pillot *et al.*, 2010; Ibarra *et al.*, 2012; Schmid *et al.*, 2012; Schmidt *et al.*, 2012). This expression is proposed to produce small-RNA information that each nurse cell type loads into the corresponding, adjacent gamete(s) to impose genome-wide silencing of TE activity and thereby protect the next generation against mutagenic effects. For pollen, a mechanistic basis involving small interfering RNAs has been demonstrated (Martínez *et al.*, 2016). TE silencing could be especially critical in maize, where the genome is far richer in both TE types and sheer numbers of TEs. The genome consists of more than 85% transposable elements, most of which are transcriptionally silent in the sporophyte (Schnable *et al.*, 2009). Interestingly, TE transcripts accumulate to higher levels in both gametophytes prior to fertilization than in diploid plant tissues (Chettoor *et al.*, 2014). Future work will be needed to determine in which cell types and developmental stages this TE expression occurs, and how it relates to epigenetic programming after fertilization. Once TE expression associated with the female gametophyte has been resolved to particular cell types and stages, including examination of the surrounding diploid nucellus (e.g. by reporter gene and transcriptomics approaches), the stage will be set to elucidate the relationship of expression of different TE family types to the timing and location of their silencing.

## 2.6  Future Prospects

Many aspects of maize gametophyte interactions are currently unresolved. Some of these can be addressed with currently available technologies, such as RNA-seq, while others could require increased sensitivity in imaging or proteomics technologies. For example, there are a multitude of research opportunities related to elucidating interactions between the pollen grain/tube and the silk. The directional cues guiding the pollen tube through the transmitting tract are currently unknown, and analysis of the changes that occur in the pollen tube (transcriptomic, proteomic, etc.) during its interactions with the silk is in its infancy. Loci involved in the decisions to accept or reject a pollen grain are known, but the underlying genes have not been identified. The biochemical nature of tube growth arrest in rejected pollen is also unknown. We have begun to understand the molecular mechanisms of interaction between the male and female gametophytes at the time of pollen tube arrival and fertilization in maize, such as pollen tube guidance by EA1 and pollen tube rupture by ZmES4, but the story is far from complete. For example, what are the signaling cascades inside the pollen tube and synergid during their interactions? Some fertilization-related pathways will perhaps be inferred when work done in other species is applied to maize, but it is equally likely that some of these pathways and mechanisms will not apply to maize at all. Other, more general research questions also remain. In the female gametophyte, there is no experimental evidence for the function of the antipodal cells of maize, or in any other species for that matter. Whether they primarily function before or after fertilization is not even known at this point. Many of the details of how the female gametophyte sets up the epigenetic landscape of the egg in preparation for the next generation are also unknown. Transcripts that correspond to many maize TEs are expressed in or near the female gametophyte. They are presumed to assist in reducing TE activity in the next generation, but that hypothesis has not yet been tested. It is also not known which cells express these TE messages, (e.g. the egg cell itself, the central cell, the antipodal cells, or the nucellar cells adjacent to the embryo sac). New insights will come from dynamic models integrating the full transcriptomes of these cells, the identification of the proteins and small molecules involved in their interactions, their locations, and the visualization of these interactions in real time.

### Acknowledgment

EV and MMSE are funded jointly by the U.S. National Science Foundation, award number 1340050.

## References

Alandete-Saez, M., Ron, M. and McCormick, S. (2008) GEX3, expressed in the male gametophyte and in the egg cell of *Arabidopsis thaliana*, is essential for micropylar pollen tube guidance and plays a role during early embryogenesis. *Molecular Plant* 1, 586–598.

Amien, S., Kiwer, I., Marton, M.L., Debener, T., Geiger, D., Becker, D. and Dresselhaus, T. (2010) Defensin-like ZmES4 mediates pollen tube burst in maize via opening of the potassium channel KZM1. *PLOS Biology* 8, e1000388.

Arthur, K.M., Vejlupkova, Z., Meeley, R.B. and Fowler, J.E. (2003) Maize ROP2 GTPase provides a competitive advantage to the male gametophyte. *Genetics* 165, 2137–2151.

Baker, S.C., Robinson-Beers, K., Villanueva, J.M., Gaiser, J.C. and Gasser, C.S. (1997) Interactions among genes regulating ovule development in *Arabidopsis thaliana*. *Genetics* 145, 1109–1124.

Barnabas, B. and Fridvalsky, L. (1984) Adhesion and germination of differently treated maize pollen grains on the stigma. *Acta Botanica Hungarica* 30, 329–332.

Bayer, M., Nawy, T., Giglione, C., Galli, M., Meinnel, T. and Lukowitz, W. (2009) Paternal control of embryonic patterning in *Arabidopsis thaliana*. *Science* 323, 1485–1488.

Bloom, J.C. and Holland, J.B. (2011) Genomic localization of the maize cross-incompatibility gene, Gametophyte factor 1 (ga1). *Maydica* 56, 379–387.

Boisson-Dernier, A., Frietsch, S., Kim, T.H., Dizon, M.B. and Schroeder, J.I. (2008) The peroxin loss-of-function mutation abstinence by mutual consent disrupts male-female gametophyte recognition. *Current Biology* 18, 63–68.

Boisson-Dernier, A., Roy, S., Kritsas, K., Grobei, M.A., Jaciubek, M., Schroeder, J.I. and Grossniklaus, U. (2009) Disruption of the pollen-expressed FERONIA homologs ANXUR1 and ANXUR2 triggers pollen tube discharge. *Development* 136, 3279–3288.

Capron, A., Gourgues, M., Neiva, L.S., Faure, J.E., Berger, F., *et al.* (2008) Maternal control of male-gamete delivery in *Arabidopsis* involves a putative GPI-anchored protein encoded by the *LORELEI* gene. *Plant Cell* 20, 3038–3049.

Chen, Y.H., Li, H.J., Shi, D.Q., Yuan, L., Liu, J., *et al.* (2007) The central cell plays a critical role in pollen tube guidance in *Arabidopsis*. *Plant Cell* 19, 3563–3677.

Chettoor, A.M. and Evans, M.M.S. (2015) Correlation between a loss of auxin signaling and a loss of proliferation in maize antipodal cells. *Frontiers in Plant Science* 6, 187.

Chettoor, A.M., Givan, S.A., Cole, R.A., Coker, C.T., Unger-Wallace, E., *et al.* (2014) Discovery of novel transcripts and gametophytic functions via RNA-seq analysis of maize gametophytic transcriptomes. *Genome Biology* 15, 414.

Cole, R.A. and Fowler, J.E. (2006) Polarized growth: maintaining focus on the tip. *Current Opinion in Plant Biology* 9, 579–588.

Cosgrove, D.J., Bedinger, P. and Durachko, D.M. (1997) Group I allergens of grass pollen as cell wall-loosening agents. *Proceedings of the National Academy of Science of the United States of America* 94, 6559–6564.

Dong, J., Kim, S.T. and Lord, E.M. (2005) Plantacyanin plays a role in reproduction in Arabidopsis. *Plant Physiology* 138, 778–789.

Dresselhaus, T., Lausser, A. and Marton, M.L. (2011) Using maize as a model to study pollen tube growth and guidance, cross-incompatibility and sperm delivery in grasses. *Annals of Botany* 108, 727–737.

Escobar-Restrepo, J.M., Huck, N., Kessler, S., Gagliardini, V., Gheyselinck, J., Yang, W.C. and Grossniklaus, U. (2007) The FERONIA receptor-like kinase mediates male-female interactions during pollen tube reception. *Science* 317, 656–660.

Estruch, J.J., Kadwell, S., Merlin, E. and Crossland, L. (1994) Cloning and characterization of a maize pollen-specific calcium-dependent calmodulin-independent protein kinase. *Proceedings of the National Academy of Sciences of the United States of America* 91, 8837–8841.

Evans, M.M.S. and Kermicle, J.L. (2001) Teosinte crossing barrier1, a locus governing hybridization of teosinte with maize. *Theoretical and Applied Genetics* 103, 259–265.

Gibbon, B.C., Kovar, D.R. and Staiger, C.J. (1999) Latrunculin B has different effects on pollen germination and tube growth. *Plant Cell* 11, 2349–2363.

Gutiérrez-Marcos, J.F., Costa, L.M. and Evans, M.M.S. (2006) Maternal gametophytic baseless1 is required for development of the central cell and early endosperm patterning in maize (*Zea mays*). *Genetics* 174, 317–329.

Hao, L., Liu, J., Zhong, S., Gu, H. and Qu, L.J. (2016) AtVPS41-mediated endocytic pathway is essential for pollen tube–stigma interaction in *Arabidopsis*. *Proceedings of the National Academy of Sciences of the United States of America* 113, 6307–6312.

Higashiyama, T., Kuroiwa, H., Kawano, S. and Kuroiwa, T. (1998) Guidance *in vitro* of the pollen tube to the naked embryo sac of *Torenia fournieri*. *Plant Cell* 10, 2019–2032.

Higashiyama, T., Yabe, S., Sasaki, N., Nishimura, Y., Miyagishima, S., Kuroiwa, H. and Kuroiwa, T. (2001) Pollen tube attraction by the synergid cell. *Science* 293, 1480–1483.

House, L.R. and Nelson, O.E. (1958) Tracer study of pollen-tube growth in cross-sterile maize. *Journal of Heredity* 49, 18–21.

Huck, N., Moore, J.M., Federer, M. and Grossniklaus, U. (2003) The *Arabidopsis* mutant *feronia* disrupts the female gametophytic control of pollen tube reception. *Development* 130, 2149–2159.

Hulskamp, M., Schneitz, K. and Pruitt, R.E. (1995) Genetic evidence for a long-range activity that directs pollen tube guidance in Arabidopsis. *Plant Cell* 7, 57–64.

Ibarra, C.A., Feng, X., Schoft, V.K., Hsieh, T.F., Uzawa, R., *et al.* (2012) Active DNA demethylation in plant companion cells reinforces transposon methylation in gametes. *Science* 337, 1360–1364.

Iwano, M., Entani, T., Shiba, H., Kakita, M., Nagai, T., *et al.* (2009) Fine-tuning of the cytoplasmic $Ca^{2+}$ concentration is essential for pollen tube growth. *Plant Physiology* 150, 1322–1334.

Kasahara, R.D., Portereiko, M.F., Sandaklie-Nikolova, L., Rabiger, D.S. and Drews, G.N. (2005) *MYB98* is required for pollen tube guidance and synergid cell differentiation in *Arabidopsis*. *Plant Cell* 17, 2981–2992.

Kawashima, T., Maruyama, D., Shagirov, M., Li, J., Hamamura, Y., *et al.* (2014) Dynamic F-actin movement is essential for fertilization in *Arabidopsis thaliana*. *eLife* 3, e04501.

Kermicle, J.L. (2006) A selfish gene governing pollen-pistil compatibility confers reproductive isolation between maize relatives. *Genetics* 172, 499–506.

Kermicle, J.L. and Evans, M.M.S. (2010) The *Zea mays* sexual compatibility gene *ga2*: naturally occurring alleles, their distribution, and role in reproductive isolation. *Journal of Heredity* 101, 737–749.

Kermicle, J.L., Taba, S. and Evans, M.M.S. (2006) The gametophyte-1 locus and reproductive isolation among *Zea mays* subspecies. *Maydica* 51, 219–225.

Kessler, S.A., Shimosato-Asano, H., Keinath, N.F., Wuest, S.E., Ingram, G., Panstruga, R. and Grossniklaus, U. (2010) Conserved molecular components for pollen tube reception and fungal invasion. *Science* 330, 968–971.

Kim, S., Mollet, J.C., Dong, J., Zhang, K., Park, S.Y. and Lord, E.M. (2003) Chemocyanin, a small basic protein from the lily stigma, induces pollen tube chemotropism. *Proceedings of the National Academy of Sciences of the United States of America* 100, 16125–16130.

Kindiger, B. (1994) A staining procedure for pollen grain chromosomes of maize. In: Freeling, M. and Walbot, V. (eds.) *The Maize Handbook*. Springer, New York, pp. 476–480.

Kovar, D.R., Drobak, B.K. and Staiger, C.J. (2000) Maize profilin isoforms are functionally distinct. *Plant Cell* 12, 583–598.

Lausser, A., Kliwer, I., Srilunchang, K.O. and Dresselhaus, T. (2010) Sporophytic control of pollen tube growth and guidance in maize. *Journal of Experimental Botany* 61, 673–682.

Leydon, A.R., Beale, K.M., Woroniecka, K., Castner, E., Chen, J., *et al.* (2013) Three MYB transcription factors control pollen tube differentiation required for sperm release. *Current Biology* 23, 1209–1214.

Liu, J., Zhong, S., Guo, X., Hao, L., Wei, X., *et al.* (2013) Membrane-bound RLCKs LIP1 and LIP2 are essential male factors controlling male–female attraction in *Arabidopsis*. *Current Biology* 23, 993–998.

Liu, X., Castro, C., Wang, Y., Noble, J., Ponvert, N., *et al.* (2016) The role of LORELEI in pollen tube reception at the interface of the synergid cell and pollen tube requires the modified eight-cysteine motif and the receptor-like kinase FERONIA. *Plant Cell* 28, 1035–1052.

Liu, Y., Tewari, R., Ning, J., Blagborough, A.M., Garbom, S., *et al.* (2008) The conserved plant sterility gene HAP2 functions after attachment of fusogenic membranes in *Chlamydomonas* and *Plasmodium* gametes. *Genes & Development* 22, 1051–1068.

Lu, Y., Kermicle, J.L. and Evans, M.M. (2014) Genetic and cellular analysis of cross-incompatibility in *Zea mays*. *Plant Reproduction* 27, 19–29.

Luu, D.T., Qin, X., Morse, D. and Cappadocia, M. (2000) S-RNase uptake by compatible pollen tubes in gametophytic self-incompatibility. *Nature* 407, 649–651.

Mangelsdorf, P.C. and Jones, D.F. (1926) The expression of mendelian factors in the gametophyte of maize. *Genetics* 11, 423–455.

Martínez, G., Panda, K., Kohler, C. and Slotkin, R.K. (2016) Silencing in sperm cells is directed by RNA movement from the surrounding nurse cell. *Nature Plants* 2, 16030.

Marton, M.L., Cordts, S., Broadhvest, J. and Dresselhaus, T. (2005) Micropylar pollen tube guidance by egg apparatus 1 of maize. *Science* 307, 573–576.

Marton, M.L., Fastner, A., Uebler, S. and Dresselhaus, T. (2012) Overcoming hybridization barriers by the secretion of the maize pollen tube attractant ZmEA1 from *Arabidopsis* ovules. *Current Biology* 22, 1194–1198.

Miyazaki, S., Murata, T., Sakurai-Ozato, N., Kubo, M., Demura, T., Fukuda, H. and Hasebe, M. (2009) ANXUR1 and 2, sister genes to FERONIA/SIRENE, are male factors for coordinated fertilization. *Current Biology* 19, 1327–1331.

Mori, T., Kuroiwa, H., Higashiyama, T. and Kuroiwa, T. (2006) GENERATIVE CELL SPECIFIC 1 is essential for angiosperm fertilization. *Nature Cell Biology* 8, 64–71.

Mori, T., Igawa, T., Tamiya, G., Miyagishima, S.Y. and Berger, F. (2014) Gamete attachment requires GEX2 for successful fertilization in *Arabidopsis. Current Biology* 24, 170–175.

Muller, L.M., Lindner, H., Pires, N.D., Gagliardini, V. and Grossniklaus, U. (2016) A subunit of the oligosaccharyltransferase complex is required for interspecific gametophyte recognition in *Arabidopsis. Nature Communications* 7, 10826.

Nelson, O.E. (1994) The gametophyte factors of maize. In: Freeling, M. and Walbot, V. (eds.) *The Maize Handbook.* Springer, New York, pp. 298–302.

Okuda, S., Tsutsui, H., Shiina, K., Sprunck, S., Takeuchi, H., *et al.* (2009) Defensin-like polypeptide LUREs are pollen tube attractants secreted from synergid cells. *Nature* 458, 357–361.

Ouyang, Y. and Zhang, Q. (2013) Understanding reproductive isolation based on the rice model. *Annual Review of Plant Biology* 64, 111–135.

Palanivelu, R. and Preuss, D. (2006) Distinct short-range ovule signals attract or repel *Arabidopsis thaliana* pollen tubes *in vitro. BMC Plant Biology* 6, 7.

Palanivelu, R., Brass, L., Edlund, A.F. and Preuss, D. (2003) Pollen tube growth and guidance is regulated by *POP2*, an *Arabidopsis* gene that controls GABA levels. *Cell* 114, 47–59.

Pillot, M., Baroux, C., Vazquez, M.A., Autran, D., Leblanc, O., *et al.* (2010) Embryo and endosperm inherit distinct chromatin and transcriptional states from the female gametes in *Arabidopsis. Plant Cell* 22, 307–320.

Prado, A.M., Porterfield, D.M. and Feijo, J.A. (2006) Nitric oxide is involved in growth regulation and re-orientation of pollen tubes. *Development* 131, 2707–2714.

Qin, Y., Leydon, A.R., Manziello, A., Pandey, R., Mount, D., *et al.* (2009) Penetration of the stigma and style elicits a novel transcriptome in pollen tubes, pointing to genes critical for growth in a pistil. *PLOS Genetics* 5, e1000621.

Randolph, L.F. (1936) Developmental morphology of the caryopsis in maize. *Journal of Agricultural Research* 53, 881–916.

Ray, S.M., Park, S.S. and Ray, A. (1997) Pollen tube guidance by the female gametophyte. *Development* 124, 2489–2498.

Rotman, N., Rozier, F., Boavida, L., Dumas, C., Berger, F. and Faure, J.E. (2003) Female control of male gamete delivery during fertilization in *Arabidopsis thaliana. Current Biology* 13, 432–436.

Safadi, F., Reddy, V.S. and Reddy, A.S. (2000) A pollen-specific novel calmodulin-binding protein with tetratricopeptide repeats. *Journal of Biological Chemistry* 275, 35457–35470.

Schmid, M.W., Schmidt, A., Klostermeier, U.C., Barann, M., Rosenstiel, P. and Grossniklaus, U. (2012) A powerful method for transcriptional profiling of specific cell types in eukaryotes: laser-assisted microdissection and RNA sequencing. *PLOS ONE* 7, e29685.

Schmidt, A., Schmid, M.W. and Grossniklaus, U. (2012) Analysis of plant germline development by high-throughput RNA profiling: technical advances and new insights. *Plant Journal* 70, 18–29.

Schnable, P.S., Ware, D., Fulton, R.S., Stein, J.C., Wei, F., *et al.* (2009) The B73 maize genome: complexity, diversity, and dynamics. *Science* 326, 1112–1115.

Slotkin, R.K., Vaughn, M., Borges, F., Tanurdzic, M., Becker, J.D., Feijo, J.A. and Martienssen, R.A. (2009) Epigenetic reprogramming and small RNA silencing of transposable elements in pollen. *Cell* 136, 461–472.

Song, X., Yuan, L. and Sundaresan, V. (2014) Antipodal cells persist through fertilization in the female gametophyte of *Arabidopsis. Plant Reproduction* 27, 197–203.

Sprunck, S., Rademacher, S., Vogler, F., Gheyselinck, J., Grossniklaus, U. and Dresselhaus, T. (2012) Egg cell-secreted EC1 triggers sperm cell activation during double fertilization. *Science* 338, 1093–1097.

Suen, D.F. and Huang, A.H. (2007) Maize pollen coat xylanase facilitates pollen tube penetration into silk during sexual reproduction. *Journal of Biological Chemistry* 282, 625–636.

Uebler, S., Dresselhaus, T. and Marton, M. (2013) Species-specific interaction of EA1 with the maize pollen tube apex. *Plant Signaling and Behavior* 8, e25682–1.

Valdivia, E.R., Wu, Y., Li, L.C., Cosgrove, D.J. and Stephenson, A.G. (2007) A group-1 grass pollen allergen influences the outcome of pollen competition in maize. *PLOS ONE* 2, e154.

Wang, S.S., Wang, F., Tan, S.J., Wang, M.X., Sui, N. and Zhang, X.S. (2014) Transcript profiles of maize embryo sacs and preliminary identification of genes involved in the embryo sac–pollen tube interaction. *Frontiers in Plant Science* 5, 702.

Weatherwax, P. (1926) Persistence of the antipodal tissue in the development of the seed of maize. *Bulletin of the Torrey Botanical Club* 53, 381–384.

Woriedh, M., Wolf, S., Marton, M.L., Hinze, A., Gahrtz, M., Becker, D. and Dresselhaus, T. (2013) External application of gametophyte-specific ZmPMEI1 induces pollen tube burst in maize. *Plant Reproduction* 26, 255–266.

Wu, H.M., Wong, E., Ogdahl, J. and Cheung, A.Y. (2000) A pollen tube growth-promoting arabinogalactan protein from *Nicotiana alata* is similar to the tobacco TTS protein. *Plant Journal* 22, 165–176.

Xu, Z. and Dooner, H.K. (2006) The maize aberrant pollen transmission 1 gene is a SABRE/KIP homolog required for pollen tube growth. *Genetics* 172, 1251–1261.

Zhang, H., Liu, X., Zhang, Y., Jiang, C., Cui, D., *et al.* (2012) Genetic analysis and fine mapping of the Ga1-S gene region conferring cross-incompatibility in maize. *Theoretical and Applied Genetics* 124, 459–465.

# 3 Endosperm Development and Cell Specialization

Junpeng Zhan[1], Joanne M. Dannenhoffer[2] and Ramin Yadegari[1,*]

[1]*School of Plant Sciences, University of Arizona, USA;* [2]*Department of Biology, Central Michigan University, Michigan, USA*

## 3.1 Introduction

The endosperm of angiosperms is a seed structure that provides nutrients and signals for embryo development and seedling germination (Li and Berger, 2012; Olsen and Becraft, 2013). In cereal crops, it occupies the largest portion of the mature grain, contains large amounts of storage compounds including primarily carbohydrates and storage proteins, and is an important source of biofuel (Lopes and Larkins, 1993; Sabelli and Larkins, 2009; FAO, 2015). Because of its value and relatively large size, maize endosperm has become a model system for studies of endosperm development.

Angiosperm seed development is initiated by a double fertilization during which one of two sperm cells fuses with the egg cell within the female gametophyte (embryo sac) to produce the diploid embryo (1 maternal:1 paternal) and the other fertilizes the central cell to form the triploid endosperm (2 maternal:1 paternal) (Friedman *et al.*, 2008; Hamamura *et al.*, 2012). Subsequently, in maize endosperm the nuclei undergo proliferation, creating a coenocyte that becomes cellularized and then differentiates into at least seven recognizable cell types: the basal endosperm transfer layer (BETL);

aleurone (AL); embryo-surrounding region (ESR); central starchy endosperm (CSE); subaleurone (SA); conducting zone (CZ); and basal intermediate zone (BIZ). Concurrent with cell differentiation, the endosperm undergoes two major phases of mitotic proliferation, an early period that lasts until 8–12 days after pollination (DAP) in the central region, and a late period that continues until 20–25 DAP in the outer cell layers (AL and SA). Starting around 8–10 DAP, cells in the central portion of the endosperm gradually switch from a mitotic cell cycle to endoreduplication and become filled with starch and storage proteins. They eventually undergo maturation and desiccation (Sabelli and Larkins, 2009; Becraft and Gutiérrez-Marcos, 2012). These developmental events correspond to three important physiological periods: (i) a lag period (approximately the first 2 weeks after pollination); (ii) a grain-filling period (from approximately 2 weeks after pollination until 6–7 weeks after pollination); and (iii) a final period during which grain filling ceases and the kernel matures (Johnson and Tanner, 1972; Jones *et al.*, 1985). Although most of the kernel dry weight is gained during the grain-filling period, the preceding lag period is a critical formative

*Corresponding author e-mail: yadegari@email.arizona.edu

 

phase during which kernel sink strength and storage capacity are established through cell division and plastid proliferation (Reddy and Daynard, 1983; Jones *et al.*, 1985, 1996). Moreover, the length of the lag period has recently been shown to correlate positively with kernel size (Sekhon *et al.*, 2014). This chapter provides an overview of the early period of maize endosperm development (i.e. the lag period), with an emphasis on our current knowledge of molecular mechanisms that regulate endosperm cell differentiation.

## 3.2 Coenocyte Formation and Cellularization

Upon fertilization, the central cell (primary endosperm cell) nucleus undergoes multiple rounds of division without cytokinesis, forming a multinucleate coenocyte. The coenocyte consists of a thin layer of cytoplasm surrounding a large central vacuole, and it fills the majority of the volume of the embryo sac. As coenocytic divisions proceed synchronously, the nuclei spread along the periphery of the central vacuole from the micropylar end (near the embryo) toward the antipodal end of the embryo sac (Randolph, 1936; Monjardino *et al.*, 2007; Leroux *et al.*, 2014). In maize endosperm, coenocytic nuclear proliferation takes place within about 3 DAP and results in a cell containing 128 to 512 nuclei (Fig. 3.1A), without a significant increase in the size of the unfertilized embryo sac (Randolph, 1936; Kiesselbach, 1999; Leroux *et al.*, 2014). Immunohistochemical staining of microtubules in barley and Arabidopsis show the nuclei are evenly spaced within the coenocyte by internuclear radial microtubule systems (RMSs) that emerge around nuclear-cytoplasmic domains (NCDs) (Brown *et al.*, 1994, 1999). The uncoupling of mitosis and cytokinesis contrasts with what occurs in somatic cells, where mitotic divisions involve formation of a phragmoplast that directs cell plate formation between the daughter cells (Jurgens, 2005). Likely due to this difference, nuclear divisions in the coenocytic endosperm proceed faster than in the embryo; by the time the zygote undergoes its first cell division, endosperm nuclei have already divided two to three times (Randolph, 1936). Therefore, coenocyte formation could be an evolutionary strategy to rapidly populate the central cell with nuclei and support rapid mitotic proliferation afterward (Sabelli and Larkins, 2009). In wheat and barley, a short-lived phragmoplast forms between dividing sister nuclei of the coenocytic endosperm, and cell plates are transiently deposited in wheat, suggesting phragmoplast function is suppressed after it is initiated in the coenocyte (Brown *et al.*, 1994; Tian *et al.*, 1998). Therefore, the absence of cell membrane and cell wall synthesis during coenocytic development could be due to suppression of phragmoplast formation.

Starting about 2.5 to 4 DAP, the coenocyte cellularizes within approximately one day in two consecutive but distinguishable phases: alveolation and partitioning (Monjardino *et al.*, 2007; Leroux *et al.*, 2014). Alveolation begins with formation of vesicles that deliver wall materials to the space between NCDs to form anticlinal cell walls separating peripheral nuclei. This results in tube-like alveoli that are closed at the end facing the coenocyte wall and open toward the central vacuole (Fig. 3.1B). Subsequently, nuclei in the alveoli divide periclinally to generate a second layer of nuclei that are displaced inward and separated by deposition of a periclinal wall thus forming an outer layer of cells and an inner second layer of alveoli (Brown *et al.*, 1994, 1999; Monjardino *et al.*, 2007; Leroux *et al.*, 2014). The process of alveolation in maize proceeds repeatedly until up to four layers of cells are formed, and is followed by random cellular partitioning of the central vacuole (Leroux *et al.*, 2014). Cellularization through this later partitioning appears to be a mechanism unique to maize, as in other cereals cellularization is completed by alveolation (Brown *et al.*, 1994, 1996a,b; Leroux *et al.*, 2014).

Evidence from many angiosperm model systems suggests that timing of the transition from coenocytic to cellular proliferation is a key decision in endosperm development,

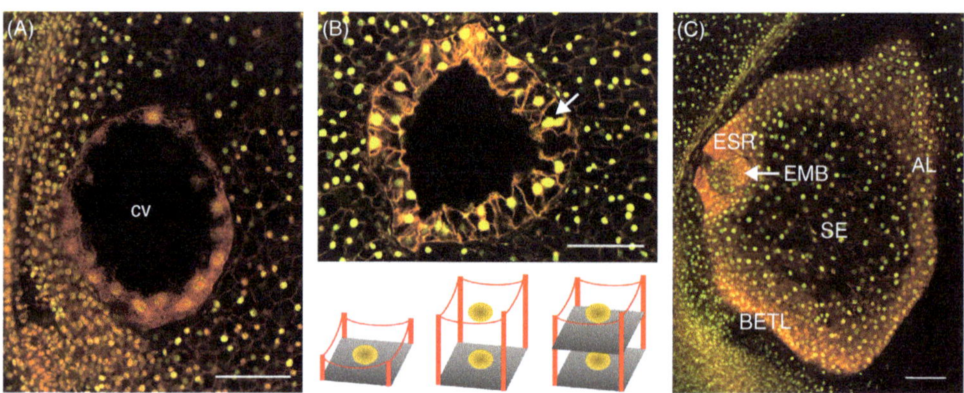



**Fig. 3.1.** Early proliferation of maize endosperm. Confocal micrographs of kernels at stages of coenocytic proliferation (A), cellularization (alveolation phase) (B), and beginning of cell differentiation upon cellularization (C). The arrow in (B) indicates a nucleus undergoing periclinal division. Abbreviations: AL, aleurone; BETL, basal endosperm transfer layer; CV, central vacuole; EMB, embryo; ESR, embryo-surrounding region; SE, starchy endosperm. Scale bars = 100 µm.

as it has a high correlation with endosperm/seed size (Li and Berger, 2012; Orozco-Arroyo *et al.*, 2015; Gehring and Satyaki, 2017). For example, in maize and Arabidopsis, seeds with an excess of the maternal genome generally show precocious cellularization and reduced endosperm size, while seeds with paternal genomic excess generally exhibit delayed cellularization and increased endosperm size (Cooper, 1951; Scott *et al.*, 1998; von Wangenheim and Peterson, 2004; Pennington *et al.*, 2008). Similar correlation between cellularization timing and endosperm size was also observed in interspecific crosses within multiple genera (Bushell *et al.*, 2003; Ishikawa *et al.*, 2011; Rebernig *et al.*, 2015; Garner *et al.*, 2016; Oneal *et al.*, 2016). In nearly all these crosses, with the *Oryza* interspecific crosses being the only exception, failure in endosperm cellularization causes seed lethality, and hence establishes interploidy or interspecific hybridization barriers. These phenomena have been interpreted using the parental conflict theory (also known as the kinship theory) of genomic imprinting (Haig and Westoby, 1989; Bushell *et al.*, 2003; Haig, 2014).

Genomic imprinting is the allele-biased expression of genes in a parent-of-origin-dependent manner (Abramowitz and Bartolomei, 2012; Peters, 2014). In plants, imprinting is observed predominantly in the endosperm

(Gehring, 2013; Rodrigues and Zilberman, 2015). The parental conflict theory hypothesizes that in situations where a mother contributes nutritional resources to offspring of multiple different fathers, the paternally inherited genes foster uptake of as many nutrients as possible to increase fitness of the fathers-derived progeny, whereas the maternal genes tend to evenly distribute nutrition to all offspring (Haig and Westoby, 1989; Haig, 2014). According to this theory, maternally-expressed imprinted genes (MEGs) tend to limit endosperm growth (size), while paternally-expressed genes (PEGs) tend to promote it. An extensive set of studies in Arabidopsis have implicated chromatin-regulatory mechanisms, including many histone modification- and DNA methylation-related processes in regulation of cellularization and seed size (Li and Berger, 2012; Gehring and Satyaki, 2017). Mutations in components of the Fertilization-Independent Seed (FIS)-Polycomb Repressive Complex 2 (PRC2), a putative H3K27 methyltransferase, have been shown to result in endosperm cellularization failure, increased coenocytic proliferation, and ultimately seed abortion (Li and Berger, 2012; Gehring and Satyaki, 2017). The gene networks regulated by the FIS-PRC2 complex during early endosperm development are beginning to be deciphered functionally. Among its direct targets is the type I MADS-box family transcription-factor (TF) gene *AGAMOUS-LIKE62 (AGL62)*; mutations in *AGL62* result in precocious endosperm cellularization and reduced seed size (Kang *et al.*, 2008; Hehenberger *et al.*, 2012). Arabidopsis DNA METHYLTRANSFERASE1 (MET1) is responsible for maintaining DNA (CpG) methylation (Finnegan and Dennis, 1993). Global hypomethylation of the paternal genome in anti-sense or loss-of-function *met1* mutants delayed endosperm cellularization and increased seed size, while hypomethylation of the maternal genome has opposite effects (Adams *et al.*, 2000; Xiao *et al.*, 2006). These observations support a model in which genomic imprinting contributes to regulation of endosperm cellularization, and establishment of interploidy/ interspecific hybridization barriers. This concept has recently been reinforced by the observations that paternal inheritance of mutations in some PEGs (*ADMETOS, SUVH7, PEG2* and *PEG9*) can lead to partial normalization of the timing of endosperm cellularization in interploidy hybridizations (Kradolfer *et al.*, 2013; Wolff *et al.*, 2015).

Genetic studies in Arabidopsis have also uncovered a maternal or sporophytic contribution to the regulation of endosperm cellularization and seed size based on the analysis of the HAIKU (IKU) pathway genes (*IKU1, IKU2, MINISEED3,* and *SHORT HYPOCOTYL UNDER BLUE1*), and the AP2 family TF gene *APETALA2 (AP2)* (Li and Berger, 2012; Orozco-Arroyo *et al.*, 2015). Loss of function of individual IKU pathway genes results in precocious endosperm cellularization and reduced seed size, while loss of function of *AP2* results in delayed endosperm cellularization and increased seed size (Orozco-Arroyo *et al.*, 2015). How these sporophytic genes regulate endosperm development remains unclear; it is likely some signaling mechanisms mediate interactions between the three major seed components, the seed coat, the embryo, and the endosperm, with the latter perhaps acting as a central hub for regulatory activities that coordinate interactions with the other two components (Berger *et al.*, 2006; Orozco-Arroyo *et al.*, 2015). Available data on the genetic and epigenetic players in Arabidopsis

support a key role for cellularization in proper endosperm development—and by extension in proper development of the whole seed. How this key step occurs and the nature of the gene networks that regulate the transition from coenocytic to cellular proliferation remain to be determined. In maize, the nature of such regulatory processes is even less well understood. Genes encoding a number of chromatin-modifying enzymes, including MET1 and two FIS-PRC2-component-related proteins, have been identified (Steward *et al.*, 2000; Danilevskaya *et al.*, 2003; Haun *et al.*, 2007); however, their roles, if any, in regulation of endosperm cellularization are elusive.

### 3.3 Pattern Formation

Endosperm patterning proceeds in two main phases. The first establishes the micropylar and chalazal domains of the endosperm along the micropylar-chalazal (MC) axis, while the second results in differentiated cell types that are arranged according to radial symmetry (Costa *et al.*, 2004; Li and Berger, 2012). The first phase is reflected by the reported micropylar/embryo-surrounding-region-specific gene expression patterns in the endosperm of Arabidopsis and maize (Opsahl-Ferstad *et al.*, 1997; Bonello *et al.*, 2000; Tanaka *et al.*, 2001; Baud *et al.*, 2005; Ingouff *et al.*, 2005; Yang *et al.*, 2008), and the distinct mitotic domains within the coenocytic endosperm of Arabidopsis (Boisnard-Lorig *et al.*, 2001; Brown *et al.*, 2003). Mutations in genes encoding components of Arabidopsis FIS-PRC2 display female-gametophytic defects, and lead to ectopic chalazal endosperm development (Sorensen *et al.*, 2001; Guitton *et al.*, 2004), suggesting female-gametophytic or maternally-encoded functions are required for establishing the early MC endosperm axis. In maize, the maternal gametophytic mutation *baseless1 (bsl1)* that disrupts central cell polarity also alters the spatial expression patterns of BETL marker genes in both coenocytic and cellularized endosperm (Gutiérrez-Marcos *et al.*, 2006), suggesting that the regulatory program controlling basal endosperm patterning is already active, at least partially, in the unfertilized central cell.

Two cytokinin biosynthetic genes [*ISO-PENTENYL TRANSFERASE-4* (*IPT-4*) and *IPT-8*] were identified as predominantly expressed in the chalazal region of the coenocytic endosperm of Arabidopsis, indicating a potential role of polarized cytokinin localization in MC axis patterning (Li *et al.*, 2013). Interestingly, cytokinin has also been shown to be critical for the establishment of the MC axis that pre-exists in the female gametophyte (Tekleyohans *et al.*, 2017). Whether this axis influences formation of the corresponding MC axis in the endosperm is unknown; however, the extensive similarities between the female gametophyte and endosperm developmental patterns, including the sequence of coenocytic profileration and cellularization, suggest both female gametophyte development and endosperm development are regulated by an ancient genetic program (Olsen *et al.*, 1999). Therefore, elucidation of the role of cytokinin is critical to understanding endosperm pattern formation and could provide insight into its evolutionary history.

## 3.4 Cell Fate Specification and Differentiation

The second phase of endosperm patterning involves cell fate specification of epidermal and inner cells, which is described below in conjunction with other cell differentiation events. Starting around 4 DAP, maize endosperm differentiates into four specialized compartments or cell types that become cytologically distinguishable by around 6 DAP, namely the BETL, AL, SE, and ESR (Fig. 3.1C and Fig. 3.2A). These main cell types then become further differentiated (by about 8 DAP), leading to the emergence of three additional recognizable cell types: the BIZ, CZ, and SA (Fig. 3.2B). Although some of these individual cell types have not been characterized fully or associated with any specialized biological function, each of the

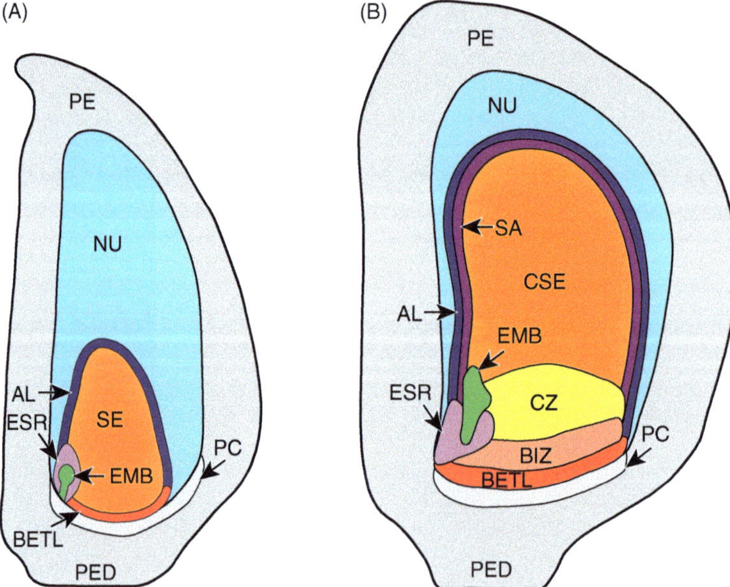


**Fig. 3.2.** Cell differentiation in maize endosperm. Schematic diagrams of the maize kernel showing the relative position of endosperm cell types and other kernel compartments at early (A) and late (B) differentiation phases. Abbreviations: AL, aleurone; BETL, basal endosperm transfer layer; CSE, central starchy endosperm; CZ, conducting zone; EMB, embryo; ESR, embryo-surrounding region; NU, nucellus; PC, placento-chalaza; PE, pericarp; PED, pedicel; SE, starchy endosperm.

resulting seven cell types occupies a specific territory within the early endosperm and possesses a unique set of cytological characteristics and gene expression programs (Leroux *et al.*, 2014; Zhan *et al.*, 2015). The BETL and AL are discussed in detail in other chapters (Chapters 5 and 6). Here we focus primarily on cell differentiation events associated with the main cell types in the context of their individual functions (if known) and their overall contribution to the development of the endosperm as a storage compartment.

The BETL contains a single layer of cells that is located at the base of the endosperm in direct contact with the maternal placento-chalazal zone (PC) (Fig. 3.2). The main function of the BETL is transport of nutrients from maternal tissue into the SE (Gunning and Pate, 1969; Pate and Gunning, 1972; Shannon *et al.*, 1986). BETL cell fate is likely specified during coenocytic proliferation of the endosperm. In barley, mRNA of the BETL marker gene, *END1*, is localized at the basal region of the coenocytic endosperm (Doan *et al.*, 1996). Consistent with this, the maize *END1* ortholog, named *BETL-9*, also exhibits a basal endosperm-specific mRNA accumulation pattern in the coenocytic and cellularized endosperm (Gutiérrez-Marcos *et al.*, 2006; Royo *et al.*, 2014). The phenotype of the maize *globby-1 (glo-1)* mutant supports the hypothesis that BETL cell fate is already specified during coenocytic proliferation (Costa *et al.*, 2003). In this mutant, BETL differentiation is disrupted to variable extents and expression of BETL marker genes is reduced at the coenocytic phase. Interestingly, the *glo-1* mutant also shows abnormal endosperm cellularization in the basal region, suggesting the *GLO-1* gene plays a broad role in both early cell proliferation and differentiation of the endosperm (Costa *et al.*, 2003).

Extensive evidence indicates BETL differentiation is regulated genetically and epigenetically. A detailed set of studies recently sought to uncover the gene regulatory networks (GRNs) of the BETL, with a particular focus on MYB-Related Protein-1 (MRP-1), a MYB-related TF family member that plays a key role in BETL differentiation (Yuan *et al.*,

2016; Doll *et al.*, 2017; see Chapter 5, this volume). MRP-1 directly regulates numerous BETL-expressed genes, including the *MATERNALLY EXPRESSED GENE-1 (MEG-1)* gene that encodes a small cysteine-rich peptide (Zhan *et al.*, 2015). The *MEG-1* gene itself was shown to be necessary and sufficient for regulation of BETL differentiation (Costa *et al.*, 2012). *MEG-1* is an imprinted gene that shows a maternally-biased expression pattern in endosperm around 4 to 6 DAP (Gutiérrez-Marcos *et al.*, 2004), strongly suggesting BETL differentiation is also epigenetically regulated. In support of this notion, the 2 maternal:1 paternal genomic ratio was shown to be critical for BETL differentiation, as both maternal and paternal genomic excess disrupt the process to varying extents (Charlton *et al.*, 1995; Pennington *et al.*, 2008). Nonetheless, the nature of MRP-1- and MEG-1-mediated regulation of BETL differentiation remains to be determined.

BIZ and CZ are two endosperm cell types located immediately internal to the BETL and differentiate later than BETL (Fig. 3.2B). The CZ contains highly elongated cells (about 3.5 times longer than wide) with tapering end walls, sparse cytoplasm, and large nuclei (Leroux *et al.*, 2014). The CZ cells are believed to transport nutrient solutes throughout the endosperm (Becraft, 2001), but different from the BETL, the CZ cells lack cell wall ingrowths and distinct vacuoles (Leroux *et al.*, 2014). The BIZ contains 2–4 layers of cells that sit between the BETL and the CZ (Leroux *et al.*, 2014). These cells are often grouped as part of the BETL (Sabelli and Larkins, 2009; Olsen and Becraft, 2013; Yuan *et al.*, 2016), but increasing evidence suggests that the BIZ cells constitute a unique cell type in the endosperm. The BIZ cells show intermediate, but largely different, characteristics from the BETL and CZ in terms of cell elongation, the extent of cell wall ingrowth, cytoplasmic density, and nuclear size, supporting their distinction as a unique cell type (Monjardino *et al.*, 2013; Leroux *et al.*, 2014). Accordingly, recent mRNA *in situ* hybridization assays detected mRNAs localized exclusively in the BETL or BIZ, as well as mRNAs that are preferentially localized in both BETL and BIZ at 6 to 8 DAP (Li *et al.*, 2014). Because the differentiation of CZ and BIZ appears to begin later than in the four main cell types, and because they are related to the main cell types either clonally or functionally, the CZ and BIZ can be viewed as specialized sub-regions of the main cell types. The delayed timing of their differentiation relative to the BETL could be related to a need for the ever-increasing rate of movement of photo-assimilates through the developing BETL for utilization and storage in the inner endosperm cells, including the SE.

The AL is the peripheral layer of cells that covers the entire surface of the endosperm, except the BETL and ESR regions (Fig. 3.2). During endosperm development, the AL stores proteins, lipids, non-starch carbohydrates, and mineral nutrients. During seed germination, the primary role of the AL is production of hydrolytic enzymes to utilize storage proteins, nucleic acids, and carbohydrates stored in the SE (Becraft, 2007; Becraft and Yi, 2011). As such, AL is the only endosperm cell type that remains alive when the other compartments of the endosperm mature and undergo programmed cell death (Young *et al.*, 1997; Young and Gallie, 2000). The AL and BETL constitute a continuous cell layer, but have distinct morphological and cytological characteristics, as described in Chapters 5 and 6 of this volume. The cell fate of AL also seems to be controlled by a regulatory program independent from that for the BETL, because endosperm in the maize *defective kernel1* (*dek1*) mutant lacks an AL but has a normal BETL (Becraft *et al.*, 2002; Lid *et al.*, 2002).

A series of genetic experiments showed the AL and SE share a common cell lineage, and they exhibit interchangeable cell fates regulated in response to positional cues (Becraft and Asuncion-Crabb, 2000; Becraft *et al.*, 2002; Lid *et al.*, 2002). These observations indicate that AL and SE are relatively plastic and maintain their differentiated states by sensing and responding to environmental cues throughout development. A number of genes are known to be involved in regulation of AL cell fate and/or differentiation (Chapter 6). These genes and the corresponding mutants are beginning to provide a clearer picture of AL differentiation and its relation to the underlying SA and SE cells.

The SE is the major endosperm cell type that accumulates starch, DNA and storage proteins. It occupies the largest, central portion of the mature grain and represents the bulk of the endosperm mass. The SE cells are derived by differentiation of the centrally localized cells formed after cellularization and periclinal divisions of the AL/SA cells (Morrison *et al.*, 1975; Lending and Larkins, 1989; Becraft and Asuncion-Crabb, 2000). Late in differentiation, as the BIZ, CZ, and SA cell types (here considered sub-regions of SE) become visible, the central region of the SE can be further delineated and is termed CSE (Fig. 3.2B). This region contains variably sized cells and nuclei that increase in size from near the SA toward the center (Leroux *et al.*, 2014). The adjacent SA cells are filled by large vacuoles, mitochondria, and proplastids, but the cytoplasm is slightly less dense than that of the AL cells, and only a small number of protein bodies and starch grains are present (Khoo and Wolf, 1970; Lending and Larkins, 1989; Leroux *et al.*, 2014).

Due to a lack of mutants that specifically alter SE cell fate, little is known about how it is controlled. Given the reversibility in cell fates of SE and AL, as discussed above, mutants with interchangeable SE to AL phenotypes would be valuable for understanding the molecular mechanisms controlling SE cell fate. Such mutants possibly exist in uncharacterized *dek* or *empty pericarp* mutant collections (Neuffer and Sheridan, 1980; Sheridan and Neuffer, 1980; Scanlon *et al.*, 1994). By contrast, more is known about regulation of the storage programs of the SE and AL, and in recent years this knowledge has begun to shed light on SE/AL differentiation. Expression of storage-protein genes has been detected in both SE and AL at the mRNA and protein levels (Reyes *et al.*, 2011). The bZIP family TF protein, Opaque-2 (O2), and the DOF family TF, Prolamin Box-binding Factor (PBF), are two of the major regulators of the storage program (Kawakatsu and Takaiwa, 2010; Thompson and Verdier, 2012; Chapter 14, this volume). The *naked endosperm (nkd)* mutant and *DOF3* RNA interference (RNAi) knockdown lines that exhibit defects in AL cell fate and differentiation also show altered storage product accumulation, and both *NKD1* and *NKD2* directly regulate storage-protein gene expression (Yi *et al.*, 2015; Gontarek *et al.*, 2016; Qi *et al.*, 2016). Furthermore, both *O2* and *PBF* are downregulated in AL of the *nkd* mutant (Gontarek *et al.*, 2016). These findings indicate extensive interplay between the gene regulatory programs that control SE/AL cell fate and differentiation, and the programs controlling the storage function of SE and AL. Therefore, characterization of the GRNs regulated by the NKDs, O2, and PBF is expected to provide valuable insight into the regulation of AL/SE cell fate and differentiation.

The ESR consists of multiple layers of small, densely cytoplasmic and thin-walled cells that surround the endosperm cavity where the embryo develops (Schel *et al.*, 1984; Opsahl-Ferstad *et al.*, 1997; Leroux *et al.*, 2014). Early in its differentiation, the ESR surrounds the entire embryo. As differentiation proceeds, ESR cells are pushed away and crushed by the developing embryo and remain exclusively in regions around the embryo suspensor (Fig. 3.2) (Leroux *et al.*, 2014). This morphological change of the ESR is supported by mRNA localization patterns and promoter activities of the ESR marker genes, *ESR-1, 2,* and *3* (Opsahl-Ferstad *et al.*, 1997; Bonello *et al.*, 2000). In both assays, marker gene expression detected in the endosperm region surrounding the entire embryo at 5 DAP becomes restricted to a small region surrounding the embryo suspensor by 7 to 9 DAP, and to only the base of the suspensor by 12 to 15 DAP.

Increasing evidence in maize and Arabidopsis suggests that the ESR (the analogous region in Arabidopsis is referred to as the micropylar endosperm) is involved in nurturing and defense of the embryo, and also mediates signaling between the endosperm and embryo. The nutritive function is in part suggested by the finding that an invertase inhibitor gene (*INVINH-1*) is expressed in the maize ESR and likely functions in modulating invertase activity to regulate sugar transport into the embryo (Bate *et al.*, 2004). The ESR-specific expression patterns of two maize genes—the *ANDROGENIC*

*EMBRYO-3 (AE-3)* gene encoding a small hydrophilic protein (Magnard *et al.*, 2000; Sevilla-Lecoq *et al.*, 2003) with structural similarity to the BETL-expressed basal layer antifungal proteins (Serna *et al.*, 2001) and *ESR-6*, encoding a defensin-like protein (Balandin *et al.*, 2005)—suggest these gene products function within the ESR in defense of the embryo. The role of the ESR in signaling between the endosperm and the embryo is supported by multiple lines of evidence in both Arabidopsis and maize. The Arabidopsis *ZHOUPI (ZOU)* gene is expressed predominantly in the ESR, and loss of function of *ZOU* results in retarded endosperm breakdown and impaired epidermal development of the embryo. The latter indicates a critical role of the ESR region in control of embryo epidermal development through a signaling pathway (Yang *et al.*, 2008; Xing *et al.*, 2013). RNAi knockdown of *ZOU* in maize endosperm also results in retarded breakdown of the ESR and the adjacent suspensor (Grimault *et al.*, 2015). In addition, the Arabidopsis EMBRYO SURROUNDING FACTOR 1 (ESF1) accumulates in the central cell before fertilization and in the ESR after fertilization, and is required for early embryo patterning, likely through a non-cell-autonomous pathway (Costa *et al.*, 2014). These observations support a conserved linkage between embryonic and endosperm cell fates in both eudicots and monocots.

The mechanisms controlling ESR cell fate specification are unknown, but the *ESR-1, 2,* and *3* genes are possibly involved in the process. These genes encode small secreted proteins that show partial homology to Arabidopsis CLAVATA3 (CLV3) (Cock and McCormick, 2001; Bonello *et al.*, 2002), a signaling peptide that functions in the maintenance of the stem cell population in shoot apical meristems (Fletcher *et al.*, 1999; Ogawa *et al.*, 2008). This suggests ESR proteins could be involved in an equivalent fashion in signaling cell fate specification and/or differentiation of the ESR region itself, or they may mediate signaling events required for epidermal and/or suspensor development in the embryo. It is also possible that differentiation of the ESR region could require signals from the embryo.

In support of embryo-to-endosperm signaling, mutant embryo-less kernels form an embryo cavity within the endosperm, yet no mRNAs of the *ESR-1, 2,* and *3* genes are detectable in cells surrounding the cavity and, furthermore, these cells lack any ESR cell morphology (Opsahl-Ferstad *et al.*, 1997). Therefore, published data support a role for an extensive set of signaling processes between the endosperm and the embryo that may underlie proper seed development (Chapter 8).

## 3.5 Key Questions and Future Directions

An understanding of the nature of regulatory programs that dictate endosperm development will ultimately contribute to the improvement of yield and quality of the maize kernel. As in other multi-cellular eukaryotic systems, cell fate specification and differentiation in the maize endosperm is likely regulated by a combination of endogenous cues and positional information, including signals within the developing endosperm as well as from surrounding kernel compartments. These cell-differentiation regulatory programs exhibit extensive interplay with related programs associated with early endosperm cell proliferation, development of polarity, and later programs associated with storage product accumulation and endosperm maturation. Therefore, a deep mechanistic understanding of key processes regulating endosperm cell differentiation will require a holistic understanding of the temporal and spatial gene expression programs occurring during endosperm development.

In recent years, numerous endosperm transcriptome profiling studies have been carried out (Sekhon *et al.*, 2011, 2013; Lu *et al.*, 2013; Chen *et al.*, 2014; Li *et al.*, 2014; Zhan *et al.*, 2015; Qu *et al.*, 2016). In addition, many genome-wide analyses of histone modifications, DNA methylation, gene imprinting, and profiles of proteins and phosphoproteins have been published (Waters *et al.*, 2011; Zhang *et al.*, 2011,

2014; Makarevitch *et al.*, 2013; Walley *et al.*, 2013, 2016; Xin *et al.*, 2013; Dong *et al.*, 2017). Although most of these studies were performed on the whole endosperm or kernel, integration of the limited amount of spatial data within the endosperm and the available temporal data of whole endosperm/kernel can provide insight into gene regulation dynamics, as illustrated by use of the endosperm transcriptome (Li *et al.*, 2014; Zhan *et al.*, 2015). To generate a high-resolution spatio-temporal atlas of gene expression in the endosperm, laser-capture-microdissection (LCM)-based transcriptome profiling has recently been proven to be a powerful approach (Emmert-Buck *et al.*, 1996; Kerk *et al.*, 2003; Thakare *et al.*, 2014; Xiong *et al.*, 2014; Yi *et al.*, 2015; Zhan *et al.*, 2015). Other cell-type-specific genome-wide studies, such as mass spectrometry analysis of protein profiles and chromatin immunoprecipitation assay of chromatin modifications, require relatively larger amounts of tissues/cells. Therefore, alternative "single-cell omics" approaches, such as fluorescence-activated sorting of tagged cells or nuclei (Deal and Henikoff, 2010; Wang and Bodovitz, 2010; Macaulay and Voet, 2014; Handley *et al.*, 2015; Clark *et al.*, 2016), as have been carried out in other systems (Brady *et al.*, 2007; Evrard *et al.*, 2012; Slane *et al.*, 2014), can be employed to overcome such limitations. Recently, fluorescent markers of AL, BETL, and SE proved useful for tracking cell fate and differentiation of these cell types (Gruis *et al.*, 2006). However, fluorescent markers of the other cell types remain to be developed. Such markers, in turn, will likely have to be identified through LCM-assisted transcriptome profiling.

Using these approaches, the resulting spatio-temporal transcriptomic and proteomic data can be used to construct co-expression networks and infer GRNs (Kang *et al.*, 2011; Downs *et al.*, 2013; Xue *et al.*, 2013; Zhan *et al.*, 2015; Walley *et al.*, 2016). The hubs of the networks, particularly TFs, are likely to play important roles in the regulation of the associated biological processes and functions related to differentiated states. The function of the hubs can be further studied by generating loss-of-function mutants using RNAi or genome-editing tools (Hannon, 2002; Townsend *et al.*, 2009; Zhang *et al.*, 2013; Bortesi and Fischer, 2015), which are expected to uncover novel functional information for endosperm-expressed genes.

## Acknowledgment

Research in the Dannenhoffer lab and the Yadegari lab is currently funded by the U.S. National Science Foundation Award IOS-1444568.

## References

Abramowitz, L.K. and Bartolomei, M.S. (2012) Genomic imprinting: recognition and marking of imprinted loci. *Current Opinion in Genetics and Development* 22, 72–78.

Adams, S., Vinkenoog, R., Spielman, M., Dickinson, H.G. and Scott, R.J. (2000) Parent-of-origin effects on seed development in *Arabidopsis thaliana* require DNA methylation. *Development* 127, 2493–2502.

Balandin, M., Royo, J., Gómez, E., Muniz, L.M., Molina, A. and Hueros, G. (2005) A protective role for the embryo surrounding region of the maize endosperm, as evidenced by the characterisation of *ZmESR-6*, a defensin gene specifically expressed in this region. *Plant Molecular Biology* 58, 269–282.

Bate, N.J., Niu, X., Wang, Y., Reimann, K.S. and Helentjaris, T.G. (2004) An invertase inhibitor from maize localizes to the embryo surrounding region during early kernel development. *Plant Physiology* 134, 246–254.

Baud, S., Wuilleme, S., Lemoine, R., Kronenberger, J., Caboche, M., Lepiniec, L. and Rochat, C. (2005) The AtSUC5 sucrose transporter specifically expressed in the endosperm is involved in early seed development in Arabidopsis. *The Plant Journal* 43, 824–836.

Becraft, P.W. (2001) Cell fate specification in the cereal endosperm. *Seminars in Cell & Development Biology* 12, 387–394.

Becraft, P.W. (2007) Aleurone cell development. In: Olsen, O.-A. (ed.) *Endosperm.* Springer, Heidelberg, pp. 45–56.

Becraft, P.W. and Asuncion-Crabb, Y. (2000) Positional cues specify and maintain aleurone cell fate in maize endosperm development. *Development* 127, 4039–4048.

Becraft, P.W. and Gutiérrez-Marcos, J.F. (2012) Endosperm development: dynamic processes and cellular innovations underlying sibling altruism. *Wiley Interdisciplinary Reviews: Developmental Biology* 1, 579–593.

Becraft, P.W. and Yi, G. (2011) Regulation of aleurone development in cereal grains. *Journal of Experimental Botany* 62, 1669–1675.

Becraft, P.W., Li, K., Dey, N. and Asuncion-Crabb, Y. (2002) The maize *dek1* gene functions in embryonic pattern formation and cell fate specification. *Development* 129, 5217–5225.

Berger, F., Grini, P.E. and Schnittger, A. (2006) Endosperm: an integrator of seed growth and development. *Current Opinion in Plant Biology* 9, 664–670.

Boisnard-Lorig, C., Colon-Carmona, A., Bauch, M., Hodge, S., Doerner, P., *et al.* (2001) Dynamic analyses of the expression of the HISTONE:YFP fusion protein in Arabidopsis show that syncytial endosperm is divided in mitotic domains. *Plant Cell* 13, 495–509.

Bonello, J.F., Opsahl-Ferstad, H.G., Perez, P., Dumas, C. and Rogowsky, P.M. (2000) *Esr* genes show different levels of expression in the same region of maize endosperm. *Gene* 246, 219–227.

Bonello, J.F., Sevilla-Lecoq, S., Berne, A., Risueno, M.C., Dumas, C. and Rogowsky, P.M. (2002) Esr proteins are secreted by the cells of the embryo surrounding region. *Journal of Experimental Botany* 53, 1559–1568.

Bortesi, L. and Fischer, R. (2015) The CRISPR/Cas9 system for plant genome editing and beyond. *Biotechnology Advances* 33, 41–52.

Brady, S.M., Orlando, D.A., Lee, J.Y., Wang, J.Y., Koch, J., *et al.* (2007) A high-resolution root spatiotemporal map reveals dominant expression patterns. *Science* 318, 801–806.

Brown, R.C., Lemmon, B.E. and Olsen, O.-A. (1994) Endosperm development in barley: microtubule involvement in the morphogenetic pathway. *Plant Cell* 6, 1241–1252.

Brown, R.C., Lemmon, B.E. and Olsen, O.-A. (1996a) Polarization predicts the pattern of cellularization in cereal endosperm. *Protoplasma* 192, 168–177.

Brown, R.C., Lemmon, B.E. and Olsen, O.-A. (1996b) Development of the endosperm in rice (*Oryza sativa* L): cellularization. *Journal of Plant Research* 109, 301–313.

Brown, R.C., Lemmon, B.E., Nguyen, H. and Olsen, O.A. (1999) Development of endosperm in *Arabidopsis thaliana*. *Sexual Plant Reproduction* 12, 32–42.

Brown, R.C., Lemmon, B.E. and Nguyen, H. (2003) Events during the first four rounds of mitosis establish three developmental domains in the syncytial endosperm of *Arabidopsis thaliana*. *Protoplasma* 222, 167–174.

Bushell, C., Spielman, M. and Scott, R.J. (2003) The basis of natural and artificial postzygotic hybridization barriers in Arabidopsis species. *Plant Cell* 15, 1430–1442.

Charlton, W.L., Keen, C.L., Merriman, C., Lynch, P., Greenland, A.J. and Dickinson, H.G. (1995) Endosperm development in *zea-mays* – implication of gametic imprinting and paternal excess in regulation of transfer layer development. *Development* 121, 3089–3097.

Chen, J., Zeng, B., Zhang, M., Xie, S., Wang, G., Hauck, A. and Lai, J. (2014) Dynamic transcriptome landscape of maize embryo and endosperm development. *Plant Physiology* 166, 252–264.

Clark, S.J., Lee, H.J., Smallwood, S.A., Kelsey, G. and Reik, W. (2016) Single-cell epigenomics: powerful new methods for understanding gene regulation and cell identity. *Genome Biology* 17, 72.

Cock, J.M. and McCormick, S. (2001) A large family of genes that share homology with CLAVATA3. *Plant Physiology* 126, 939–942.

Cooper, D.C. (1951) Caryopsis development following matings between diploid and tetraploid strains of *zea-mays*. *American Journal of Botany* 38, 702–708.

Costa, L.M., Gutiérrez-Marcos, J.F., Brutnell, T.P., Greenland, A.J. and Dickinson, H.G. (2003) The *globby1-1* (*glo1-1*) mutation disrupts nuclear and cell division in the developing maize seed causing alterations in endosperm cell fate and tissue differentiation. *Development* 130, 5009–5017.

Costa, L.M., Gutiérrez-Marcos, J.F. and Dickinson, H.G. (2004) More than a yolk: the short life and complex times of the plant endosperm. *Trends in Plant Science* 9, 507–514.

Costa, L.M., Yuan, J., Rouster, J., Paul, W., Dickinson, H. and Gutiérrez-Marcos, J.F. (2012) Maternal control of nutrient allocation in plant seeds by genomic imprinting. *Current Biology* 22, 160–165.

Costa, L.M., Marshall, E., Tesfaye, M., Silverstein, K.A., Mori, M., *et al.* (2014) Central cell-derived peptides regulate early embryo patterning in flowering plants. *Science* 344, 168–172.

Danilevskaya, O.N., Hermon, P., Hantke, S., Muszynski, M.G., Kollipara, K. and Ananiev, E.V. (2003) Duplicated fie genes in maize: expression pattern and imprinting suggest distinct functions. *Plant Cell* 15, 425–438.

Deal, R.B. and Henikoff, S. (2010) A simple method for gene expression and chromatin profiling of individual cell types within a tissue. *Developmental Cell* 18, 1030–1040.

Doan, D.N., Linnestad, C. and Olsen, O.A. (1996) Isolation of molecular markers from the barley endosperm coenocyte and the surrounding nucellus cell layers. *Plant Molecular Biology* 31, 877–886.

Doll, N.M., Depege-Fargeix, N., Rogowsky, P.M. and Widiez, T. (2017) Signaling in early maize kernel development. *Molecular Plant* 10, 375–388.

Dong, X., Zhang, M., Chen, J., Peng, L., Zhang, N., Wang, X. and Lai, J. (2017) Dynamic and antagonistic allele-specific epigenetic modifications controlling the expression of imprinted genes in maize endosperm. *Molecular Plant* 10, 442–455

Downs, G.S., Bi, Y.M., Colasanti, J., Wu, W., Chen, X., *et al.* (2013) A developmental transcriptional network for maize defines coexpression modules. *Plant Physiology* 161, 1830–1843.

Emmert-Buck, M.R., Bonner, R.F., Smith, P.D., Chuaqui, R.F., Zhuang, Z., *et al.* (1996) Laser capture microdissection. *Science* 274, 998–1001.

Evrard, A., Bargmann, B.O., Birnbaum, K.D., Tester, M., Baumann, U. and Johnson, A.A. (2012) Fluorescence-activated cell sorting for analysis of cell type-specific responses to salinity stress in Arabidopsis and rice. *Methods in Molecular Biology* 913, 265–276.

FAO (2015) FAO Statistical Pocketbook 2015. Food and Agriculture Organization of the United Nations, Rome, Italy. Available at: http://www.fao.org/documents/card/en/c/383d384a-28e6-47b3-a1a2-2496a9e017b2 (accessed May 21, 2017).

Finnegan, E.J. and Dennis, E.S. (1993) Isolation and identification by sequence homology of a putative cytosine methyltransferase from *Arabidopsis thaliana*. *Nucleic Acids Research* 21, 2383–2388.

Fletcher, J.C., Brand, U., Running, M.P., Simon, R. and Meyerowitz, E.M. (1999) Signaling of cell fate decisions by *CLAVATA3* in *Arabidopsis* shoot meristems. *Science* 283, 1911–1914.

Friedman, W.E., Madrid, E.N. and Williams, J.H. (2008) Origin of the fittest and survival of the fittest: relating female gametophyte development to endosperm genetics. *International Journal of Plant Sciences* 169, 79–92.

Garner, A.G., Kenney, A.M., Fishman, L. and Sweigart, A.L. (2016) Genetic loci with parent-of-origin effects cause hybrid seed lethality in crosses between *Mimulus* species. *The New Phytologist* 211, 319–331.

Gehring, M. (2013) Genomic imprinting: insights from plants. *Annual Review of Genetics* 47, 187–208.

Gehring, M. and Satyaki, P.R. (2017) Endosperm and imprinting, inextricably linked. *Plant Physiology* 173, 143–154.

Gontarek, B.C., Neelakandan, A.K., Wu, H. and Becraft, P.W. (2016) NKD transcription factors are central regulators of maize endosperm development. *Plant Cell* 28, 2916–2936.

Grimault, A., Gendrot, G., Chamot, S., Widiez, T., Rabille, H., *et al.* (2015) ZmZHOUPI, an endosperm-specific basic helix-loop-helix transcription factor involved in maize seed development. *The Plant Journal* 84, 574–586.

Gruis, D.F., Guo, H., Selinger, D., Tian, Q. and Olsen, O.-A. (2006) Surface position, not signaling from surrounding maternal tissues, specifies aleurone epidermal cell fate in maize. *Plant Physiology* 141, 898–909.

Guitton, A.E., Page, D.R., Chambrier, P., Lionnet, C., Faure, J.E., Grossniklaus, U. and Berger, F. (2004) Identification of new members of Fertilisation Independent Seed Polycomb Group pathway involved in the control of seed development in *Arabidopsis thaliana*. *Development* 131, 2971–2981.

Gunning, B.E.S. and Pate, J.S. (1969) Transfer cells plant cells with wall ingrowths, specialized in relation to short distance transport of solutes – their occurrence, structure, and development. *Protoplasma* 68, 107–133.

Gutiérrez-Marcos, J.F., Costa, L.M., Biderre-Petit, C., Khbaya, B., O'Sullivan, D.M., *et al.* (2004) *Maternally expressed gene1* is a novel maize endosperm transfer cell-specific gene with a maternal parent-of-origin pattern of expression. *Plant Cell* 16, 1288–1301.

Gutiérrez-Marcos, J.F., Costa, L.M. and Evans, M.M. (2006) Maternal gametophytic baseless1 is required for development of the central cell and early endosperm patterning in maize (*Zea-mays*). *Genetics* 174, 317–329.

Haig, D. (2014) Coadaptation and conflict, misconception and muddle, in the evolution of genomic imprinting. *Heredity* 113, 96–103.

Haig, D. and Westoby, M. (1989) Parent-specific gene-expression and the triploid endosperm. *The American Naturalist* 134, 147–155.

Hamamura, Y., Nagahara, S. and Higashiyama, T. (2012) Double fertilization on the move. *Current Opinion in Plant Biology* 15, 70–77.

Handley, A., Schauer, T., Ladurner, A.G. and Margulies, C.E. (2015) Designing cell-type-specific genome-wide experiments. *Molecular Cell* 58, 621–631.

Hannon, G.J. (2002) RNA interference. *Nature* 418, 244–251.

Haun, W.J., Laoueille-Duprat, S., O'Connell, M.J., Spillane, C., Grossniklaus, U., *et al.* (2007) Genomic imprinting, methylation and molecular evolution of maize enhancer of zeste (Mez) homologs. *The Plant Journal* 49, 325–337.

Hehenberger, E., Kradolfer, D. and Kohler, C. (2012) Endosperm cellularization defines an important developmental transition for embryo development. *Development* 139, 2031–2039.

Ingouff, M., Haseloff, J. and Berger, F. (2005) Polycomb group genes control developmental timing of endosperm. *The Plant Journal* 42, 663–674.

Ishikawa, R., Ohnishi, T., Kinoshita, Y., Eiguchi, M., Kurata, N. and Kinoshita, T. (2011) Rice interspecies hybrids show precocious or delayed developmental transitions in the endosperm without change to the rate of syncytial nuclear division. *The Plant Journal* 65, 798–806.

Johnson, D.R. and Tanner, J.W. (1972) Calculation of the rate and duration of grain filling in corn (*Zea-mays* L.). *Crop Science* 12, 485–486.

Jones, R.J., Roessler, J. and Ouattar, S. (1985) Thermal environment during endosperm cell-division in maize – effects on number of endosperm cells and starch granules. *Crop Science* 25, 830–834.

Jones, R.J., Schreiber, B.M.N. and Roessler, J.A. (1996) Kernel sink capacity in maize: genotypic and maternal regulation. *Crop Science* 36, 301–306.

Jurgens, G. (2005) Cytokinesis in higher plants. *Annual Review of Plant Biology* 56, 281–299.

Kang, H.J., Kawasawa, Y.I., Cheng, F., Zhu, Y., Xu, X., *et al.* (2011) Spatio-temporal transcriptome of the human brain. *Nature* 478, 483–489.

Kang, I.H., Steffen, J.G., Portereiko, M.F., Lloyd, A. and Drews, G.N. (2008) The AGL62 MADS domain protein regulates cellularization during endosperm development in *Arabidopsis*. *Plant Cell* 20, 635–647.

Kawakatsu, T. and Takaiwa, F. (2010) Cereal seed storage protein synthesis: fundamental processes for recombinant protein production in cereal grains. *Plant Biotechnology Journal* 8, 939–953.

Kerk, N.M., Ceserani, T., Tausta, S.L., Sussex, I.M. and Nelson, T.M. (2003) Laser capture microdissection of cells from plant tissues. *Plant Physiology* 132, 27–35.

Khoo, U. and Wolf, M.J. (1970) Origin and development of protein granules in maize endosperm. *American Journal of Botany* 57, 1042–1050.

Kiesselbach, T.A. (1999) *The Structure and Reproduction of Corn*. Cold Spring Harbor Laboratory Press, Cold Spring Harbor, New York.

Kradolfer, D., Wolff, P., Jiang, H., Siretskiy, A. and Kohler, C. (2013) An imprinted gene underlies postzygotic reproductive isolation in *Arabidopsis thaliana*. *Developmental Cell* 26, 525–535.

Lending, C.R. and Larkins, B.A. (1989) Changes in the zein composition of protein bodies during maize endosperm development. *Plant Cell* 1, 1011–1023.

Leroux, B.M., Goodyke, A.J., Schumacher, K.I., Abbott, C.P., Clore, A.M., *et al.* (2014) Maize early endosperm growth and development: from fertilization through cell type differentiation. *American Journal of Botany* 101, 1259–1274.

Li, G., Wang, D., Yang, R., Logan, K., Chen, H., *et al.* (2014) Temporal patterns of gene expression in developing maize endosperm identified through transcriptome sequencing. *Proceedings of the National Academy of Sciences of the United States of America* 111, 7582–7587.

Li, J. and Berger, F. (2012) Endosperm: food for humankind and fodder for scientific discoveries. *The New Phytologist* 195, 290–305.

Li, J., Nie, X., Tan, J.L. and Berger, F. (2013) Integration of epigenetic and genetic controls of seed size by cytokinin in *Arabidopsis*. *Proceedings of the National Academy of Sciences of the United States of America* 110, 15479–15484.

Lid, S.E., Gruis, D., Jung, R., Lorentzen, J.A., Ananiev, E., *et al.* (2002) The *defective kernel 1* (*dek1*) gene required for aleurone cell development in the endosperm of maize grains encodes a membrane protein of the calpain gene superfamily. *Proceedings of the National Academy of Sciences of the United States of America* 99, 5460–5465.

Lopes, M.A. and Larkins, B.A. (1993) Endosperm origin, development, and function. *Plant Cell* 5, 1383–1399.

Lu, X., Chen, D., Shu, D., Zhang, Z., Wang, W., *et al.* (2013) The differential transcription network between embryo and endosperm in the early developing maize seed. *Plant Physiology* 162, 440–455.

Macaulay, I.C. and Voet, T. (2014) Single cell genomics: advances and future perspectives. *PLOS Genetics* 10, e1004126.

Magnard, J.L., Le Deunff, E., Domenech, J., Rogowsky, P.M., Testillano, P.S., *et al.* (2000) Genes normally expressed in the endosperm are expressed at early stages of microspore embryogenesis in maize. *Plant Molecular Biology* 44, 559–574.

Makarevitch, I., Eichten, S.R., Briskine, R., Waters, A.J., Danilevskaya, O.N., *et al.* (2013) Genomic distribution of maize facultative heterochromatin marked by trimethylation of H3K27. *Plant Cell* 25, 780–793.

Monjardino, P., Machado, J., Gil, F.S., Fernandes, R. and Salema, R. (2007) Structural and ultrastructural characterization of maize coenocyte and endosperm cellularization. *Canadian Journal of Botany* 85, 216–223.

Monjardino, P., Rocha, S., Tavares, A.C., Fernandes, R., Sampaio, P., Salema, R. and da Camara Machado, A. (2013) Development of flange and reticulate wall ingrowths in maize (*Zea mays* L.) endosperm transfer cells. *Protoplasma* 250, 495–503.

Morrison, I.N., Kuo, J. and O'Brien, T.P. (1975) Histochemistry and fine structure of developing wheat aleurone cells. *Planta* 123, 105–116.

Neuffer, M.G. and Sheridan, W.F. (1980) Defective kernel mutants of maize. I. Genetic and lethality studies. *Genetics* 95, 929–944.

Ogawa, M., Shinohara, H., Sakagami, Y. and Matsubayashi, Y. (2008) *Arabidopsis* CLV3 peptide directly binds CLV1 ectodomain. *Science* 319, 294.

Olsen, O.-A. and Becraft, P.W. (2013) Endosperm development. In: Becraft, P.W. (ed.) *Seed Genomics*. Wiley-Blackwell, Oxford, UK, pp. 43–62.

Olsen, O.-A., Linnestad, C. and Nichols, S.E. (1999) Developmental biology of the cereal endosperm. *Trends in Plant Science* 4, 253–257.

Oneal, E., Willis, J.H. and Franks, R.G. (2016) Disruption of endosperm development is a major cause of hybrid seed inviability between *Mimulus guttatus* and *Mimulus nudatus*. *The New Phytologist* 210, 1107–1120.

Opsahl-Ferstad, H.G., Le Deunff, E., Dumas, C. and Rogowsky, P.M. (1997) *ZmEsr*, a novel endosperm-specific gene expressed in a restricted region around the maize embryo. *The Plant Journal* 12, 235–246.

Orozco-Arroyo, G., Paolo, D., Ezquer, I. and Colombo, L. (2015) Networks controlling seed size in Arabidopsis. *Plant Reproduction* 28, 17–32.

Pate, J.S. and Gunning, B.E.S. (1972) Transfer cells. *Annual Review of Plant Physiology* 23, 173–196.

Pennington, P.D., Costa, L.M., Gutiérrez-Marcos, J.F., Greenland, A.J. and Dickinson, H.G. (2008) When genomes collide: aberrant seed development following maize interploidy crosses. *Annals of Botany* 101, 833–843.

Peters, J. (2014) The role of genomic imprinting in biology and disease: an expanding view. *Nature Review Genetics* 15, 517–530.

Qi, X., Li, S., Zhu, Y., Zhao, Q., Zhu, D. and Yu, J. (2016) *ZmDof3*, a maize endosperm-specific Dof protein gene, regulates starch accumulation and aleurone development in maize endosperm. *Plant Molecular Biology* 93, 7–20.

Qu, J., Ma, C., Feng, J., Xu, S., Wang, L., *et al.* (2016) Transcriptome dynamics during maize endosperm development. *PLOS ONE* 11, e0163814.

Randolph, L.F. (1936) *Developmental Morphology of the Caryopsis in Maize*. United States Department of Agriculture, Washington, DC.

Rebernig, C.A., Lafon-Placette, C., Hatorangan, M.R., Slotte, T. and Kohler, C. (2015) Non-reciprocal interspecies hybridization barriers in the capsella genus are established in the endosperm. *PLOS Genetics* 11, e1005295.

Reddy, V.M. and Daynard, T.B. (1983) Endosperm characteristics associated with rate of grain filling and kernel size in corn. *Maydica* 28, 339–355.

Reyes, F.C., Chung, T., Holding, D., Jung, R., Vierstra, R. and Otegui, M.S. (2011) Delivery of prolamins to the protein storage vacuole in maize aleurone cells. *Plant Cell* 23, 769–784.

Rodrigues, J.A. and Zilberman, D. (2015) Evolution and function of genomic imprinting in plants. *Genes & Development* 29, 2517–2531.

Royo, J., Gómez, E., Sellam, O., Gerentes, D., Paul, W. and Hueros, G. (2014) Two maize *END-1* orthologs, *BETL9* and *BETL9like*, are transcribed in a non-overlapping spatial pattern on the outer surface of the developing endosperm. *Frontiers in Plant Science* 5, 180.

Sabelli, P.A. and Larkins, B.A. (2009) The development of endosperm in grasses. *Plant Physiology* 149, 14–26.

Scanlon, M.J., Stinard, P.S., James, M.G., Myers, A.M. and Robertson, D.S. (1994) Genetic analysis of 63 mutations affecting maize kernel development isolated from Mutator stocks. *Genetics* 136, 281–294.

Schel, J.H.N., Kieft, H. and Vanlammeren, A.A.M. (1984) Interactions between embryo and endosperm during early developmental stages of maize caryopses (*Zea-Mays*). *Canadian Journal of Botany* 62, 2842–2853.

Scott, R.J., Spielman, M., Bailey, J. and Dickinson, H.G. (1998) Parent-of-origin effects on seed development in *Arabidopsis thaliana. Development* 125, 3329–3341.

Sekhon, R.S., Lin, H., Childs, K.L., Hansey, C.N., Buell, C.R., de Leon, N. and Kaeppler, S.M. (2011) Genome-wide atlas of transcription during maize development. *The Plant Journal* 66, 553–563.

Sekhon, R.S., Briskine, R., Hirsch, C.N., Myers, C.L., Springer, N.M., *et al.* (2013) Maize gene atlas developed by RNA sequencing and comparative evaluation of transcriptomes based on RNA sequencing and microarrays *PLOS ONE* 8, e61005.

Sekhon, R.S., Hirsch, C.N., Childs, K.L., Breitzman, M.W., Kell, P., *et al.* (2014) Phenotypic and transcriptional analysis of divergently selected maize populations reveals the role of developmental timing in seed size determination. *Plant Physiology* 165, 658–669.

Serna, A., Maitz, M., O'Connell, T., Santandrea, G., Thevissen, K. *et al.* (2001) Maize endosperm secretes a novel antifungal protein into adjacent maternal tissue. *The Plant Journal* 25, 687–698.

Sevilla-Lecoq, S., Deguerry, F., Matthys-Rochon, E., Perez, P., Dumas, C. and Rogowsky, P.M. (2003) Analysis of *ZmAE3* upstream sequences in maize endosperm and androgenic embryos. *Sexual Plant Reproduction* 16, 1–8.

Shannon, J.C., Porter, G.A. and Knievel, D.P. (1986) Phloem unloading and transfer of sugars into developing corn endosperm. In: Cronshaw, J., Lucas, W.J. and Guiaquinta, R.T. (eds.) *Phloem Transport*. Alan Liss, Inc., New York, pp. 265–277.

Sheridan, W.F. and Neuffer, M.G. (1980) Defective kernel mutants of maize II. Morphological and embryo culture studies. *Genetics* 95, 945–960.

Slane, D., Kong, J., Berendzen, K.W., Kilian, J., Henschen, A., *et al.* (2014) Cell type-specific transcriptome analysis in the early *Arabidopsis thaliana* embryo. *Development* 141, 4831–4840.

Sorensen, M.B., Chaudhury, A.M., Robert, H., Bancharel, E. and Berger, F. (2001) Polycomb group genes control pattern formation in plant seed. *Current Biology* 11, 277–281.

Steward, N., Kusano, T. and Sano, H. (2000) Expression of ZmMET1, a gene encoding a DNA methyltransferase from maize, is associated not only with DNA replication in actively proliferating cells, but also with altered DNA methylation status in cold-stressed quiescent cells. *Nucleic Acids Research* 28, 3250–3259.

Tanaka, H., Onouchi, H., Kondo, M., Hara-Nishimura, I., Nishimura, M., Machida, C. and Machida, Y. (2001) A subtilisin-like serine protease is required for epidermal surface formation in Arabidopsis embryos and juvenile plants. *Development* 128, 4681–4689.

Tekleyohans, D.G., Nakel, T. and Gross-Hardt, R. (2017) Patterning the female gametophyte of flowering plants. *Plant Physiology* 173, 122–129.

Thakare, D., Yang, R., Steffen, J.G., Zhan, J., Wang, D., *et al.* (2014) RNA-Seq analysis of laser-capture microdissected cells of the developing central starchy endosperm of maize. *Genomics Data* 2, 242–245.

Thompson, R.D. and Verdier, J. (2012) Networks of seed storage protein regulation in cereals and legumes at the dawn of the omics era. In: Agrawal, G.K. and Rakwal, R. (eds.) *Seed Development: OMICS Technologies toward Improvement of Seed Quality and Crop Yield.* Springer, Dordrecht, The Netherlands, pp. 187–210.

Tian, G.W., You, R.L., Guo, F.L. and Wang, X.C. (1998) Microtubular cytoskeleton of free endosperm nuclei during division in wheat. *Cytologia* 63, 427–433.

Townsend, J.A., Wright, D.A., Winfrey, R.J., Fu, F., Maeder, M.L., Joung, J.K. and Voytas, D.F. (2009) High-frequency modification of plant genes using engineered zinc-finger nucleases. *Nature* 459, 442–445.

von Wangenheim, K.H. and Peterson, H.P. (2004) Aberrant endosperm development in interploidy crosses reveals a timer of differentiation. *Developmental Biology* 270, 277–289.

Walley, J.W., Shen, Z., Sartor, R., Wu, K.J., Osborn, J., Smith, L.G. and Briggs, S.P. (2013) Reconstruction of protein networks from an atlas of maize seed proteotypes. *Proceedings of the National Academy of Sciences of the United States of America* 110, E4808–E4817.

Walley, J.W., Sartor, R.C., Shen, Z., Schmitz, R.J., Wu, K.J., *et al.* (2016) Integration of omic networks in a developmental atlas of maize. *Science* 353, 814–818.

Wang, D. and Bodovitz, S. (2010) Single cell analysis: the new frontier in "omics." *Trends in Biotechnology* 28, 281–290.

Waters, A.J., Makarevitch, I., Eichten, S.R., Swanson-Wagner, R.A., Yeh, C.T., *et al.* (2011) Parent-of-origin effects on gene expression and DNA methylation in the maize endosperm. *Plant Cell* 23, 4221–4233.

Wolff, P., Jiang, H., Wang, G., Santos-Gonzalez, J. and Kohler, C. (2015) Paternally expressed imprinted genes establish postzygotic hybridization barriers in *Arabidopsis thaliana. eLife* 4.

Xiao, W., Brown, R.C., Lemmon, B.E., Harada, J.J., Goldberg, R.B. and Fischer, R.L. (2006) Regulation of seed size by hypomethylation of maternal and paternal genomes. *Plant Physiology* 142, 1160–1168.

Xin, M., Yang, R., Li, G., Chen, H., Laurie, J., *et al.* (2013) Dynamic expression of imprinted genes associates with maternally controlled nutrient allocation during maize endosperm development. *Plant Cell* 25, 3212–3227.

Xing, Q., Creff, A., Waters, A., Tanaka, H., Goodrich, J. and Ingram, G.C. (2013) ZHOUPI controls embryonic cuticle formation via a signalling pathway involving the subtilisin protease ABNORMAL LEAF-SHAPE1 and the receptor kinases GASSHO1 and GASSHO2. *Development* 140, 770–779.

Xiong, Y., Mei, W., Kim, E.D., Mukherjee, K., Hassanein, H., *et al.* (2014) Adaptive expansion of the maize *maternally expressed gene* (*Meg*) family involves changes in expression patterns and protein secondary structures of its members. *BMC Plant Biology* 14, 204.

Xue, Z., Huang, K., Cai, C., Cai, L., Jiang, C.Y., *et al.* (2013) Genetic programs in human and mouse early embryos revealed by single-cell RNA sequencing. *Nature* 500, 593–597.

Yang, S., Johnston, N., Talideh, E., Mitchell, S., Jeffree, C., Goodrich, J. and Ingram, G. (2008) The endosperm-specific ZHOUPI gene of *Arabidopsis thaliana* regulates endosperm breakdown and embryonic epidermal development. *Development* 135, 3501–3509.

Yi, G., Neelakandan, A.K., Gontarek, B.C., Vollbrecht, E. and Becraft, P.W. (2015) The naked endosperm genes encode duplicate INDETERMINATE domain transcription factors required for maize endosperm cell patterning and differentiation. *Plant Physiology* 167, 443–456.

Young, T.E. and Gallie, D.R. (2000) Regulation of programmed cell death in maize endosperm by ab-scisic acid. *Plant Molecular Biology* 42, 397–414.

Young, T.E., Gallie, D.R. and DeMason, D.A. (1997) Ethylene-mediated programmed cell death during maize endosperm development of wild-type and *shrunken2* genotypes. *Plant Physiology* 115, 737–751.

Yuan, J., Bateman, P. and Gutiérrez-Marcos, J.F. (2016) Genetic and epigenetic control of transfer cell development in plants. *Journal of Genetics and Genomics* 43, 533–539.

Zhan, J., Thakare, D., Ma, C., Lloyd, A., Nixon, N.M., *et al.* (2015) RNA sequencing of laser-capture microdissected compartments of the maize kernel identifies regulatory modules associated with endosperm cell differentiation. *Plant Cell* 27, 513–531.

Zhang, M., Zhao, H., Xie, S., Chen, J., Xu, Y., *et al.* (2011) Extensive, clustered parental imprinting of protein-coding and noncoding RNAs in developing maize endosperm. *Proceedings of the National Academy of Sciences of the United States of America* 108, 20042–20047.

Zhang, M., Xie, S., Dong, X., Zhao, X., Zeng, B., *et al.* (2014) Genome-wide high resolution parental-specific DNA and histone methylation maps uncover patterns of imprinting regulation in maize. *Genome Research* 24, 167–176.

Zhang, Y., Zhang, F., Li, X., Baller, J.A., Qi, Y., *et al.* (2013) Transcription activator-like effector nucle-ases enable efficient plant genome engineering. *Plant Physiology* 161, 20–27.

# 4 What Can We Learn from Maize Kernel Mutants?

Donald R. McCarty*

*Department of Horticultural Sciences, University of Florida, USA*

## 4.1 Introduction

Maize kernel mutants have provided insight into the mechanisms of embryo and endosperm formation for more than a century (Neuffer and Sheridan, 1980; Sheridan and Neuffer, 1980; Clark and Sheridan 1991; Sheridan and Clark, 1993). Advances in genomics technologies revolutionized our ability to learn from them, and recent application of transposon mutagenesis enabled their genome-wide analysis (McCarty *et al.*, 2005, 2013; Hunter *et al.*, 2014). With current gene discovery and genome editing technologies, there is no longer a distinction between forward and reverse genetics approaches to linking genes and phenotypes. Moreover, genetic and phenotypic analyses can be integrated with other types of genomic data that place genes in networks, providing even deeper insight into their functions. Broadly speaking, maize seed mutants fall into three categories: (i) defective kernel (*dek*) mutations that affect both endosperm and embryo (Fig. 4.1, top); (ii) embryo-specific (*emb*) mutations with more or less normal endosperm formation (Fig. 4.1, bottom); and (iii) endosperm-specific mutations. Important members of the last group include the opaque and shrunken mutants that affect storage protein and starch biosynthesis, respectively, and those that affect aleurone differentiation and pigmentation. Many of these mutations and pathways are described in other chapters of this book; here I focus on *dek* and *emb* mutants.

## 4.2 Defective Kernel Mutants

In their classic 1980 papers, Neuffer and Sheridan made insightful observations about *dek* mutants that are pertinent to our understanding of the genetics and genomics of seed formation: (i) there are many *deks*; although only rough estimates of the potential number of genes in the three categories can be obtained from systematic mutagenesis studies, it is likely *deks* (broadly defined) outnumber the endosperm- and embryo-specific categories by a substantial margin; (ii) in spite of a tendency for *deks* to more severely impact embryo than endosperm development, in many cases *dek* mutant embryos can be rescued by tissue culture with a basal medium of nutrients; (iii) genetically non-concordant seeds generated using B-A chromosome translocations suggest that non-autonomous

*Corresponding author e-mail: drm@ufl.edu

 

dek mutants

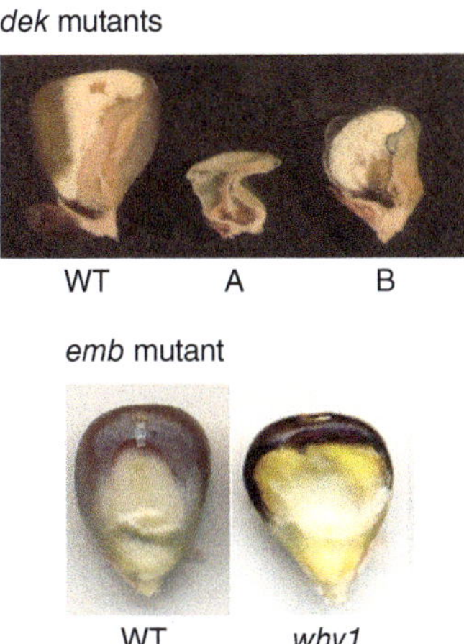


**Fig. 4.1.** *dek* and *emb* mutants. Top: *dek* mutants vary in their impacts on embryo and endosperm. Mutant "A" has a severe empty pericarp phenotype with profound effects on both embryo and endosperm and "B" illustrates a common pattern of the embryo being affected more severely than endosperm. Bottom: In W22 *why1* exhibits a classic *emb* phenotype revealing the empty embryo cavity in the endosperm (Shen *et al.*, 2013). The transition stage *why1* embryo is not visible. (Photo credits: Karen E. Koch, Jonathan Saunders, Jiani Yang and Masaharu Suzuki, Horticultural Sciences Department, University of Florida)

cross-feeding interactions between endosperm and embryo are relatively common; i.e. it is sometimes the case among *deks* that a mutant embryo can be at least partially rescued by a normal endosperm, and vice versa. It is instructive to revisit these observations in light of our current understanding of gene expression and metabolism in the developing seed (see Chapter 7).

One likely reason for the abundance of *deks* is the extensive overlap of gene expression in the embryo and endosperm. Roughly 90% of seed-expressed genes have significant mRNA expression in both embryo and endosperm (Lu *et al.*, 2013; Zhan *et al.*, 2015). Preferential expression in one organ or another seems the rule rather than strict tissue/organ specificity. Consequently, it is not surprising that many mutations in essential genes affect both organs, giving rise to early seed-lethal or *dek* phenotypes. These include mutations in so-called "housekeeping" genes which have essential functions, e.g. protein synthesis, central metabolism, DNA replication, etc. These genes have the ironic status of being simultaneously important and (to some) boring! Technologies that enable genome-wide analyses of genetic networks allow us to approach this complex class of genes with a less biased perspective, and in the process perhaps learn that they are interesting.

Nevertheless, the "housekeeping" framework is not sufficient for understanding *deks*. A paradox is that even though *deks* typically impact embryo development at least as strongly as endosperm development, in many cases *dek* embryos can be rescued in tissue culture (Sheridan and Neuffer, 1980). This can be rationalized in a metabolic framework if *dek* mutants correspond to genes required for synthesis of essential metabolites, such as vitamins, amino acids, etc. Indeed, Neuffer and Sheridan were motivated in part to identify auxotrophs in such pathways. While some *deks* fit this category, in many cases their mutant embryos can be rescued by culturing on a basal media. In these mutants, embryo growth is not restricted by an obvious metabolic lesion. Rather, embryo growth is constrained by an environment determined by the endosperm.

### 4.2.1 Embryo dependence on a functional endosperm

A second conceptual framework for understanding *dek* mutants comes from the long-standing observation that a functional endosperm is required for embryo development (Brink and Cooper, 1947; Pennington *et al.*, 2008). For example, geneticists have long known that the balance of maternal and paternal genomes in the triploid endosperm is a significant barrier to interploidy crosses in maize and other species (Lin *et al.*, 1984; Chapter 9, this volume). An analysis of maize seed formed by heterofertilization indicates that embryo size is affected by genetic relatedness of filial embryo and endosperm genomes, whereas endosperm size is not (Wu *et al.*, 2013). Hence, the endosperm genotype can influence allocation of maternal resources to the embryo. Although the underlying roles of gene imprinting (Lin *et al.*, 1984; Costa *et al.*, 2012) and genome dosage balance in endosperm development have been studied extensively (Birchler and Hart, 1987; Birchler, 1993), we still know surprisingly little about precisely how and why embryo development is dependent on the status of the endosperm (Chapter 9, this volume). While endosperm–embryo signaling interactions during early seed development have received appropriate attention (Costa *et al.*, 2014; Chapter 8, this volume), it is likely the maize embryo's dependence on endosperm extends into the grain-filling phase of seed development.

The dependence of the embryo on the endosperm has a structural manifestation in the maize seed. The endosperm extends beneath the embryo, physically separating it from the maternal vasculature at the base of the kernel (Leroux *et al.*, 2014). Thus, the endosperm is positioned to mediate and perhaps control transfer of nutrients to the embryo throughout its development. Prior to about 7 days after pollination (DAP), a distinct set of endosperm cells form an embryo-surrounding region (ESR) with features suggesting active secretion of nutrients to the embryo (Schel *et al.*, 1984). By the time of rapid grain-filling, *c.*12 DAP, the ESR is replaced by cells that line the embryo-cavity in the endosperm (Cossegal *et al.*, 2007; Leroux *et al.*, 2014). The extent and significance of nutrient transport across this interface is an open question. Recent studies highlighting the importance of this region suggest that transport across it is physiologically important to embryo development.

Of particular interest are genes expressed specifically in the endosperm, that when mutated create *dek* phenotypes affecting both the endosperm and embryo. Striking examples are *meg1* (Costa *et al.*, 2012) and *sweet4c* (Sosso *et al.*, 2015). The *meg1* gene encodes a peptide required for differentiation of the basal endosperm transfer cell layer (BETL), whereas *sweet4c* encodes a hexose transporter specifically expressed in the BETL (Sosso *et al.*, 2015; Chapter 5, this volume). The *meg1* gene is imprinted and expressed preferentially from the maternal genome. RNAi mutants that inhibit

*meg1* expression in the basal endosperm reduce endosperm and embryo size (Costa *et al.*, 2012). Consistent with Neuffer and Sheridan (1980), the impact of reduced *meg1* expression on embryo size is greater than the effect on endosperm. Similarly, loss-of-function mutants in *sweet4c* severely impact development of both embryo and endosperm, producing a severe *dek* (empty pericarp subtype) phenotype (Sosso *et al.*, 2015). However, if rescued from the mutant kernel and supplied sugars in culture, the embryo is capable of developing into a seedling. The implication is that whereas *in vivo* the embryo has limited capability for importing sugars from the maternal transfer zone, this can be bypassed in culture. Hence, there is intimate coupling between metabolic signals and BETL differentiation during endosperm development (Sosso *et al.*, 2015). Together, these results suggest BETL differentiation is a critical point of failure in *dek* mutants that exhibit endosperm-dominant phenotypes. Consistent with this hypothesis, BETL differentiation is also disrupted in seeds that abort due to endosperm genome dosage imbalance (Pennington *et al.*, 2008; Chapter 9, this volume).

Other mutants that point to a strong connection between embryo growth and BETL differentiation include plastid *6-phosphogluconate dehydrogenase* (*6-pgd*) (Spielbauer *et al.*, 2013), the *de18* auxin biosynthetic mutant (Bernardi *et al.*, 2012) and the *miniature 1* (*mn1*) cell wall invertase mutant (Cheng *et al.*, 1996). Analysis of the *6-pgd* mutant indicates that activity of the oxidative pentose phosphate pathway in endosperm is required for BETL formation and embryo development (Spielbauer *et al.*, 2013). Its starch-deficient phenotype is an interesting contrast to the shrunken/brittle mutants that specifically ablate starch biosynthesis while having little impact on BETL formation and embryo development. The impacts of *de18* and *mn1* on embryo growth are intermediate in severity. Although the primary lesion in *de18* is disruption of auxin synthesis in endosperm, the mutation causes a roughly proportional reduction in endosperm and embryo size (Bernardi *et al.*, 2012). The *mn1* mutant (Cheng *et al.*, 1996) is deficient in an endosperm-localized cell wall invertase that normally works in concert with the *sweet4c* hexose transporter to promote the import of sugars from the maternal phloem unloading zone (Sosso *et al.*, 2015). While *mn1* sharply reduces endosperm size, mutant embryos are viable and capable of germination (Cheng *et al.*, 1996).

The relationship between BETL function and embryo growth is worthy of further investigation, since the BETL's importance implies the endosperm's capacity for nutrient uptake is critical for embryo growth. In Arabidopsis, sucrose secreted by the inner integuments of the seed coat traverses the endosperm on its way to the embryo, and several SWEET transporters expressed in endosperm are implicated in secretion of sugars into the apoplast surrounding the embryo (Chen *et al.*, 2015). If "pass-through" transport predominates in maize, then nutrient flux to the embryo would be directly proportional to that across the BETL. Alternatively, if the endosperm's primary role in supporting the embryo during grain-filling is to establish "sink strength" of the kernel, flux across the BETL might be only indirectly related to flux to the embryo. The concept of "sink strength" is not well defined at the molecular level, but it is plausible that, due to its position in the seed, the endosperm has a unique capacity to elicit acquisition of sugars and other nutrients from maternal tissues (Chapter 15). Once released into the seed apoplastic space, nutrients would be available to both organs.

While the BETL has received much attention as a critical component of the endosperm–maternal plant interface, we lack a comparable understanding of what transpires at the endosperm–embryo interface. Interactions across that interface likely occur throughout development, up to and including the interplay of gibberellic acid (GA) and abscisic acid (ABA) hormone signals that control re-mobilization of starch reserves during seed germination (Hoecker *et al.*, 1995, 1999). I will revisit the formation of this interface below in the discussion of mutants that alter partitioning of resources between embryo and endosperm.

## 4.2.2 Genetic dissection of embryo–endosperm interactions

To systematically dissect genetic interactions between embryo and endosperm that underlie *dek* phenotypes, ideally one would manipulate the genotypes of filial organs independently, so a mutant embryo can be studied in a seed with a wild-type endosperm, and vice versa. The B-A chromosome translocations of maize provide this capability (see Chapter 9, this volume). B-A translocation chromosomes non-disjoin frequently at the second mitotic division of pollen development, giving rise to pollen grains with non-identical sperm: one is deficient for a specific chromosome arm (hypoploidy), while the other carries an additional copy of that arm (hyperploidy). When used to pollinate a plant carrying a *dek* mutation, pollen from a B-A translocation stock produces two non-concordant seed genotypes, hypoploid (mutant) embryo/ hyperploid (wild-type) endosperm and hyperploid (wild-type) embryo/hypoploid (mutant) endosperm, depending on which of the two sperm fertilizes the egg and the central cell, respectively.

B-A translocation experiments confirm that "endosperm dominant" phenotypes are not limited to endosperm-specific genes, such as *meg1* (Costa *et al.*, 2012) and *sweet4c* (Sosso *et al.*, 2015). A common pattern for *dek* mutants is that a wild-type endosperm is minimally affected by the presence of a mutant embryo, whereas a mutant endosperm is sufficient to block development of both organs. For example, the *empty pericarp 6* (*emp6*) gene encodes an RNA-binding protein that is essential for both embryo and endosperm formation (Chettoor *et al.*, 2015). While B-A translocation experiments confirmed *emp6* function is required independently in embryo and endosperm, a wild-type endosperm will develop normally, regardless of whether the embryo is mutant or wild type. By contrast, an emp6 mutant endosperm blocks development of a wild type embryo. Similar patterns are exhibited by the *rough endosperm 3* (*rgh3*) RNA splicing factor mutant (Fouquet *et al.*, 2011) and the aforementioned *6-pgd* mutant (Speilbauer *et al.*, 2013).

In some cases, a wild-type endosperm can be shown to at least partially rescue a mutant embryo. This category of mutants is especially interesting, because the interaction phenotype suggests enrichment for functions specifically involved in endosperm support of the embryo. Neuffer and Sheridan (1980) used B-A translocations to study interactions of embryo and endosperm genotypes in 19 *dek* mutants. In 4 of 19 cases, a normal endosperm could at least partially rescue a mutant embryo. A similar pattern of wild-type endosperm partially rescuing a mutant embryo was described for *dek24* (Chang and Neuffer, 1994).

Given the relatively small number of *deks* analyzed by B-A translocations, it is likely we have barely sampled mutants that exhibit significant non-autonomous endosperm–embryo interactions. Moreover, it is likely such interactions escape detection, except in favorable cases where the endosperm–embryo interaction is particularly strong due to the confounding effects of segmental aneuploidy, particularly in the endosperm (Chang and Neuffer, 1994; see Chapter 9, this volume). These limitations notwithstanding, a systematic B-A translocation analysis of *dek* mutants would likely provide insight into the developmental and metabolic pathways that underlie endosperm–embryo interactions. Inclusion of appropriate controls can enhance the sensitivity of the analysis, especially if the experiment is performed with an inbred background. Systematic analysis of embryo–endosperm interactions using B-A translocations would be greatly facilitated if the identities and genome locations of the mutations of interest are known, because: (i) one then knows which of the standard set of 19 B-A stocks is needed to uncover each mutant, thus minimizing the number of crosses; and (ii) if the mutant genes have been identified, the results can be integrated with other genomics analyses, including networks and models developed from transcriptomics, proteomics and metabolomics data. A large collection of transposon tagged seed mutants with precisely mapped locations in an inbred genome is ideal for this purpose (McCarty *et al.*, 2005, 2013; Hunter *et al.*, 2014).

### 4.2.3 Embryo-lethal (*emb*) mutants

The *embs* are a large class of mutants with defective embryos, but more-or-less normal endosperm size and morphology (Chapter 7). The latter property distinguishes them from *deks*. This class of maize mutants is unique in the sense that the status of the endosperm is usually less obvious in comparable mutants of other organisms, such as Arabidopsis (Tzafrir *et al.*, 2004; Bryant *et al.*, 2011). Hence, maize *emb* mutants have the potential to identify novel genes required for embryogenesis in grasses and possibly other plants.

Maize *emb* mutants are especially important, because the mechanisms of embryogenesis are less well understood in grasses than Arabidopsis. Zhao *et al.* (2017) provided a review of early embryogenesis in Arabidopsis and maize, including a cogent set of open questions. Key conserved processes include: (i) establishment of apical–basal polarity; (ii) elaboration of the apical–basal axis via patterning of polar auxin transport; (iii) differentiation of shoot and root apical meristems along the apical–basal axis; and (iv) activation of embryo-specific gene expression and hormone signaling programs by the LEC1-AFL-B3 (LAFL) transcription factor network (Suzuki *et al.*, 2008; Jia *et al.*, 2014). In Arabidopsis, apical–basal polarity of the zygote is acquired prior to the first cell division and is maintained by polar auxin transport (Zhao *et al.*, 2017). Patterning of polar auxin transport is regulated by localization of PIN proteins. Apical and basal cell lineages are marked early by expression of WOX8 and WOX2, respectively. WOX genes of the WUSCHEL transcription factor family are key regulators of meristem function. By contrast, while apical–basal polarity is also apparent by the first zygotic cell division in maize; patterns of PIN protein localization and WOX expression (specifically WUS1 and WOX5 orthologs) are not evident until the transition stage, by which time the embryo contains at least 100 cells (Forestan *et al.*, 2010; Chen *et al.*, 2014; Zhao *et al.*, 2017). Hence, there is a substantial gap in our understanding of events between the initial polarization of the zygote and meristem differentiation. Once meristems have formed, specialized gene expression and hormone signaling programs activated by the LAFL network control deposition of protein, lipid and starch reserves in cotyledons (scutellum), culminating in formation of a desiccation-tolerant, dormant (quiescent) embryo.

Arabidopsis embryo mutants are enriched for genes involved in plastid genome expression (Tzafrir *et al.*, 2004), such as plastid ribosomal protein genes (Romani *et al.*, 2012). Such mutants are typically blocked at the globular embryo stage, immediately prior to differentiation of the shoot apical meristem (SAM). In dicots, the plastid genome contains several genes essential for cell viability, suggesting a basis for embryo lethality of mutations that block plastid gene expression (Bryant *et al.*, 2011). Because the plastid genome in grasses does not include the aforementioned essential genes, maize plastid mutants typically produce albino seedling phenotypes rather than embryo-lethality (Stern *et al.*, 2004). Or so we thought...

## 4.3 The Curious Role of Plastid Signaling in Maize Embryogenesis

Given the fundamental importance of meristem differentiation, Clark and Sheridan (1991) and Sheridan and Clark (1993) focused on a substantial subset of *embs* that arrest development at the transition stage (see Chapter 7, this volume). A number of this type of *emb* mutant share a common, unexpected feature: most disrupt plastid gene expression (Ma and Dooner, 2004; Magnard *et al.*, 2004; Sosso *et al.*, 2012; Shen *et al.*, 2013; Li *et al.*, 2015; Yang *et al.*, 2016). This finding was counter to expectations based on prior evidence showing maize plastid mutants typically survive embryogenesis (Stern *et al.*, 2004). Resolution of the paradox is that the phenotype of plastid gene expression mutants (*emb* vs. *albino*) is strictly dependent on genetic background (Zhang *et al.*, 2013; Yang *et al.*, 2016). In the W22 inbred background, plastid mutants have a characteristic *emb* phenotype in which

embryo development arrests at the transition stage, whereas the same mutant in a B73 background segregates with an albino seedling phenotype (Yang *et al.*, 2016). These results have three surprising implications: (i) plastid gene expression is evidently required during early embryogenesis in W22. The maize embryo, in contrast to that of Arabidopsis, does not form photosynthetically active chloroplasts at any stage of development. Moreover, proplastids in the transition stage embryo show no evidence of differentiation that entails expression of plastid-encoded genes (Shen *et al.*, 2013); (ii) the requirement for plastid gene expression is suppressed in B73, indicating that a small number of genes determine whether or not plastid gene expression is required during maize embryogenesis; and (iii) in many respects, the *emb* phenotype in W22 is analogous to the *emb* phenotypes of Arabidopsis plastid mutants, which prevent SAM formation (Romani *et al.*, 2012). This suggests that a reassessment of the broader role of plastid signaling in plant embryogenesis could be in order (Bryant *et al.*, 2011).

*emb* mutants in W22 are readily tested for background suppression by crossing into B73 and screening for segregation of albino seedlings in the $F_2$. Indeed, we used background dependence to infer that *duf177a*, a gene encoding a highly conserved protein of unknown function, is likely to influence plastid gene expression (Yang *et al.*, 2016). This was confirmed by demonstrating that *duf177a* is required for accumulation of plastid 23S rRNA. $F_2$ segregation data indicate that several genes differentiate W22 and B73 in this respect (Yang *et al.*, 2016).

In retrospect, we could have anticipated the conditional role of plastid gene expression in maize embryogenesis. As is so often the case in maize genetics, the phenomenon underlying suppression of *embs* was very likely observed previously in a different context. Coe and co-workers (1988) showed that genetic background strongly modifies leaf striping phenotypes of the *iojap* (Han *et al.*, 1992) and *striate2* (Williams and Kermicle, 1974) mutants, where subsets of leaf cells are deficient in plastid ribosomes. Indeed, it was known that in some backgrounds *iojap* has an *emb* phenotype (Coe *et al.*, 1988). One of the genetic modifiers, *inhibitor of striate 1* (ISR), was shown to encode a plastid-localized, hydrolase-like protein of unknown function (Park *et al.*, 2000). In developing leaves, ISR inhibits proliferation of cells that lack plastid ribosomes, thus narrowing the albino stripes in leaves of *striate2* plants (Park *et al.*, 2000). Plausibly, a similar inhibition of cell proliferation during embryogenesis in backgrounds that carry dominant ISR alleles could account for the *emb* phenotype of plastid mutants.

## 4.4 Embryo Cavity Formation and Coordination of Embryo and Endosperm Growth

Another intriguing phenomenon revealed by examination of *emb* mutants is formation of the embryo cavity in the endosperm (Clark and Sheridan, 1991; Fig. 4.1, bottom). In *emb* mutant seed, the endosperm autonomously forms an empty cavity in the space that would normally be occupied by the embryo. The embryo cavity is typically somewhat larger than would be needed to accommodate a normal embryo. Moreover, the surface of the cavity is rough and disorganized in appearance, in contrast to the smooth interface that normally forms between the developing embryo and endosperm. These features suggest formation of the embryo–endosperm interface is coordinated by mutual interactions between the two organs. As noted above, this interface potentially plays a crucial role mediating transfer of nutrients and hormone signals during development and germination. Several mutants disrupt this interface, causing loss of coordination between embryo and endosperm growth. In W22 (ACR) (Brink's color converted W22, i.e. A1, C1 and R1 alleles introgressed into standard W22), which carries genes required for aleurone anthocyanin accumulation, the *viviparous-8* (*vp8*) mutant has a distinctive "widow's peak" phenotype characterized by absence of anthocyanin in the aleurone cells surrounding the lateral margins of the embryo cavity

(Suzuki *et al.*, 2008). In this background, development of the *vp8* embryo is retarded, whereas the embryo cavity is enlarged, accentuating the anthocyanin pattern. *Vp8* encodes a membrane-localized carboxypeptidase, and is the ortholog of the *Amp1* gene of Arabidopsis (Helliwell *et al.*, 2001). *Vp8* regulates the abscisic acid level during embryo development, at least in part through regulation of the LAFL transcription factor network (Suzuki *et al.*, 2008; Suzuki and McCarty, 2008). The molecular basis for its independent role in embryo cavity formation in the endosperm is unknown.

Another class of mutants that disrupts coordination of the interface of embryo and endosperm growth are a group of "big embryo" mutants that increase embryo size at the expense of the endosperm (Suzuki *et al.*, 2015). The *bige1* mutant has an enlarged embryo that over-grows the endosperm cavity, causing disruption of the interface at the margin of the scutellum. In the disturbed region at the margin of the cavity, aleurone cells are less pigmented and less clearly differentiated compared to normal. The BigE1 gene encodes a conserved MATE transporter that is localized to the trans-Golgi apparatus (Suzuki *et al.*, 2015). Because *bige1* organ size and meristem phenotypes have striking similarities to phenotypes of CYP78A P450 mutants of Arabidopsis (Anastasiou *et al.*, 2007) and rice (Nagasawa *et al.*, 2013), we speculate BIGE1 might transport an as yet unidentified product or intermediate in the still enigmatic CYP78A signaling pathway.

## 4.5 LAFL Network Coordination of Hormone Signaling Associated with Transition from Embryogenesis to Germination and Vegetative Development

Both *vp8* and *bige1* can be classified as heterochronic mutants that alter the timing or rate of developmental processes. Although a functional connection between *vp8* and *bige1* is not yet established, both alter the vegetative meristem and cause accelerated production of lateral organs (Suzuki *et al.*,

2008, 2015). However, the genes have opposing effects on regulation of the LAFL B3 transcription factor network that has a central role in embryo development. LEC1 expression is inhibited in the *vp8* mutant, whereas in the *bige1* mutant increased embryo size is associated with prolonged expression of LEC1 during development. Studies in Arabidopsis reveal the LAFL B3 network is autoregulated, such that overexpression of one gene in the network can be sufficient to activate the entire network (Jia *et al.*, 2013). During seed maturation, VAL type B3 transcription factors (Suzuki *et al.*, 2007; Suzuki and McCarty, 2008) act as repressors that counter the autoactivation cycle, leading to network inactivation in the mature seed (Jia *et al.*, 2014). VAL and AFL type B3 DNA binding proteins function as repressors and activators, respectively, of embryonic gene expression and recognize the same RY/SPH cis-element (Suzuki *et al.*, 1997, 2007). Autoregulation is enabled, because the principal genes in the LAFL network contain multiple RY/SPH regulatory elements in upstream as well as internal locations. VAL-mediated repression of the network late in embryo development is necessary for transition from seed to seedling development (Jia *et al.*, 2013, 2014).

While key elements of the LAFL B3 network are conserved in grasses, there are intriguing differences. Grasses do not have an ortholog of the LEC2 B3 protein, suggesting organization of the network could be simplified somewhat in maize. Orthologs ABI3 and VP1 in Arabidopsis and maize, respectively, play a special role in the network by coupling activity of the B3 network to ABA signaling during the maturation phase of embryo development (Suzuki *et al.*, 2007, 2008, 2014). Integration of B3 network activity and ABA signaling is mediated by the unique domain architecture of VP1/ABI3 proteins (McCarty *et al.*, 1991; Carson *et al.*, 1997; Suzuki *et al.*, 2014). An ABA-regulated co-activator/co-repressor domain (COAR) located in the N-terminal region is coupled with a B3 DNA binding domain that binds specifically to RY/SPH cis-elements that are also recognized by LEC2, FUS3, and VAL B3 proteins (Carson *et al.*, 1997; Suzuki *et al.*, 2014).

## 4.6 Signaling across the Embryo–Endosperm Interface

In addition to mediating activation of maturation-related gene expression in the embryo, VP1 is also required for repression of germination-specific gene expression in the aleurone (Hoecker *et al.*, 1995, 1999). Although *Vp1*-mediated repression of germination-specific genes, including α-amylase, is cell-autonomous within the aleurone (Hoecker *et al.*, 1995), B-A translocation experiments and double mutant analyses using GA- and ABA-deficient mutants reveal that de-repression of α-amylase genes in *vp1* aleurone cells is conditioned by non-cell-autonomous signals from the embryo (Hoecker *et al.*, 1999). With respect to de-repression of the germination response, ablation of the embryo (by introducing an early-acting *emb* mutant) is equivalent to the presence of a viviparous (germinating) embryo in the seed, indicating that the embryo is normally a source of signals that interact with *Vp1* to inhibit germination during seed maturation. ABA synthesized in the developing embryo can only partially account for the effect of embryo status on VP1-mediated repression in aleurone cells, implying that the embryo is a source of unidentified non-ABA signals (Hoecker *et al.*, 1999).

## 4.7 Open Questions and Research Opportunities in Kernel Development

The insights gained thus far from an analysis of seed mutants highlight outstanding questions and research opportunities. The following description is by no means comprehensive and is intended to illuminate questions related to a few key topics.

### 4.7.1 Metabolic, structural, and signaling mechanisms that coordinate growth and development of embryo and endosperm

The following are some of the challenges: (i) delineating mechanisms that form and support the embryo–endosperm interface throughout seed development; (ii) describing the molecular interface, including metabolite forms and nutrients that can cross this interface; (iii) testing the metabolite "pass through" hypothesis by identifying predicted transporters that mediate transfer of carbon (sugars) and nitrogen (amino acids) to the embryo; (iv) defining the concept of "sink strength" at the molecular level (see Chapter 15); and (v) identifying mechanisms that partition resources and regulate growth of the embryo and endosperm.

Formation of the embryo cavity in the endosperm is an important process for establishing the embryo–endosperm interface. Virtually nothing is known about how this structure is created. The rough surface of the cavity in *emb* mutants highlights disruption of interactions between endosperm and embryo that create a normally smooth interface.

What role does the endosperm–embryo interface play in embryo growth during grain-filling? If there is significant "pass-through" transfer of sugars, amino acids and other molecules across the endosperm–embryo interface throughout development, we might expect to find an array of transporters in the epithelial tissues that line both sides of the interface. Alternatively, if the endosperm indirectly affects embryo uptake of nutrients by determining overall "sink strength," then the embryo need not be dependent on transfer of nutrients from the endosperm. In that case, how is relative partitioning of resources between endosperm and embryo controlled? Are there circumstances in which the endosperm and embryo compete for resources? If not, then why not? For example, endosperms of starch-deficient (shrunken) mutants accumulate high levels of sugar and assimilate much less carbon than wild-type, but this seems to have little impact on embryo growth.

### 4.7.2 Applying the power of comparative genetics and genomics to seed development

An important principle of developmental genetics is that due to the contingencies of evolution and environment, different organisms often utilize different genetic programs. For this reason, there is much to be learned from comparative genetic and genomic analyses of fundamental processes, such as seed development. In addition to the open questions highlighted by Zhao *et al.* (2017), we note that in comparing maize and Arabidopsis embryogenesis one obvious contrast is the vast difference in scale of the embryo, beginning at the earliest stages of development when patterns are established. What mechanisms pattern the maize embryo and did they evolve to accommodate its large size, or are they inherently insensitive to changes in scale?

### 4.7.3 Understanding the role of plastid gene expression in maize embryogenesis

Do the novel genetics revealed in maize embryogenesis identify a fundamental role for plastid signaling in plant embryogenesis? The background-dependent phenotypes of *emb* mutants implicate plastid gene expression in early embryogenesis. This is in spite of the fact that proplastids in the maize embryo do not discernibly differentiate at the transition stage, and do not form chloroplasts at any stage prior to germination. Nevertheless, parallels with corresponding plastid mutants in Arabidopsis suggest a fundamental reassessment of the role of plastid signaling in plant embryogenesis could be in order. Robust genetic and genomic tools will allow identification of genes that differentiate W22 and B73 in this respect. This is, on one hand, a straightforward genome mapping and analysis problem, and its solution will likely provide insight into the underlying pathway. On the other hand, the full extent of genetic variation in maize for this phenotype has not yet been explored. Maize inbreds can be genetically classified into permissive (B73-like) and non-permissive (W22-like) groups, depending on whether reference mutants such as *duf177a* have *emb* or albino seedling phenotypes. Our *duf177a* segregation data (Yang *et al.*, 2016) suggest that multiple loci in B73 contribute to suppression of *emb* phenotypes, so it is possible that a broad survey of inbreds will identify additional genes.

### 4.7.4 Understanding how the LAFL B3 network mediates the transition from embryogenesis to quiescence to vegetative development

Among open questions on this topic are the following: (i) the mechanism by which VP8 regulates the LAFL network and ABA turnover; (ii) the relationship between BigE1 and CYP78A signaling; (iii) the relationship between CYP78A signaling and the LAFL network; (iv) the nature of chromatin modifications implicated in autoregulation of the LAFL B3 transcription factor network; and finally (v) identification of non-ABA signals that mediate VP1 repression of germination-specific gene expression in the aleurone.

## References


Anastasiou, E., Kenz, S., Gerstung, M., MacLean, D., Timmer, J., Fleck, C. and Lenhard, M. (2007) Control of plant organ size by *KLUH/CYP78A5*-dependent intercellular signaling. *Developmental Cell* 13, 843–856.

Bernardi, J., Lanubile, A., Li, Q.B., Kumar, D., Kladnik, A., *et al.* (2012) Impaired auxin biosynthesis in the *defective endosperm18* mutant is due to mutational loss of expression in the *ZmYuc1* gene encoding endosperm-specific YUCCA1 protein in maize. *Plant Physiology* 160, 1318–1328.

Birchler, J.A. (1993) Dosage analysis of maize endosperm development. *Annual Review of Genetics* 27, 181–204.

Birchler, J.A. and Hart, J.R. (1987) Interaction of endosperm size factors in maize. *Genetics* 117, 309–317.

Brink, R.A. and Cooper, D.C. (1947) The endosperm in seed development. *The Botanical Review* 13, 479–541.

Bryant, N., Lloyd, J., Sweeney, C., Myouga, F. and Meinke, D. (2011) Identification of nuclear genes encoding chloroplast-localized proteins required for embryo development in Arabidopsis. *Plant Physiology* 155, 1678–1689.

Carson, C.B., Hattori, T., Rosenkrans, L., Vasil, V., Vasil, I.K., Peterson, P.A. and McCarty, D.R. (1997) The quiescent/colorless alleles of *viviparous1* show that the conserved B3 domain of VP1 is not essential for ABA-regulated gene expression in the seed. *The Plant Journal* 12, 1231–1240.

Chang, M.T. and Neuffer, M.G. (1994) Endosperm–embryo interaction in maize. *Maydica* 39, 9–18.

Chen, J., Lausser, A. and Dresselhaus, T. (2014) Hormonal responses during early embryogenesis in maize. *Biochemical Society Transactions* 42, 325–331.

Chen, L.Q., Lin, I.W., Qu, X.Q., Sosso, D., McFarlane, H.E., *et al.* (2015) A cascade of sequentially expressed sucrose transporters in the seed coat and endosperm provides nutrition for the Arabidopsis embryo. *Plant Cell* 27, 607–619.

Cheng, W.H., Taliercio, E.W. and Chourey, P.S. (1996) The *Miniature1* seed locus of maize encodes a cell wall invertase required for normal development of endosperm and maternal cells in the pedicel. *Plant Cell* 8, 971–983.

Chettoor, A.M., Yi, G., Gomez, E., Hueros, G., Meeley, R.B. and Becraft, P.W. (2015) A putative plant organelle RNA recognition protein gene is essential for maize kernel development. *Journal of Integrative Plant Biology* 57, 236–246.

Clark, J.K. and Sheridan, W.F. (1991) Isolation and characterization of 51 embryo-specific mutations of maize. *Plant Cell* 3, 935–951.

Coe, E.H., Jr., Thompson, D. and Walbot, V. (1988) Phenotypes mediated by the iojap genotype in maize. *American Journal of Botany* 75, 634–644.

Cossegal, M., Vernoud, V., Depège, N. and Rogowsky, P.M. (2007) The embryo surrounding region. In: *Endosperm*. Springer, Berlin, Heidelberg, Germany, pp. 57–71.

Costa, L.M., Yuan, J., Rouster, J., Paul, W., Dickinson, H. and Gutiérrez-Marcos, J.F. (2012) Maternal control of nutrient allocation in plant seeds by genomic imprinting. *Current Biology* 22, 160–165.

Costa, L.M., Marshall, E., Tesfaye, M., Silverstein, K.A., Mori, M., *et al.* (2014) Central cell-derived peptides regulate early embryo patterning in flowering plants. *Science* 344, 168–172.

Forestan, C., Meda, S. and Varotto, S. (2010) ZmPIN1-mediated auxin transport is related to cellular differentiation during maize embryogenesis and endosperm development. *Plant Physiology* 152, 1373–1390.

Fouquet, R., Martin, F., Fajardo, D.S., Gault, C.M., Gómez, E., *et al.* (2011) Maize rough endosperm3 encodes an RNA splicing factor required for endosperm cell differentiation and has a nonautonomous effect on embryo development. *Plant Cell* 23, 4280–4297.

Han, C.D., Coe, E.H., Jr. and Martienssen, R.A. (1992) Molecular cloning and characterization of iojap (ij), a pattern striping gene of maize. *The EMBO Journal* 11, 4037.

Helliwell, C.A., Chin-Atkins, A.N., Wilson, I.W., Chapple, R., Dennis, E.S. and Chaudhury, A. (2001) The Arabidopsis *AMP1* gene encodes a putative glutamate carboxypeptidase. *Plant Cell* 13, 2115–2125.

Hoecker, U., Vasil, I.K. and McCarty, D.R. (1995) Integrated control of seed maturation and germination programs by activator and repressor functions of Viviparous-1 of maize. *Genes & Development* 9, 2459–2469.

Hoecker, U., Vasil, I.K. and McCarty, D.R. (1999) Signaling from the embryo conditions Vp1-mediated repression of α-amylase genes in the aleurone of developing maize seeds. *The Plant Journal* 19, 371–377.

Hunter, C.T., Suzuki, M., Saunders, J., Wu, S., Tasi, A., McCarty, D.R. and Koch, K.E. (2014) Phenotype to genotype using forward-genetic Mu-seq for identification and functional classification of maize mutants. *Frontiers in Plant Science* 4, 545.

Jia, H., McCarty, D.R. and Suzuki, M. (2013) Distinct roles of LAFL network genes in promoting the embryonic seedling fate in the absence of VAL repression. *Plant Physiology* 163, 1293–1305.

Jia, H., Suzuki, M. and McCarty, D.R. (2014) Regulation of the seed to seedling developmental phase transition by the LAFL and VAL transcription factor networks. *Wiley Interdisciplinary Reviews: Developmental Biology* 3, 135–145.

Leroux, B.M., Goodyke, A.J., Schumacher, K.I., Abbott, C.P., Clore, A.M., *et al.* (2014) Maize early endosperm growth and development: from fertilization through cell type differentiation. *American Journal of Botany* 101, 1259–1274.

Li, C., Shen, Y., Meeley, R., McCarty, D.R. and Tan, B.C. (2015) Embryo defective 14 encodes a plastid-targeted cGTPase essential for embryogenesis in maize. *The Plant Journal* 84, 785–799.

Lin, B.Y. (1984) Ploidy barrier to endosperm development in maize. *Genetics* 107, 103–115.

Lu, X., Chen, D., Shu, D., Zhang, Z., Wang, W., *et al.* (2013) The differential transcription network between embryo and endosperm in the early developing maize seed. *Plant Physiology* 162, 440–455.

Ma, Z. and Dooner, H.K. (2004) A mutation in the nuclear-encoded plastid ribosomal protein S9 leads to early embryo lethality in maize. *The Plant Journal* 37, 92–103.

Magnard, J.L., Heckel, T., Massonneau, A., Wisniewski, J.P., Cordelier, S., *et al.* (2004) Morphogenesis of maize embryos requires *ZmPRPL35-1* encoding a plastid ribosomal protein. *Plant Physiology* 134, 649–663.

McCarty, D.R., Hattori, T., Carson, C.B., Vasil, V., Lazar, M. and Vasil, I.K. (1991) The *Viviparous-1* developmental gene of maize encodes a novel transcriptional activator. *Cell* 66, 895–905.

McCarty, D.R., Mark Settles, A., Suzuki, M., Tan, B.C., Latshaw, S., *et al.* (2005) Steady-state transposon mutagenesis in inbred maize. *The Plant Journal* 44, 52–61.

McCarty, D.R., Latshaw, S., Wu, S., Suzuki, M., Hunter, C.T., Avigne, W.T. and Koch, K.E. (2013) Mu-seq: sequence-based mapping and identification of transposon induced mutations. *PLOS ONE* 8, e77172.

Nagasawa, N., Hibara, K.I., Heppard, E.P., Vander Velden, K.A., Luck, S., *et al.* (2013) *GIANT EMBRYO* encodes CYP78A13, required for proper size balance between embryo and endosperm in rice. *The Plant Journal* 75, 592–605.

Neuffer, M.G. and Sheridan, W.F. (1980) Defective kernel mutants of maize. I. Genetic and lethality studies. *Genetics* 95, 929–944.

Park, S.H., Chin, H.G., Cho, M.J., Martienssen, R.A. and Han, C.D. (2000) Inhibitor of striate conditionally suppresses cell proliferation in variegated maize. *Genes & Development* 14, 1005–1016.

Pennington, P.D., Costa, L.M., Gutiérrez-Marcos, J.F., Greenland, A.J. and Dickinson, H.G. (2008) When genomes collide: aberrant seed development following maize interploidy crosses. *Annals of Botany* 101, 833–843.

Romani, I., Tadini, L., Rossi, F., Masiero, S., Pribil, M., *et al.* (2012) Versatile roles of Arabidopsis plastid ribosomal proteins in plant growth and development. *The Plant Journal* 72, 922–934.

Schel, J.H.N., Kieft, H. and Lammeren, A.V. (1984) Interactions between embryo and endosperm during early developmental stages of maize caryopses (*Zea mays*). *Canadian Journal of Botany* 62, 2842–2853.

Shen, Y., Li, C., McCarty, D.R., Meeley, R. and Tan, B.C. (2013) Embryo defective12 encodes the plastid initiation factor 3 and is essential for embryogenesis in maize. *The Plant Journal* 74, 792–804.

Sheridan, W.F. and Clark, J.K. (1993) Mutational analysis of morphogenesis of the maize embryo. *The Plant Journal* 3, 347–358.

Sheridan, W.F. and Neuffer, M.G. (1980) Defective kernel mutants of maize II. Morphological and embryo culture studies. *Genetics* 95, 945–960.

Sosso, D., Canut, M., Gendrot, G., Dedieu, A., Chambrier, P., *et al.* (2012) *PPR8522* encodes a chloroplast-targeted pentatricopeptide repeat protein necessary for maize embryogenesis and vegetative development. *Journal of Experimental Botany* 63, 5843–5857.

Sosso, D., Luo, D., Li, Q.B., Sasse, J., Yang, J., *et al.* (2015) Seed filling in domesticated maize and rice depends on SWEET-mediated hexose transport. *Nature Genetics* 47, 1489–1493. DOI:10.1038/ng.3422

Spielbauer, G., Li, L., Römisch-Margl, L., Do, P.T., Fouquet, R., *et al.* (2013) Chloroplast-localized *6-phosphogluconate dehydrogenase* is critical for maize endosperm starch accumulation. *Journal of Experimental Botany* 64, 2231–2242.

Stern, D.B., Hanson, M.R. and Barkan, A. (2004) Genetics and genomics of chloroplast biogenesis: maize as a model system. *Trends in Plant Science* 9, 293–301.

Suzuki, M. and McCarty, D.R. (2008) Functional symmetry of the B3 network controlling seed development. *Current Opinion in Plant Biology* 11, 548–553.

Suzuki, M., Kao, C.Y. and McCarty, D.R. (1997) The conserved B3 domain of VIVIPAROUS1 has a cooperative DNA binding activity. *Plant Cell* 9, 799–807.

Suzuki, M., Wang, H.H.Y. and McCarty, D.R, (2007) Repression of the *LEAFY COTYLEDON 1/B3* regulatory network in plant embryo development by *VP1/ABSCISIC ACID INSENSITIVE 3-LIKE B3* genes. *Plant Physiology* 143, 902–911.

Suzuki, M., Latshaw, S., Sato, Y., Settles, A.M., Koch, K.E., *et al.* (2008) The maize *Viviparous8* locus, encoding a putative ALTERED MERISTEM PROGRAM1-like peptidase, regulates abscisic acid accumulation and coordinates embryo and endosperm development. *Plant Physiology* 146, 1193–1206.

Suzuki, M., Wu, S., Li, Q. and McCarty, D.R. (2014) Distinct functions of COAR and B3 domains of maize VP1 in induction of ectopic gene expression and plant developmental phenotypes in Arabidopsis. *Plant Molecular Biology* 85, 179–191.

Suzuki, M., Sato, Y., Wu, S., Kang, B.H. and McCarty, D.R. (2015) Conserved functions of the MATE Transporter BIG EMBRYO1 in regulation of lateral organ size and initiation rate. *Plant Cell* 27, 2288–2300.

Tzafrir, I., Pena-Muralla, R., Dickerman, A., Berg, M., Rogers, R., *et al.* (2004) Identification of genes required for embryo development in Arabidopsis. *Plant Physiology* 135, 1206–1220.

Williams, E. and Kermicle, J.L. (1974) Fine structure of plastids in maize leaves carrying the striate-2 gene. *Protoplasma* 79, 401–408.

Wu, C.C., Diggle, P.K. and Friedman, W.E. (2013) Kin recognition within a seed and the effect of genetic relatedness of an endosperm to its compatriot embryo on maize seed development. *Proceedings of the National Academy of Sciences of the United States of America* 110, 2217–2222.

Yang, J., Suzuki, M. and McCarty, D.R. (2016) Essential role of conserved DUF177A protein in plastid 23S rRNA accumulation and plant embryogenesis. *Journal of Experimental Botany* 67, 5447–5460.

Zhan, J., Thakare, D., Ma, C., Lloyd, A., Nixon, N.M., *et al.* (2015) RNA sequencing of laser-capture microdissected compartments of the maize kernel identifies regulatory modules associated with endosperm cell differentiation. *Plant Cell* 27, 513–531.

Zhang, Y.F., Hou, M.M. and Tan, B.C. (2013) The requirement of WHIRLY1 for embryogenesis is dependent on genetic background in maize. *PLOS ONE* 8, e67369.

Zhao, P., Begcy, K., Dresselhaus, T. and Sun, M.X. (2017) Does early embryogenesis in eudicots and monocots involve the same mechanism and molecular players? *Plant Physiology* 173, 130–142.

# 5 The Basal Endosperm Transfer Layer (BETL): Gateway to the Maize Kernel

**Prem S. Chourey[1] and Gregorio Hueros[2,*]**
*[1]U.S. Department of Agriculture, Agricultural Research Service, and University of Florida, USA; [2]Departamento de Biomedicina y Biotecnología, Universidad de Alcalá, Madrid, Spain*

## 5.1 Introduction

The maize basal endosperm transfer layer (BETL), with its unique location at the juncture of maternal and filial tissues (Fig. 5.1), plays a critical role in grain-filling and defense. Symplastic discontinuity between the mother plant and BETL is elaborated through programmed cell death (PCD) in the placenta–chalaza (P–C) region. Early in development, cells in the BETL undergo structural modification through development of wall ingrowths (WIGs), which facilitate transport of sugars, nutrients, and water into the kernel. WIG development is an evolutionarily conserved trait, as it occurs in other cereal and plant species, including the maize precursor, teosinte. The BETL partitions the current and subsequent plant generation and creates an antimicrobial barrier between them with cytotoxic peptides. Our insight into the structure, function, and signaling roles of the BETL will foster future research into the development and function of this important seed tissue.

The BETL in most maize varieties is a dome-shaped layer composed of two to three strata of highly specialized, elongated transfer cells (TC) (Kiesselbach, 1949; McCurdy and Hueros, 2014). The BETL in maize and its relatives is morphologically distinctive (Jain *et al.*, 2008a; Dermastia *et al.*, 2009); it is contiguous with, but very different from, the aleurone layer. The TCs are characterized by a labyrinth wall with flange type (ridge-like) projections—the WIGs—that increase the plasma membrane surface area. WIGs are a defining hallmark of many types of plant transfer cells (Offler *et al.*, 2003) and are postulated to confer enhanced solute transport and anchor numerous functions, including receptor sites for various signaling molecules. Robust WIG growth is essential for normal seed development, as shriveled, miniature, defective/aborted kernel mutants are not only associated with but also potentially caused by retarded WIG formation (Kang *et al.*, 2009).

The most cited function of the BETL is transport of photo-assimilates and nutrients from phloem termini in the pedicel to filial tissues in the kernel; there are no plasmodesmata connecting these cells (Felker and Shannon, 1980; Kang *et al.*, 2009). The P-C region that separates the BETL and the vascular elements is composed of several layers of dead, empty cells (Figure 5.1). These cells undergo PCD about 4 days after pollination

*Corresponding author e-mail: gregorio.hueros@uah.es

 

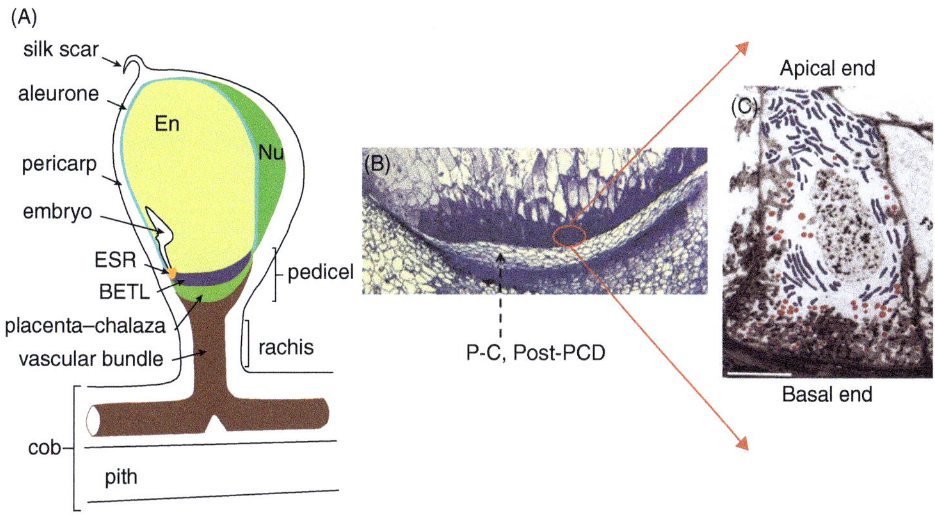



**Fig. 5.1.** (A) Schematic diagram of a maize kernel about 10 DAP. En, endosperm; Nu, nucellus; ESR, embryo-surrounding region; BETL, basal endosperm transfer cell layer. (B) Light microscopy image of the pedicel area. The section was stained with toluidine blue and transfer cells are dark blue; P-C, placenta–chalaza. (C) Electron microscopy image of a single transfer cell. ER cisternae are digitally colored in blue, and mitochondria are colored in red.

(DAP), a stage that coincides with the initiation of BETL differentiation (Kladnik *et al.*, 2004). The loss of nuclei in the P-C layers occurs in a coordinated fashion, indicating PCD is non-cell autonomous and very likely responds to a signal originated in the BETL (Kladnik *et al.*, 2004). PCD is speculated to increase hydraulic conductance to developing kernels (Kladnik *et al.*, 2004; Dermastia *et al.*, 2009). PCD is also seen in teosinte (Dermastia *et al.*, 2009) and sorghum (Jain *et al.*, 2008a); however unlike maize, the P-C region in sorghum is not a mass of dead and empty cells, but rather is composed of a sac or a cavity filled with sap, which may confer greater tolerance to drought through increased uptake of hexose and water by turgor sensing (Jain *et al.*, 2008a; Chapter 15, this volume).

## 5.2 Cellular Changes during BETL Development

In the maize caryopsis, vascular terminals in the pedicel form a cup-shaped cushion filled with several layers of crushed, dead maternal cells in the P-C area (Kladnick

*et al.*, 2004). Epithelial cells committed to differentiate into aleurone at any other area of the endosperm surface differentiate into TCs in the area facing the P-C. The entire endosperm, except for the TC region, is surrounded by a cuticular layer (Davis *et al.*, 1990), suggesting the BETL is the only site for metabolic exchange between filial and maternal tissues.

Ultrastructural studies (Davis *et al.*, 1990; Monjardino *et al.*, 2013) revealed that TCs elongate along the apico-basal axis and are enclosed in a very thick cell wall. The cell wall first grows homogeneously, but later develops a large number of inward projections (WIGs) that in maize are of the flange-type; in the most differentiated cells, large rims emerge from the inner surface of the cell wall (predominantly from the basal side of the cell) until the intracellular space is almost filled with cell wall material. Close examination of WIG ultrastructure (Monjardino *et al.*, 2013) showed accumulation of reticulate-type (papillate projections) ingrowths in basal, anticlinal cell walls. The reticulate WIGs eventually contact, forming a sponge-like network that embeds organelles on the basal side of the TCs.

The BETL is far from homogeneous (Thompson *et al.*, 2001). It contains cells with variable shapes and WIG accumulation. At 12–16 DAP, a gradient of TC development is observed along two axes (Royo *et al.*, 2007), with the most differentiated cells, including the prototypical type described above, positioned at the basal cell layer near the adgerminal side of the endosperm. Towards the abgerminal side and the inner layers, cells show progressively less WIG development and a more cubic shape. Indeed, cells at the abgerminal pole of the BETL are morphologically indistinguishable from aleurone cells, but they express TC-specific genes. These gradients of cell development suggest TC differentiation might be regulated by a combination of position effect and signaling substances diffusing from the lower part of the kernel.

Using cryofixation and low temperature embedding, Kang *et al.*, (2009) showed that the BETL in normal maize endosperm differs in multiple ways from that in the *miniature1* (*mn1*) mutant (Miller and Chourey, 1992; Cheng *et al.*, 1996). *Mn1* encodes Cell Wall Invertase-2 (INCW2), an enzyme closely associated with WIGs. Mitochondria (mt) in TCs have a unique polar spatial distribution, i.e. they are more abundant at the basal pole of these cells, adjacent to WIGs, than at the apical end (Fig. 5.1), which has little or no WIG formation. Further, the polarity of mt distribution appears to be established as early as 7 DAP, prior to WIG formation. Once WIGs are formed, mt become associated with them, suggesting a potential functional linkage. In the *mn1* mutant, these cells show greatly stunted and reduced density of WIGs and mt. This is attributable to lower levels of hexoses (see below), which are critical precursors for cell wall biosynthesis (Kang *et al.*, 2009). Similar reductions in mt density were also reported for TCs in maize *emp4*, an *empty pericarp* lethal seed mutant that lacks *EMP4*, a constitutive mt-localized protein. Although mt deficiency is the causal basis for the *emp4* seed phenotype, there is sparse growth of WIGs and reduced INCW2, as in *mn1* (Gutiérrez-Marcos *et al.*, 2007; Kang *et al.*, 2009). Simultaneous reduction of mitochondria and WIGs in these mutants, where both affect diverse aspects of energy metabolism, is interesting and suggests a potential relationship of the BETL in energy metabolism. Interestingly, WIG-deficiency is also seen in *Zmsweet4* (Sosso *et al.*, 2015), presumably due to, among other factors, reduced *Mn1* expression. An alternative and somewhat complementary interpretation of these results is that differentiation of the BETL and the development of WIGs are high energy-demanding processes and mutations that perturb nutrient transport and metabolism have an impact on BETL differentiation.

Kang *et al.* (2009) reported a significant increase in Golgi density in normal TCs between 7 and 12 DAP; during this period *mn1* cells have nearly 51% fewer Golgi. Given that Golgi cisternae are a key site for protein glycosylation, this could reflect a feed-forward response to reduced hexose levels in *mn1* (LeClere *et al.*, 2010; Chourey *et al.*, 2012). Not surprisingly, a glycoproteome profile of the two genotypes showed decreased glycosylation in *mn1* endosperm (Silva-Sanchez *et al.*, 2014, see below). TCs are also enriched in endoplasmic reticulum (ER), which is altered in *mn1* as evident from a swollen and dilated morphology.

## 5.3 Role of the BETL in Sugar Transport

As the first filial cell layer in maize endosperm, the BETL plays a pivotal role in sugar transport and physiology (see Chapters 11, 12, and 15, this volume). A substantial proportion of sucrose and hexoses enter these cells from the post-phloem P-C cells (Shannon 1972; Schmalstig and Hitz, 1987). Studies of the *mn1* mutation show a dramatic reduction in endosperm mass due to the loss of INCW2 (Miller and Chourey, 1992; Cheng *et al.*, 1996). Further, it is clear from *in vitro* kernel culture studies that the metabolic release of hexoses is critical for normal seed development, because feeding hexose sugars fails to rescue the *mn1* phenotype (Cheng and Chourey, 1999). As expected from the INCW2-deficiency, the *mn1* basal endosperm has greatly reduced levels of glucose and fructose and elevated levels of sucrose compared to wild type (*Mn1*) seed (LeClere *et al.*, 2010; Chourey *et al.*, 2012). This could be the causal factor for the pleiotropic effects of the *mn1* mutation, including reduced auxin levels (LeClere *et al.*, 2010) and co-ordinated downregulation of genes critical to sucrose and hexose metabolism (Chourey *et al.*, 2012). Among these is a significant downregulation of aldolase, the branch point controlling fructose flux towards gly-colysis and respiration, and a pivotal point for control of carbon partitioning.

The increase of sucrose in *mn1* endo-sperm supports earlier data suggesting that, although sucrose hydrolysis is not essential, its uptake (Schmalstig and Hitz, 1987) and cleavage to hexoses is critical for its utiliza-tion and growth of the endosperm. Shannon *et al.* (1986) suggested sucrose is resynthe-sized in the BETL, creating a "futile" cycle of sucrose cleavage and resynthesis that provides a regulatory force for maintaining a physio-logical sucrose gradient between phloem termini and the endosperm. Indeed, this hy-pothesis is consistent with results showing the presence of several key enzymes for re-versible cleavage and synthesis of sucrose. These include a major sucrose synthesizing enzyme, sucrose phosphate synthase (Cheng *et al.*, 1996), a *sucrose non-fermenting-related-kinase*, *SnRK1*, known to control the sucrose–starch transition (Jain *et al.*, 2008b), and an isozyme of sucrose synthase, SUS1 (Chen and Chourey, 1989). Recently, Sosso *et al.* (2015) identified a hexose transporter, Sucrose <u>Wi</u>ll <u>E</u>ventually be <u>E</u>xported <u>T</u>rans-porters (ZmSWEET4c), that is BETL-specific and postulated to function in a linear pathway with INCW2 in sugar translocation to starch accumulating cells in the endo-sperm (see Chapter 12). Loss of the transporter protein in *zmsweet4c-umu1* is associated with reduced levels of INCW2, retarded WIG growth, and ultimately shriveled mature ker-nels. A similar seed phenotype is observed in the SWEET-deficient mutant of rice, *oss-weet4-1* (Sosso *et al.*, 2015).

## 5.4 Regulation of BETL Differentiation and Development

A prominent role of the BETL is to facilitate transport of nutrients into the developing kernel, and transcriptomic and proteomic analyses document a large number of me-tabolite transporters and nutrient process-ing enzymes expressed in this tissue. Most genes involved in BETL transport functions, however, do not predominate nor are they exclusively expressed in this tissue. Inter-estingly, transcriptome analysis also iden-tified a large and diverse set of genes exclu-sively and abundantly expressed in the BETL, suggesting overlooked functions of endosperm TCs.

Since the discovery of *BETL-1*, the first TC-specific gene identified (Hueros *et al.*, 1995), it was evident that endosperm TCs dis-play highly specific gene expression patterns. *BETL-1* was identified by screening genes expressed around 10 DAP, just before the start of grain filling. This can be considered a hallmark of TC-specific genes. Subsequent efforts to identify TC-specific genes were based on a simple experimental design: 10 DAP kernels were dissected into upper (top, TC-lacking) and lower (bottom or basal, TC-containing) halves, and the RNA popu-lations compared by differential expression in the cDNA libraries. This led to identifica-tion of BETL-specific genes and identified a prominent class of genes, including *BETL-1*, potentially associated with microbial defense. These genes are highly expressed, TC-specific, and encode small (lower than 10 kDa), cysteine-rich precursor proteins containing a hydro-phobic signal peptide of variable length that targets them for secretion. In all cases ana-lyzed, the mature protein accumulates on the maternal side of the TC layer, suggesting polarized secretion (reviewed in Royo *et al.*, 2007; McCurdy and Hueros, 2014). The se-quence and expression features suggest these genes might be part of an innate defense mechanism protecting filial tissues against pathogens (fungi and bacteria) originating in the mother plant. Indeed, a potent effect on fungi was demonstrated for the mature form of BETL-2 and its related isoforms (re-named basal layer antifungal proteins (BAPs), Serna *et al.*, 2001). BAPs undergo complex post-translational proteolytic processing that first eliminates the hydrophobic signal peptide and then the protein's N-terminus. Mature BAPs are highly cytotoxic and accu-mulate in the P-C region during PCD.

In a recent study (Lopato *et al.*, 2014, and references therein), it was shown that these TC-specific small peptides are structurally related and comprise a family of non-specific lipid transfer proteins (nsLTPs). Plant nsLTPs have low levels of sequence identity, but are associated with functions related to cell–cell interactions, notably resistance against pathogens. Lopato *et al.* (2014) identified four classes of nsLTPs among cereal TC-specific genes, with the main class comprising defensin-related molecules, like BETL-1, BETL-3, BETL-10 and the large family of MEG-1 (maternally expressed gene-1) related proteins. A second group includes BETL-4, which shares homology with the Bowman-Birk family of α-amylase/trypsin inhibitors. The third group of nsLTPs includes BETL-2 related antifungal proteins, and the fourth group is represented by BETL-9, a gene of unknown function, but the only member of the family that shows orthologs (End-1 and TaPR60) in barley and wheat TCs, and Arabidopsis, where up to four orthologous genes mark boundary regions at the base of branches, the stem, flowers and fruits.

An alternative role for at least some of these proteins in cell signaling has been proposed. Members of the MEG-1 family, for instance, are associated with TC differentiation and efficiency of grain filling (Gutiérrez-Marcos *et al.*, 2004; Costa *et al.*, 2012).

Top versus bottom endosperm expression screenings also identified TC-specific genes with potential regulatory roles. Among them, *ZmMRP-1* (Gómez *et al.*, 2002), was suggested as a master regulator of the TC gene expression program. *ZmMRP-1* encodes a transcription factor containing a single MYB-like DNA binding domain. The gene is expressed in cytoplasmic domains at the base of the endosperm coenocyte as early as 3 DAP, a period that precedes cellularization and, consequently, the appearance of TC morphology. This protein was found to effectively transactivate promoters of *BETL-1* and *BETL-2* (Gómez *et al.*, 2002); subsequent studies using transient and transgenic approaches identified five additional transfer cell-specific genes regulated

by ZmMRP-1: *BETL-9*, *BETL-10*, *MEG-1*, *ZmTCRR-1*, and *ZmTCRR-2* (Gómez *et al.*, 2002, 2009; Gutiérrez-Marcos *et al.*, 2004; Muñiz *et al.*, 2006, 2010).

Dissection of ZmMRP-1 interaction with *BETL-1* and *2* promoters revealed a 12 base-pair (bp) motif containing two direct repeats of the sequence TATCTA/C, defined as the TC-box, located within 100 bp upstream of the TATA box (Barrero *et al.*, 2006). In a recent study (Zhan *et al.*, 2015), a combination of bioinformatics and yeast 1-hybrid approaches extended the number of genes controlled by *ZmMRP-1* to a minimum of 93, highlighting the importance of this transcription factor in the function of the tissue. A thorough analysis of the promoters of these genes helped to refine the TC-box, which might consist of a single TATCTA/C repeat with some degeneracy in the core sequence (Zhan *et al.*, 2015).

Unfortunately, ectopic expression of *ZmMRP-1* with ubiquitously expressed promoters (i.e. ubiquitin, 35S) produced no viable transformants of maize, Arabidopsis or tobacco (Gómez *et al.*, 2009), suggesting that the protein is possibly toxic outside the BETL. Expression under control of tissue-specific promoters or inducible ones allowed recovery of transgenic plants, but they expressed only trace amounts of the transcription factor. In one case, expression with an aleurone-specific promoter, *BETL-9like* (Royo *et al.*, 2014), resulted in partial differentiation of the abgerminal region of the aleurone into TCs (Gómez *et al.*, 2009). These cells acquired TC morphology and expressed TC-specific molecular markers, suggesting expression of *ZmMRP-1* initiated a cascade of events through which epidermal cells fated to be aleurone differentiated into TCs. There are some facts, however, that do not support this hypothesis. Since the cell transformation was only partial, observed temporarily, and limited to the abgerminal side of the aleurone (Gómez *et al.*, 2009), it is very likely *ZmMRP-1* is not sufficient to promote complete differentiation of TCs and requires additional factors, possibly signals from maternal tissues that are ineffective in the aleurone after a certain developmental stage.

The signals regulating *ZmMRP-1* expression were investigated by expression of promoter:GUS constructs in maize and heterologous systems (Barrero *et al.*, 2009). In maize, GUS expression was confined to TCs, indicating that the promoter provides tissue-specific regulation. In tobacco, Arabidopsis, and barley, however, the *ZmMRP-1* promoter was active in several boundary regions where exchange of solutes takes place, such as branch points and the base of flowers and fruits. The promoter was shown to be responsive to the concentration of sugars in systems as diverse as yeast and germinating Arabidopsis seeds (Barrero *et al.*, 2009), perhaps indicating a general sensitivity towards the metabolic status of the cells.

In addition to the previously described roles of the BETL in transport and microbial protection, there is evidence for a third role in that the BETL could serve as a sensor of the metabolic state of the plant. Specifically, two elements of signal transduction pathways known as two-component-systems (TCS) were identified among the genes regulated by *ZmMRP-1* (Muñiz *et al.*, 2006, 2010). The prototypical plant TCS is dedicated to cytokinin sensing (Hwang and Sheen, 2001). It involves a plasma membrane receptor with histidine kinase activity that auto-phosphorylates in response to the hormone. Subsequently, a chain of phosphate transfer reactions moves the signal through a histidine phosphotransfer protein (HPT) to a type-B response regulator (type B-RR), a MYB-related transcription factor, that regulates genes responsive to cytokinin. Among genes regulated by the type B-RR are type-A response regulators (type A-RR), which are negative modulators of the pathway that compete with the type B-RR for phosphate transfer. Type A-RRs can also be regulated by other signals, thus allowing crosstalk between cytokinin signaling and other regulatory pathways. For instance, type A-RRs have been found to be repressed by auxin or ethylene in different signaling contexts (reviewed in El-Showk *et al.*, 2013).

The TC-specific elements identified within this signaling pathway, *ZmTCRR-1* and *ZmTCRR-2*, belong to the type-A RR class of molecules. Surprisingly, the transcriptional

regulator of both genes, *ZmMRP-1*, contains a single MYB-like DNA binding domain related to that in the type-B RRs, but lacks any sign of an RR domain. *ZmTCRR-1* and *-2* might have been co-opted by the *ZmMRP-1* regulated pathway, possibly to interact with hormone-regulated signal transduction pathways. Although transcripts of these genes are exclusively localized in TCs between 8 and 14 DAP, the ZmTCRR proteins accumulate in the inner endosperm at the region above the BETL known as the conducting zone (Muñiz *et al.*, 2006, 2010). The conductive zone is thought to facilitate symplastic transport of nutrients towards metabolite accumulation regions (Becraft, 2001). This suggests *ZmMRP-1* is involved in an intercellular signal transduction pathway by promoting expression of proteins migrating into the endosperm, presumably to integrate kernel development with external inputs. The signaling pathways interacting with *ZmTCRR-1* and *-2* in the conducting zone are currently unknown.

These findings imply a potential role of *ZmMRP-1* in BETL development and grain filling. Circumstantial evidence supporting this hypothesis is based on gene expression studies in small (KSS) and large kernel (KLS) inbreds (Zhang *et al.*, 2016). KLS lines accumulate dry matter faster than KSS inbreds, indicating more efficient grain filling; *ZmMRP-1* was found to be significantly upregulated in KLS kernels (see Chapter 16, this volume).

RNAseq technologies have produced an almost complete picture of the BETL transcriptome at developmental stages most relevant to its differentiation. Xiong *et al.* (2011) identified BETL genes expressed at the initial stage of the grain filling (10 DAP) when transfer cells are most active. More recently, Zhan *et al.* (2015) analyzed the transcriptome of kernel domains at 8 DAP based on tissue capture by laser dissection. The depth of sequencing, the developmental stage, and spatial distribution of tissues, including 10 maternal and filial compartments, made this an excellent resource for TC studies. Principal component and hierarchical clustering analyses clearly differentiated the BETL from other endosperm compartments.

Among filial tissues, only the embryo produced a comparably distinct expression profile. Inter-compartment comparisons identified 912 BETL-specific genes. The co-regulated Module-18, corresponding to the BETL expression signature, contained all previously reported TC-specific genes. BETL functions are clearly represented in Module-18, including transport, defense, and signaling. As discussed above, *in silico* analyses and yeast 1-hybrid experiments revealed that *ZmMRP-1* is directly regulating at least 93 genes, representing a large proportion (15%) of those in this module (616 genes). Interestingly, a few highly expressed aleurone marker genes, for instance *BETL-9like* (Royo *et al.* 2014), had small but significant expression levels in the BETL. These data suggest that BETL differentiation is not complete by 8 DAP, and overlap may exist in expression of TC and aleurone genes.

One of the puzzling observations from high-throughput transcriptomic analyses of cereals is the lack of conservation among gene networks thought to regulate TC differentiation. Thiel *et al.* (2012) conducted an extensive RNAseq analysis of the barley TC transcriptome at 3, 5, and 7 days after flowering (DAF), focusing on identification of TCS elements. They identified nearly 40 genes associated with TCS signal transduction pathways expressed in the barley endosperm TC layer. Comparison with public EST collections suggested nearly half of these could be TC-specific. Furthermore, specific combinations of TCS elements are associated with different developmental stages (cellularization, differentiation, and maturation); however, no *ZmMRP-1* ortholog was identified in the barley TC transcriptome. Furthermore, the type-A RR molecules identified in maize clearly separated from the barley genes in this class. These results indicate that TC signaling in barley could be based on a more canonical use of TCS elements, suggesting that the maize ZmMRP-1/ZmT-CRR1,2 pathway could be a later acquisition, perhaps for modulation of some tropical cereal-specific features. Certainly, there are important anatomical differences in the TCs of barley and maize, but both species develop the flange-type cell wall architecture

and are relatively close in terms of evolutionary distance. Thus, conservation in the basic regulatory circuits determining TCs differentiation would have been expected.

A TC-specific transcriptome has also been studied in *Vicia faba* cotyledons. *V. faba* embryos are covered by an epidermal layer of TCs on their abaxial surface. The adaxial side contains no TCs, but when the cotyledons are dissected and cultured, the epidermal cells spontaneously undergo trans-differentiation into TCs (Andriunas *et al.*, 2013). This system has value for studies on TC differentiation (Zhang *et al.*, 2015c). Contrary to the situation in barley, TCS elements were found not to be correlated with TC induction, and the authors suggested this might uncover the molecular basis for morphological differences between flange (present in barley, wheat, and maize) and reticulate (present in *Vicia*) TC wall architecture. In another study, also in *V. faba*, Arun-Chinnappa and McCurdy (2016) characterized the expression of transcription factors during TC differentiation and found an association between the initiation of TC differentiation and the expression of WRKY and ethylene-responsive transcription factors. The expression of transcription factors of the MYB-family, associated with the synthesis of secondary cell wall in Arabidopsis, was also found to be significantly linked with the process, but again no role could be assigned to any *ZmMRP-1* ortholog in *Vicia*.

In contrast to the situation with transcriptome analysis, much remains to be done in proteomic analyses during BETL differentiation in maize. Two studies were carried out using soluble and cell wall-associated proteins (Silva-Sanchez *et al.*, 2013) and glycoproteins (Silva-Sanchez *et al.*, 2014) from the lower third of the 12 DAP kernel report proteome profiles enriched for BETL proteins. These papers provide a useful study of a proteome resource influenced by the INCW-deficiency in the maize *mn1* mutant. Overall, the data provide insight on how a loss-of-function mutation of a BETL-specific metabolic gene, *Mn1*, can lead to the pleiotropic changes in the *mn1* mutant (Chourey *et al.*, 2012).

## 5.5 Hormonal Regulation of BETL Differentiation and Development

Differentiation of epithelial TCs has been thoroughly studied in *V. faba* by exploiting induction in culture of the TC trans-differentiation process (Andriunas *et al.*, 2013). These studies led to a model explaining the induction and maturation of this tissue through interaction between the cotyledon abgerminal epidermis and the maternal seed coat, uncovering the role of ethylene and reactive oxygen species (ROS) signaling in the process. Early in development, a high level of glucose produced by seed coat extracellular invertase blocks ethylene signaling in the cotyledon epidermal cells. At the onset of cotyledon growth, the inner cell layers of the seed coat are crushed, resulting in a reduction in invertase production, and hence a decrease in glucose repression and generation of an ethylene signal. In the epidermal cells, ethylene signaling is amplified following an auxin-regulated burst in ethylene production. Ethylene induces expression of respiratory burst oxidases that migrate to the external surface of the epidermis, causing a polarized ROS and $H_2O_2$ signaling cascade that ultimately induces synthesis of uniform walls and then of cell wall ingrowths. The model for ingrowth production has subsequently been improved to include the role of $H_2O_2$-regulated $Ca^{2+}$ channels and a localized intracellular increase in $Ca^{2+}$ (Zhang *et al.*, 2015a, b).

Unfortunately, knowledge of hormone-mediated regulation of TC differentiation in maize kernels is far less complete. There are indications that mechanisms similar to those found in *Vicia* might also operate in the maize BETL, and the model could be used as a framework to organize diverse observations seen in maize. However, the transcriptome comparisons have thus far failed to identify common regulatory pathways.

Several studies in maize analyzing hormone and sugar regulation at the whole kernel level focused on grain filling and late maturation stages. It is known, for instance, that cytokinins dominate signaling during early seed developmental phases, being gradually replaced by auxin after 10 DAP (Lur and Setter, 1993). Forestan *et al.* (2010) reported an initial burst of auxin connected to BETL differentiation; specifically, immature TCs were found to accumulate auxin and high levels of the auxin efflux carrier, ZmPIN1. Furthermore, auxin accumulation and expression of ZmPIN1 and the TC marker BETL-1 were greatly reduced in TCs of the auxin-deficient mutant, *defective endosperm-18* (*de18*). Cells at the BETL showed a very weak TC morphology in this mutant (Forestan *et al.*, 2010). Bernardi *et al.* (2012) found that developing endosperm in maize has the highest indole-3-acetic acid (IAA) levels relative to any tissue in the plant. Additionally, their data suggest that IAA biosynthesis in endosperm cells is local, including the TCs (Forestan *et al.*, 2010; LeClere *et al.*, 2010; Bernardi *et al.*, 2012). Molecular analyses also suggest the IAA-deficiency in the *de18* mutant is due to changes in expression of *ZmYuc1*, a gene encoding an enzyme (YUC1) critical in IAA biosynthesis.

The earliest evidence of possible sugar–auxin (IAA) cross-talk in developing maize endosperm was seen through reduced levels of *ZmYuc1* transcript abundance and the IAA levels in basal regions of *mn1* endosperm (LeClere *et al.*, 2008, 2010). Further analyses based on expression of a ZmYUC promoter:GUS fusion in Arabidopsis revealed that glucose promoted, while sucrose reduced, GUS expression (LeClere *et al.*, 2010). Indeed, glucose was shown to affect transcription of many auxin-responsive genes, including the *YUC* genes of Arabidopsis (Mishra *et al.*, 2009). Overall, these studies indicate that sugar influx into sink tissues may influence sink size and strength by regulating auxin levels.

There are indications suggesting that, as reported in the *Vicia* system, PCD in maternal tissues might have a signaling role, perhaps through production of ROS during BETL differentiation. It was found that PCD in the nucellar P-C, a fundamental event in the construction of a functional transport system, is fertilization-dependent and controlled by zygotic tissues (Kladnik *et al.*, 2004). Furthermore, PCD at the P-C has been suggested to be promoted in a fertilization-dependent way by accumulation of cytokinins in the pedicel (Rijavec *et al.*, 2011). In this way, and resembling the situation described in *Vicia* cotyledons, the proto-transfer cells at the filial tissues would trigger the production of a polarizing signal in the maternal tissues, which in turn would induce the cell wall modifications that characterize transfer cells.

## 5.6 Future Directions

The pleiotropic phenotypes of BETL development in the maize *mn1* mutant provide insight into the BETL's complex and important functions, and should provide a starting point for future research to better understand the complexity of sugar physiology in relation to seed development in maize and sorghum. Clearly, cell wall invertase-catalyzed sucrose hydrolysis in the BETL is of critical importance; however, sucrose biosynthesis also occurs in these cells, as evidenced by the abundance of transcripts for sucrose biosynthetic enzymes. Subcellular localization of proteins related to sucrose biosynthesis is of interest in the context of WIG function. If these enzymes co-localize with WIGs, as is the case of INCW2, new insights regarding sugar physiology and WIG function could be learned. The maize BETL is a model for studying polarized secretion (Kang *et al.*, 2009); specifically, WIG formation requires massive amounts of cell wall proteins and polysaccharides delivered from Golgi stacks. Some evidence suggests that signals for PCD in the maternal P-C region originate in the BETL. It is now possible to investigate the transcriptional regulation of polarity establishment and PCD (Xiong *et al.*, 2011). Finally, much remains to be learned about sugar level and auxin regulation of endosperm development. It is

noteworthy that the developing endosperm shows the highest concentration of IAA recorded in any plant tissue. Molecular genetic analyses are needed to characterize IAA biosynthetic genes, specifically those that are sugar-responsive, to understand the auxin–sugar relationship in developing endosperm. The role of *ZmMRP-1* and the signaling pathway by which it regulates the transfer cell differentiation process and/or the construction of a barrier against pathogens can be investigated using *zmmrp-1* mutants. Unfortunately, the existence of high levels of genetic redundancy for this pathway hampers this research. Valuable information on the possible roles of *ZmMRP-1* could be obtained through examination of the biochemical functions of the peptide cocktail produced as a result of the *ZmMRP-1* activity. The critical roles of hormone (auxin, ethylene) and ROS signaling in TC differentiation is well established for *Vicia faba* cotyledons (Andriunas *et al.*, 2013) and similar insight is anticipated from investigating maize BETL differentiation. The inaccessibility of the maize BETL for hormonal manipulation for *in vitro* experiments is a technical obstacle, but knowledge of the BETL transcriptome, in combination with gene editing technologies for targeted mutagenesis, provides an alternative approach to these investigations.

## Acknowledgment

We would like to thank Drs. Alan Myers (Iowa State University) and Charles Hunter (USDA ARS, Gainesville, FL) for critical reading of the manuscript. We thank Dr. B. Larkins for his careful editing of the early versions of the chapter.

## References

Andriunas, F.A., Zhang, H.M., Xia, X., Patrick, J.W. and Offler, C.E. (2013) Intersection of transfer cells with phloem biology: broad evolutionary trends, function, and induction. *Frontiers in Plant Science* 4, 221.

Arun-Chinnappa, K.S. and McCurdy, D.W. (2016) Identification of candidate transcriptional regulators of epidermal transfer cell development in *Vicia faba* cotyledons. *Frontiers in Plant Science* 7, 717.

Barrero, C., Muñiz, L.M., Gómez, E., Hueros, G. and Royo, J. (2006) Molecular dissection of the interaction between the transcriptional activator *ZmMRP-1* and the promoter of *BETL-1*. *Plant Molecular Biology* 62, 655–668.

Barrero, C., Royo, J., Grijota-Martinez, C., Faye, C., Paul, W., *et al.* (2009) The promoter of *ZmMRP-1*, a maize transfer cell-specific transcriptional activator, is induced at solute exchange surfaces and responds to transport demands. *Planta* 229, 235–247.

Becraft, P.W. (2001) Cell fate specification in the cereal endosperm. *Seminars in Cell Development Biology* 12, 387–394.

Bernardi, J., Lanubile, A., Li, Q.-B., Kumar, D., Kladnik, A., *et al.* (2012) Impaired auxin biosynthesis in the *defective endosperm18* mutant is due to mutational loss of expression in the *ZmYuc1* gene encoding endosperm-specific YUCCA1 protein in maize. *Plant Physiology* 160, 1318–1328.

Chen, Y.-C. and Chourey, P.S. (1989) Spatial and temporal expression of two sucrose synthase genes in maize: immunohistological evidence. *Theoretical and Applied Genetics* 78, 553–559.

Cheng, W.-H. and Chourey, P.S. (1999) Genetic evidence that the invertase-mediated release of hexoses is a critical function for the normal development of seed in maize. *Theoretical and Applied Genetics* 98, 485–495.

Cheng, W.-H., Taliercio, E.W. and Chourey, P.S. (1996) The *Miniature1* seed locus of maize encodes a cell wall invertase required for normal development of endosperm and maternal cells in the pedicel. *Plant Cell* 8, 971–983.

Chourey, P.S., Li, Q.-B. and Cevallos-Cevallos, J. (2012) Pleiotropy and its dissection through a metabolic gene Miniature1 (Mn1) that encodes a cell wall invertase in developing seeds of maize. *Plant Science* 184, 45–53.

Costa, L.M., Yuan, J., Rouster, J., Paul, W., Dickinson, H. and Gutiérrez-Marcos, J.F. (2012) Maternal control of nutrient allocation in plant seeds by genomic imprinting. *Current Biology* 22, 160–165.

Davis, R.W., Smith, J.D. and Cobb, B.G. (1990) A light and electron microscope investigation of the transfer cell region of maize caryopses. *Canadian Journal of Botany* 68, 471–479.

Dermastia, M., Kladnik, A., Koce, J.D. and Chourey, P.S. (2009) A cellular study of teosinte *Zea mays* subsp. *parviglumis* (Poaceae) caryopsis development showing several processes conserved in maize. *American Journal of Botany* 96, 1798–1807.

El-Showk, S., Ruonala, R. and Helariutta, Y. (2013) Crossing paths: cytokinin signaling and crosstalk. *Development* 140, 1373–1383.

Felker, F.C. and Shannon, J.C. (1980) Movement of $^{14}$C-labeled assimilate into kernels of *Zea mays* L. III. An anatomical examination and microautoradiographic study of assimilate transfer. *Plant Physiology* 55, 864–870.

Forestan, C., Meda, S. and Varotto, S. (2010) ZmPIN1-mediated auxin transport is related to cellular differentiation during maize embryogenesis and endosperm development. *Plant Physiology* 152, 1373–1390.

Gómez, E., Royo, J., Guo, Y., Thompson, R. and Hueros, G. (2002) Establishment of cereal endosperm expression domains: identification and properties of a maize transfer cell-specific transcription factor, *ZmMRP-1*. *Plant Cell* 14, 599–610.

Gómez, E., Royo, J., Muñiz, L.M., Sellam, O., Paul, W., *et al.* (2009) The maize transcription factor myb-related protein-1 is a key regulator of the differentiation of transfer cells. *Plant Cell* 21, 2022–2035.

Gutiérrez-Marcos, J.F., Costa, L.M., Biderre-Petit, C., Khbaya, B., O'Sullivan, D.M., *et al.* (2004) *Maternally expressed gene1* is a novel maize endosperm transfer cell-specific gene with a maternal parent-of-origin pattern of expression. *Plant Cell* 16, 1288–1301.

Gutiérrez-Marcos, J.F., Dal Prà, M., Giulini, A., Costa, L.M., Gavazzi, G., *et al.* (2007) *Empty pericarp4* encodes a mitochondrion-targeted pentatricopeptide repeat protein necessary for seed development and plant growth in maize. *Plant Cell* 19, 196–210.

Hueros, G., Varotto, S., Salamini, F. and Thompson, R.D. (1995) Molecular characterization of BET1, a gene expressed in the endosperm transfer cells of maize. *Plant Cell* 7, 747–757.

Hwang, I. and Sheen, J. (2001) Two-component circuitry in *Arabidopsis* cytokinin signal transduction. *Nature* 413, 383–389.

Jain, M., Chourey, P.S., Li, Q.-B. and Pring, D.R. (2008a) Expression of cell wall invertase and several other genes of sugar metabolism in relation to seed development in sorghum (*Sorghum bicolor*). *Journal of Plant Physiology* 165, 331–344.

Jain, M., Li, Q.-B. and Chourey, P.S. (2008b) Cloning and expression analyses of *sucrose non-fermenting-1-related kinase* (*SnRK1b*) gene during development of sorghum and maize endosperm and its implicated role in sugar-to-starch metabolic transition. *Physiologia Plantarum* 134, 161–173.

Kang, B.-H., Xiong, Y., Williams, D.S., Pozueta-Romero, D. and Chourey, P.S. (2009) *Miniature1*-encoded cell wall invertase is essential for assembly and function of wall-in-growth in the maize endosperm transfer cell. *Plant Physiology* 151, 1366–1376.

Kiesselbach, T.A. (1949) *The Structure and Reproduction of Corn*. University of Nebraska Press, Lincoln, Nebraska.

Kladnik, A., Chamusco, K., Dermastia, M. and Chourey, P. (2004) Evidence of programmed cell death in post-phloem transport cells of the maternal pedicel tissue in developing caryopsis of maize. *Plant Physiology* 136, 3572–3581.

LeClere, S., Schmelz, E.A. and Chourey, P.S. (2008) Cell wall invertase-deficient *miniature1* kernels have altered phytohormone levels. *Phytochemistry* 69, 692–699.

LeClere, S., Schmelz, E.A. and Chourey, P.S. (2010) Sugar levels regulate tryptophan-dependent auxin biosynthesis in developing maize kernels. *Plant Physiology* 153, 306–318.

Lopato, S., Borisjuk, N., Langridge, P. and Hrmova, M. (2014) Endosperm transfer cell-specific genes and proteins: structure, function and applications in biotechnology. *Frontiers in Plant Science* 5, 64.

Lur, H.S. and Setter, T.L. (1993) Role of auxin in maize endosperm development (timing of nuclear DNA endoreduplication, zein expression, and cytokinin). *Plant Physiology* 103, 273–280.

McCurdy, D.W. and Hueros, G. (2014) Transfer cells. *Frontiers in Plant Science* 5, 672. https://doi.org/10.3389/fpls.2014.00672

Miller, M.E. and Chourey, P.S. (1992) The maize invertase-deficient miniature1 seed mutant is associated with aberrant pedicel and endosperm development. *Plant Cell* 4, 297–305.

Mishra, B.S., Singh M., Aggrawal, P. and Laxmi, A. (2009) Glucose and auxin signaling in controlling *Arabidopsis thaliana* seedling root growth and development. *PLOS ONE* 4, E4502.

Monjardino, P., Rocha, S., Tavares, A.C., Fernandes, R., Sampaio, P., Salema, R. and da Câmara Machado, A. (2013) Development of flange and reticulate wall ingrowths in maize (*Zea mays* L.) endosperm transfer cells. *Protoplasma* 250, 495–503.

Muñiz, L.M., Royo, J., Gómez, E., Barrero, C., Bergareche, D. and Hueros, G. (2006) The maize transfer cell-specific type-A response regulator *ZmTCRR-1* appears to be involved in intercellular signaling. *Plant Journal* 48, 17–27.

Muñiz, L.M., Royo, J., Gómez, E., Baudot, G., Paul, W. and Hueros, G. (2010) Atypical response regulators expressed in the maize endosperm transfer cells link canonical two component systems and seed biology. *BMC Plant Biology* 10, 84.

Offler, C.E., McCurdy, D.W., Patrick, J.W. and Talbot, M.J. (2003) Transfer cells: cells specialized for a special purpose. *Annual Review of Plant Biology* 54, 431–454.

Rijavec, T., Jain, M., Dermastia, M. and Chourey, P.S. (2011) Spatial and temporal profiles of cytokinin biosynthesis and accumulation in developing caryopses of maize. *Annals of Botany* 107, 1235–1245.

Royo, J., Gómez, E. and Hueros, G. (2007) Transfer cells. In: Olsen, O.-A. (ed.) *Endosperm*. Springer, Berlin, Heidelberg, Germany, pp. 73–89.

Royo, J., Gómez, E., Sellam, O., Gerentes, D., Paul, W. and Hueros, G. (2014) Two maize *END-1* orthologs, *BETL9* and *BETL9like*, are transcribed in a non-overlapping spatial pattern on the outer surface of the developing endosperm. *Frontiers in Plant Science* 5, 180.

Schmalstig, J.D. and Hitz, W.D. (1987) Transport and metabolism of a sucrose analog (1'-fluorosucrose) into *Zea mays* L. endosperm without invertase hydrolysis. *Plant Physiology* 85, 902–905.

Shannon, J.C. (1972) Movement of [14]C-labeled assimilates into kernels of *Zea mays* L. I. Pattern and rate of sugar movement. *Plant Physiology* 49, 198–202.

Shannon, J.C., Porter, G.A. and Knievel, D.P. (1986) Phloem unloading and transfer of sugars into developing corn endosperm. In: Cronshaw, J., Lucas, W.J. and Guiquinta, R.T. (eds.) *Phloem Transport*. Alan Liss, Inc., New York, pp. 265–277.

Serna, A., Maitz, M., O'Connell, T., Santandrea, G., Thevissen, K., *et al.* (2001) Maize endosperm secretes a novel antifungal protein into adjacent maternal tissue. *Plant Journal* 25, 687–698.

Silva-Sanchez, C., Chen, S., Zhu, N., Li, Qi.-B. and Chourey, P.S. (2013) Proteomic comparison of basal endosperm in maize *miniature1* mutant and its wild-type *Mn1*. *Frontiers in Plant Science* 4, 211.

Silva-Sanchez, C., Chen, S., Zhu, N., Li, Q.-B. and Chourey, P.S. (2014) A comparative glycoproteome study of developing endosperm in the hexose-deficient *miniature1* (*mn1*) seed mutant and its wild type *Mn1* in maize. *Frontiers in Plant Science* 5, 63.

Sosso, D., Luo, D., Li, Q.-B., Sasse, J., Yang, J., *et al.* (2015) Seed filling in domesticated maize and rice depends on SWEET-mediated hexose transport. *Nature Genetics* 47, 1489–1493.

Thiel, J., Hollmann, J., Rutten, T., Weber, H., Scholz, U. and Weschke, W. (2012) 454 Transcriptome sequencing suggests a role for two-component signalling in cellularization and differentiation of barley endosperm transfer cells. *PLOS ONE* 7, e41867.

Thompson, R.D., Hueros, G., Becker, H. and Maitz, M. (2001) Development and functions of seed transfer cells. *Plant Science* 160, 775–783.

Xiong, Y., Li, Q.-B., Kang, B. and Chourey, P.S. (2011) Discovery of genes expressed in basal endosperm transfer cells in maize using 454 transcriptome sequencing. *Plant Molecular Biology Reporter* 29, 835.

Zhan, J., Thakare, D., Ma, C., Lloyd, A., Nixon, N.M., *et al.* (2015) RNA sequencing of laser-capture microdissected compartments of the maize kernel identifies regulatory modules associated with endosperm cell differentiation. *Plant Cell* 27, 513–531.

Zhang, H.M., Imtiaz, M.S., Laver, D.R., McCurdy, D.W., Offler, C.E., van Helden, D.F. and Patrick, J.W. (2015a) Polarized and persistent $Ca^{2+}$ plumes define loci for formation of wall ingrowth papillae in transfer cells. *Journal of Experimental Botany* 66, 1179–1190.

Zhang, H.M., van Helden, D.F., McCurdy, D.W., Offler, C.E. and Patrick, J.W. (2015b) Plasma membrane $Ca^{2+}$-permeable channels are differentially regulated by ethylene and hydrogen peroxide to generate persistent plumes of elevated cytosolic $Ca^{2+}$ during transfer cell trans-differentiation. *Plant Cell Physiology* 56, 1711–1720.

Zhang, H.M., Wheeler, S., Xia, X., Radchuk, R., Weber, H., Offler, C.E. and Patrick, J.W. (2015c) Differential transcriptional networks associated with key phases of ingrowth wall construction in trans-differentiating epidermal transfer cells of *Vicia faba* cotyledons. *BMC Plant Biology* 15, 103.

Zhang, X., Hirsch, C.N., Sekhon, R.S., Leon, N. and Kaeppler, S.M. (2016) Evidence for maternal control of seed size in maize from phenotypic and transcriptional analysis. *Journal of Experimental Botany* 67, 1907–1917.

# 6  Aleurone

**Bryan C. Gontarek and Philip W. Becraft***
*Department of Genetics, Development and Cell Biology,
Iowa State University, Iowa, USA*

## 6.1  Introduction

The aleurone cell layer forms at the surface of the endosperm and is present in seeds of most flowering plants. It has epidermal-like characteristics, except that it is not directly exposed to the atmosphere; rather, it is covered by maternally derived testa and pericarp. Maize aleurone has a rich history, being instrumental in fundamental discoveries by pioneering geneticists, including Barbara McClintock. Anthocyanin pigmentation of aleurone provides a convenient genetic marker that has led to the discovery of genes that regulate anthocyanin biosynthesis and endosperm development. Anthocyanin pigmentation in the aleurone has also been utilized to study the inheritance patterns and behaviors of genes. Transposable elements, imprinting and paramutation are among the significant discoveries facilitated by anthocyanin in the aleurone (McClintock, 1950; Brink, 1956; Kermicle, 1970). More recently, attention has focused on the aleurone per se, due to its important biological functions, implications for agronomic performance and industrial applications, and healthful properties.

## 6.2  Biological Functions of Aleurone

Major functions of aleurone in cereals include storage, defense, and hydrolysis. Aleurone is a major storage site where minerals, particularly phosphorus, are bound to phytic acid. Within the endosperm, essentially all the phytate and stored phosphorus occurs in globoid structures contained within protein storage vacuoles (PSVs) in the aleurone. Other minerals such as iron, magnesium, and zinc also associate with phytate-containing globoids (Regvar *et al.*, 2011). Aleurone also accumulates high levels of proteins, lipids, and vitamins (Brouns *et al.*, 2012).

The endosperm, with its high concentration of stored nutrients, is a target for pathogens and herbivorous insects and the aleurone provides an active defense response to pathogen infection. Aleurone expresses several defense-related proteins, including PR-4, CHITINASE, XYLINASE INHIBITOR PROTEIN-1 (XIP-1), and 7S GLOBULIN (Jerkovic *et al.*, 2010). It was recently shown that NKD transcription factors regulate aleurone defense-related genes, including defensins, *mlo* family members, *xylanase*

*Corresponding author e-mail: becraft@iastate.edu

 

*inhibitor protein1*, and genes involved in producing reactive oxygen species (ROS), jasmonate, and ethylene signaling (Gontarek *et al.*, 2016). Remarkably little work has been done on aleurone function in pathogen responses (Casacuberta *et al.*, 1991).

The most well-known function of aleurone is hydrolysis of storage polymers during germination. At imbibition, the aleurone secretes amylases and other hydrolases into dead starchy endosperm cells, where they break down starch, storage proteins, and nucleic acids, mobilizing free sugars, amino acids, and nucleotides to nourish the germinating seedling. Expression of genes encoding these enzymes is induced by gibberellin (GA) from the embryo and inhibited by abscisic acid (ABA) (Zentella *et al.*, 2002). Aleurone hydrolytic function culminates in a novel mode of programmed cell death (PCD) (Bethke and Jones, 2001).

## 6.3   Practical Properties of Aleurone

Defense functions of aleurone are obviously important because fungal infections of seeds decrease grain yield, quality and market value, and pose health risks from aflatoxin. Bacterial and fungal infections can inhibit germination and lead to seedling infection, negatively impacting crop stands (Stuckey *et al.*, 1985).

Rapid remobilization of storage reserves during germination is also essential for seedling emergence and crop establishment. Coordination of metabolic activities in the embryo and aleurone ensures adequate production of hydrolases to supply metabolites for seedling growth. This hydrolytic activity converts non-fermentable starch into fermentable sugars; a process exploited by the malting industry. Malting consists of the partial germination of grains, followed by heating to kill and dry them. When malted grains are steeped in warm water, the amylases produced during germination convert starch to sugar for fermentation by yeast. The rapid and uniform conversion of starch is critical for malting and depends directly on aleurone physiology.

A litany of dietary health benefits are derived from cereal bran and most can be attributed to aleurone. These include promoting cardiovascular health, anti-cancer and anti-diabetic properties, and many more (Brouns *et al.*, 2012; Lillioja *et al.*, 2013). Aleurone contains high concentrations of many healthful constituents, including fiber (arabinoxylans, β-glucans) (Lafiandra *et al.*, 2014), vitamins (tocopherols, folate) (Henderson *et al.*, 2012), antioxidants (ferulic acid and other phenolic compounds, polyamine conjugates, anthocyanins) (Islam *et al.*, 2011), minerals (P, K, Mg, Fe, Na, Al, and Zn) (Regvar *et al.*, 2011; Lillioja *et al.*, 2013), and healthful lipids (phytosterols; esters of ferulate and p-coumarate) (Iwatsuki *et al.*, 2003; Jain *et al.*, 2008).

## 6.4   Aleurone Ontogeny and Differentiation

As described in Chapters 3 and 10, endosperm initially undergoes a coenocytic phase followed by cellularization. During this process, alveoli at the endosperm periphery undergo a periclinal division to produce an outer cell layer and an inner layer of alveoli. This repeats in the inner alveoli until the endosperm is completely cellularized. The outermost layer is the founder cell population for the aleurone, while internal cells produce starchy endosperm (Becraft and Yi, 2011; Becraft and Gutiérrez-Marcos, 2012). Cells in the outer layer immediately assume a behavior distinct from internal cells; the outer cells divide primarily in anticlinal and periclinal orientations, whereas internal cells divide in random planes.

Developmental timing can vary substantially due to environment and genetics, but by 5 days after pollination (DAP) the outer endosperm cell layer shows specific expression of *BETL9-like* (formerly *al9*), indicating that these cells have already begun to differentiate as aleurone (Royo *et al.*, 2014). At 8 DAP, aleurone cells are rectangular in section, thin-walled, and highly vacuolated. By 12 DAP, the cells begin accumulating cytoplasmic inclusions, which continues until maturity, giving aleurone cells their densely staining characteristics (Becraft and Asuncion-Crabb, 2000; Leroux *et al.*, 2014).

Mature aleurone cells have a regular cuboidal shape with thick autofluorescent cell walls; they form a tessellation of variably shaped cells when viewed from the surface (Fig. 6.1A). Their dense cytoplasm results from accumulation of various inclusions, including protein-carbohydrate bodies and aleurone grains, which are globoid bodies containing phytic acid and protein surrounded by lipid droplets (Jakobsen *et al.*, 1971). Unlike starchy endosperm cells that accumulate prolamin storage proteins in ER-derived protein bodies, aleurone loads prolamins into specialized PSVs containing multilayered membranes and engulfed cytoplasm; they form in a process similar to autophagosomes (Reyes *et al.*, 2011). Aleurone cells do not accumulate starch granules and in particular genotypes, anthocyanin pigmentation specifically accumulates in the aleurone at late stages of development.

The aleurone is the only endosperm cell type to remain alive in fully mature dry seeds. ABA signaling is required to promote quiescence and desiccation tolerance, a highly specialized state involving complex physiological adaptations including the accumulation of dehydrin proteins that protect membranes and proteins (Goyal *et al.*, 2005; Vicente-Carbajosa and Carbonero, 2005).

## 6.5 Regulation of Aleurone Development

As described, cells of the outer endosperm layer, produced during cellularization, are the aleurone founder cells. However, this outer layer also contributes cells internally that will differentiate as starchy endosperm cells; position is what ultimately determines cell identity. This was revealed by a cell lineage study where cells were simultaneously marked with *C1*, which confers anthocyanin pigmentation in the aleurone, and *wx1*, a mutation causing amylose deficiency in starchy endosperm cells (Becraft and Asuncion-Crabb, 2000). Clonal sectors

arising throughout endosperm development contained both aleurone and starchy endosperm cells, indicating that the same progenitor cells continually produce both cell types. This implies that position specifies aleurone versus starchy endosperm cell fate. This hypothesis was further supported by analysis of connated kernels, showing that when abutting endosperms grow together and fuse, internalized aleurone cells transdifferentiate to starchy endosperm cells (Geisler-Lee and Gallie, 2005). Thus, cell fates are specified by as yet unknown positional cues that act throughout kernel development.

That a single lineage can produce both aleurone and starchy endosperm cell types indicates a degree of developmental plasticity, which was further supported by analysis of *defective kernel 1* (*dek1*) genetic mosaics. Loss-of-function *dek1* mutant endosperms lack aleurone and instead contain starchy endosperm cells in the outer layer (Becraft and Asuncion-Crabb, 2000; Becraft *et al.*, 2002; Lid *et al.*, 2002). This indicates the DEK1 gene product is required for aleurone cell fate; that is, *dek1* mutant cells cannot perceive or respond to cues that specify aleurone identity. Revertant sectors of a transposon-induced *dek1* allele produce wild-type aleurone cells in a mutant background. Single-celled sectors reflect reversion events that occur at the time of the last cell division during endosperm development (Becraft and Asuncion-Crabb, 2000). Hence, the peripheral layer of endosperm cells remains competent to differentiate into aleurone and the requisite positional cues are present throughout the entirety of development. Transposon-induced *dek1* loss-of-function events produce sectors of starchy endosperm cells, some as small as a single cell, in a background of wild-type aleurone (Fig. 6.1B). Aleurone cell identity is clearly recognizable midway through kernel development, thus the DEK1-mediated response is required for aleurone cells to maintain their identity; disruption of this signaling causes aleurone cells to re-differentiate as starchy endosperm.

DEK1 is an integral membrane protein with a cytoplasmic calpain protease domain

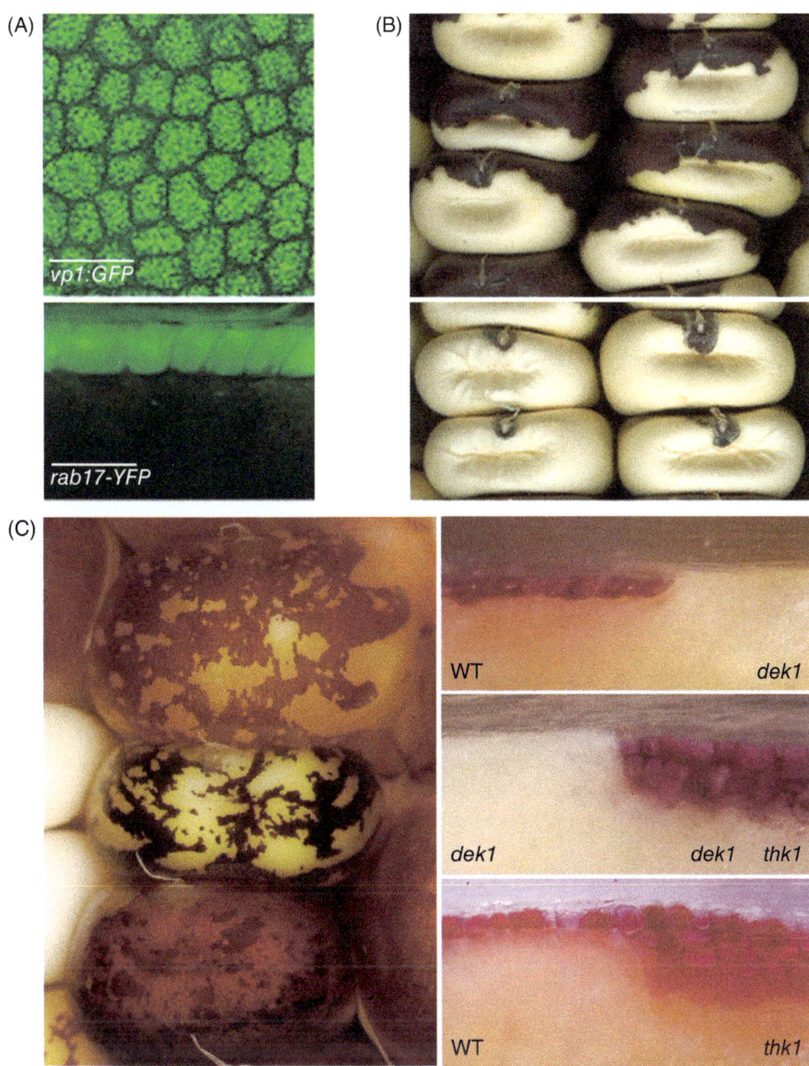



**Fig. 6.1.** (A) Marker expression in aleurone cells. Top: a top-down view of aleurone cells expressing a *vp1-promoter-GFP* transgene (Cao *et al.*, 2007). Bottom: sectional view from the side with aleurone cells expressing a *RAB17:RAB17-YFP* transgene (Gontarek *et al.*, 2016). Size bar=100 μm. (B) Examples of *dek1-D* kernels showing variable expressivity (Top: moderate expressivity; Bottom: strong expressivity) and illustrating the pattern evident in several mosaic aleurone mutants. The abgerminal face (lower side in the images) has a greater propensity for disrupted aleurone differentiation, while the silk scar region is most likely to develop aleurone (Becraft and Asuncion-Crabb, 2000; Becraft *et al.*, 2002). (C) Genetic mosaics of *dek1* and *thk1*. The panel on the left shows kernels with different sectoring patterns on a segregating ear. Each kernel corresponds to the adjacent section shown on the right. In all cases, aleurone cells are marked by anthocyanin accumulation. The top right image shows sectors of *dek1* loss in a wild-type background. Mutant *dek1* sectors are unpigmented due to lack of aleurone cells, because surface cells differentiate as starchy endosperm. The center right image shows sectors of *thk1* loss in a *dek1* mutant background. The pigmented patches are sectors of *thk1 dek1* double mutant tissue in a *dek1* single mutant unpigmented background lacking aleurone. The bottom right image shows *thk1* loss in a wild-type background. The darker purple patches are *thk1* mutant sectors with multiple aleurone layers (Yi *et al.*, 2011).

and a predicted extracellular loop region hypothesized to function as a receptor (Lid *et al.*, 2002; Kumar *et al.*, 2010). Presumably, the activity of the calpain protease is regulated and functions to cleave downstream protein(s) (Johnson *et al.*, 2008). The identity of the protein(s) that DEK1 directly regulates is unknown; however, a recent transcriptomic analysis indicated that DEK1 controls cell wall orientation and the cell division plane via cell cycle regulation (Liang *et al.*, 2015).

The *crinkly4* (*cr4*) mutant has a mosaic aleurone phenotype with sporadic areas of aleurone and starchy endosperm cell identities in the peripheral endosperm layer (Becraft *et al.*, 1996; Jin *et al.*, 2000). This indicates CR4, a receptor-like kinase, functions to positively regulate aleurone cell fate. Both CR4 and DEK1 have been proposed to function in the perception of the positional cue that specifies aleurone cell fate (Becraft *et al.*, 1996; Olsen *et al.*, 1999).

Analysis of *supernumerary aleurone layer 1* (*sal1*) suggests it is an upstream regulator of DEK1 and CR4. The *sal1* mutant has multiple aleurone layers, indicating that *Sal1+* negatively regulates aleurone cell fate (Shen *et al.*, 2003). SAL1 is a CHMP1 protein involved in multivesicle sorting and retrograde vesicle trafficking at the plasma membrane. SAL1, DEK1, and CR4 co-localize in vesicles, and it was proposed SAL1 promotes the retrograde transport of DEK1 and CR4 from the plasma membrane, thereby restricting their action and the extent of cells that differentiate as aleurone (Shen *et al.*, 2003; Tian *et al.*, 2007).

The *thick aleurone1* (*thk1*) gene, whose molecular identity remains elusive, produces a recessive mutant phenotype with multiple aleurone layers, suggesting that the *Thk1+* gene product is another negative regulator of aleurone cell layer number (Yi *et al.*, 2011). Double mutant *thk1 dek1* endosperm shows a *thk1* phenotype, indicating that *thk1* is epistatic to *dek1*. This was further confirmed with genetic mosaics, where *thk1 dek1* double mutant sectors showed the thick aleurone phenotype in a background of *dek1* mutant lacking aleurone (Fig. 6.1C) (Yi *et al.*, 2011). These data suggest that *thk1*

functions downstream of *dek1* and that the *dek1* gene product negatively regulates *thk1* function (perhaps indirectly) in the aleurone development pathway.

The *nkd1* and *nkd2* genes are duplicate factors encoding indeterminate domain (IDD) zinc finger transcription factors that regulate not only aleurone formation, but also many other aspects of endosperm development (Becraft and Asuncion-Crabb, 2000; Yi *et al.*, 2015; Gontarek *et al.*, 2016). The *nkd1 nkd2* mutant is pleotropic, affecting both the aleurone and starchy endosperm (Fig. 6.2). The mutant has multiple layers of aleurone-like cells, indicating that normal NKD functions are required to restrict the number of aleurone cell layers. The cells in these layers show sporadic lack of normal aleurone characteristics, such as anthocyanin accumulation or expression of aleurone identity markers, indicating NKD functions are also required for proper aleurone cell differentiation. In the starchy endosperm, *nkd1 nkd2* mutants are impaired in storage protein and starch deposition and carotenoid accumulation, while disrupted grain maturation causes occasional vivipary (Yi *et al.*, 2015; Gontarek *et al.*, 2016). Triple mutant *thk1 nkd1 nkd2* endosperm has more aleurone layers than either single *thk1* or *nkd1 nkd2* mutants, and the cells have impaired aleurone cell identity. This suggests *thk1* and *nkd1 nkd2* function independently for conferring aleurone cell layer number, but that *nkd1 nkd2* is downstream of *thk1* in controlling aleurone cell differentiation (Yi *et al.*, 2015).

Transcriptomic analysis explained many of the above phenotypes, revealing that NKD1 and NKD2 regulate expression of genes involved in controlling cell division, cell differentiation, anthocyanin biosynthesis, maturation, and hormone systems (Fig. 6.2) (Gontarek *et al.*, 2016). Repression of genes that promote cell division and proliferation, such as *cyclin 3B-like* and *retinoblastoma* related1, might explain the extra cell layers in the mutant (Gontarek *et al.*, 2016). NKD1 and NKD2 promote grain maturation and anthocyanin biosynthesis via transcriptional activation of the *viviparous1* (*vp1*) and *r1* genes. VP1 is necessary for

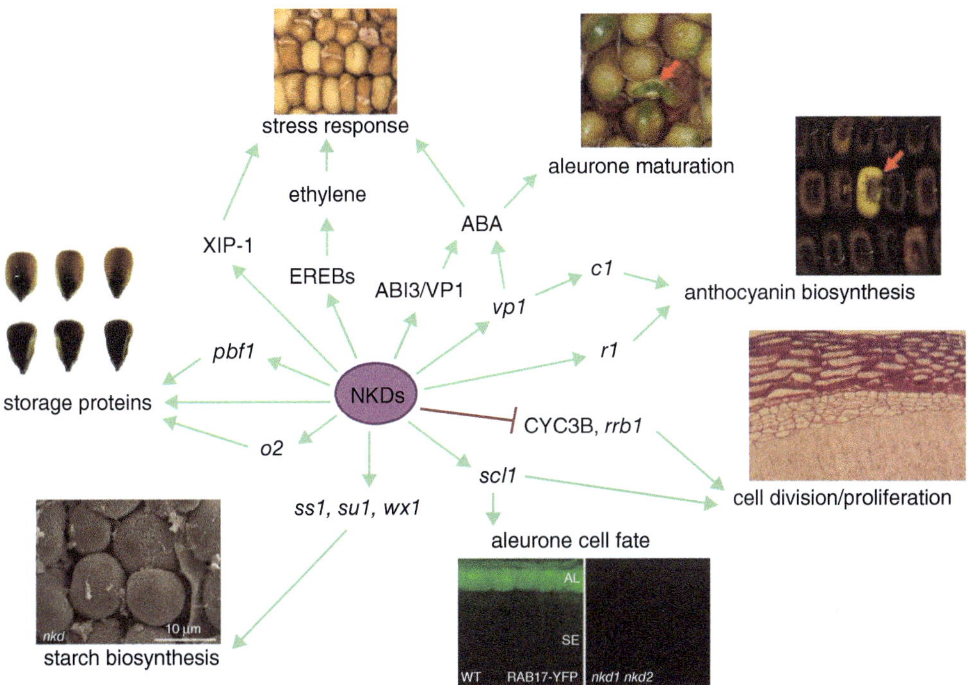



**Fig. 6.2.** Conceptual framework for the regulation of diverse aspects of endosperm development by the NKD1 and NKD2 transcription factors (Gontarek *et al.*, 2016).

ABA responses and promoting seed and aleurone maturation. VP1 also activates the transcription of *c1*, a myb transcription factor; C1 heterodimerizes with R1, a bHLH protein, to activate expression of anthocyanin biosynthetic genes (Goff *et al.*, 1992; Hattori *et al.*, 1992). NKD1 and NKD2 are required for regulation of genes in several hormone systems, including signaling and response to ABA and ethylene. These hormones promote seed maturation and are important stress regulators. Regulation of stress responses is further indicated by direct transcriptional activation of *xylanase inhibitor protein-1* (XIP-1), a pathogen defense gene (Jerkovic *et al.*, 2010); this could help to explain the propensity for *nkd1 nkd2* mutant kernels to develop infections (Fig. 6.2).

## 6.6  Remaining Questions

Despite recent advances in understanding aleurone development, many questions of both fundamental and practical importance remain. Some of these have been previously described in the literature, while others have not. The following list of questions is neither exhaustive nor necessarily in the order of significance.

### 6.6.1  What are the positional cues that specify aleurone cell fate?

Casual inspection of endosperm histology makes it apparent that positional information is integrated into establishment of cell fate because different cell types occur in particular positions of the endosperm. As described above, cells in the outermost endosperm layer interpret and respond to their position throughout the duration of development.

What is the nature of this positional information? This is an open question although several reports provide important clues. The differentiation of aleurone on isolated endosperm grown in culture strongly suggests the positional cues arise from within the endosperm per se, rather than through contact with surrounding maternal tissue (Gruis *et al.*, 2006). In this experiment, endosperms were excised from developing kernels at 6 DAP, so there was the formal possibility that early interactions with the nucellus somehow marked the surface cells in a stable manner; however, this seems unlikely, particularly given the developmental plasticity of aleurone cells discussed above. There are indications of hormonal involvement, because treatment with auxin transport inhibitors produced a multilayered aleurone, whereas overproduction of cytokinin disrupted aleurone differentiation (Geisler-Lee and Gallie, 2005; Forestan *et al.*, 2010). In recent years, patterns of biomechanical stress have returned to the fore in discussions of mechanisms that control patterns of cell division, morphogenesis and cell differentiation (Lin *et al.*, 2016; Louveaux *et al.*, 2016). Such a mechanism could well function in the surface (aleurone) layer and this model would fit well with the proposed role of DEK1 in maintaining cell–cell contact within the epidermal layer (Galletti and Ingram, 2015). Many other potential mechanisms are possible and, as such, the nature of the cue(s) that allow aleurone cells to perceive their position within endosperm tissue and differentiate appropriately remains one of the biggest unanswered questions for aleurone development.

### 6.6.2   What determines aleurone competency?

This question derives from the previous one: How do aleurone cells perceive and respond to positional cues? As discussed above, several known regulatory genes are required for this process, yet none are strictly aleurone-specific. The known aleurone mutants also affect starchy endosperm cells, often resulting in opaque and/or carotenoid-deficient endosperm. In addition, there appears to be a regional competency. Mutations in negative regulators of aleurone cell fate, such as *thk1, sal1,* and *nkd,* increase the number of aleurone cell layers (Becraft and Asuncion-Crabb, 2000; Shen *et al.*, 2003; Yi *et al.*, 2011, 2015). Yet, the effects of these mutations on aleurone cell fate are limited to the peripheral cell layers of the endosperm, indicating they are acting within a larger patterning framework that limits aleurone competency to the outer regions of the endosperm. The factors that determine this competency are unknown at present. The *globby1* mutant could be the only manifestation of a phenotype that includes scattered aleurone cells positioned internally in the endosperm (Costa *et al.*, 2003); thus, identification of this gene could be informative to understanding an important aspect of overall endosperm organization.

### 6.6.3   What is the significance of endosperm patterning reflected by mosaic aleurone mutants and how does it relate to aleurone development?

Many mutants that disrupt aleurone differentiation do so asymmetrically, with the abgerminal side of kernels much more prone to disruption than the germinal side (Fig. 6.1B). Described mutants that display this pattern include *cr4, dek1-D,* and *nkd1 nkd2* (Becraft *et al.*, 1996, 2002; Becraft and Asuncion-Crabb, 2000; Yi *et al.*, 2015), along with several others in our mutant collection. The *vp8* mutant produces the opposite mosaic pattern, with the germinal face most likely to display aleurone disruption (Suzuki *et al.*, 2008), suggesting that different developmental domains of the kernel are under distinct genetic regulation. This recurring phenotypic pattern reflects some feature of the underlying biology. Two aspects of this phenomenon remain obscure and merit discussion: (i) what is the significance of this pattern in relation to overall endosperm development; and (ii) what is the nature of the signal?

The former question is perhaps more fundamental: Is this pattern important for some aspect of endosperm development? Gradients have been described for cell division, gene expression, storage protein accumulation, endoreduplication and programmed cell death (Randolph, 1936; Kiesselbach, 1949; Kowles and Phillips, 1988; Young *et al.*, 1997; Woo *et al.*, 2001). Are the spatial distributions of some or all these activities related to the aleurone mosaic pattern? Are the spatial distributions of these activities functionally critical? Or is there an unknown aspect of endosperm development being patterned, and if so what is its significance? Learning the importance of this patterning phenomenon will require disrupting it, which is a conundrum given that we don't know the nature of the signals that establish the pattern, and if the pattern is essential, desired mutants might manifest as aborted grains.

Unknowns regarding the signal(s) include physical identity as well as the source. Presumably, it is most likely the signal has a chemical basis, though a mechanical mechanism is possible. Chemical signals could include any of the plant hormones, a small peptide, small RNA or a trafficked protein. In fact, given that the early endosperm is a coenocyte, a molecule would not necessarily even need to be mobile but could be asymmetrically distributed in the early endosperm to establish a pattern that is propagated later in the cellular endosperm. In the coenocytic Arabidopsis embryo sac, asymmetric distribution of CKI1 protein specifies central cell fate (Yuan *et al.*, 2016).

The source of the signal is also unknown, although the pattern itself provides potential clues. The embryo was an obvious candidate, but it was experimentally discounted by analysis with the *wandering embryo* (*wem*) mutant that produces embryos on random faces of the kernel. The pattern of aleurone mosaicism in *dek1 wem* and *cr4 wem* double mutants did not change with altered embryo position (Becraft and Asuncion-Crabb, 2000). However, the pattern does appear to center around the position of the silk scar. Several structures would be consistent with this pattern, including the silk itself or carpels at the silk attachment site. This is also the region of the kernel where the persistent antipodal cells reside, which could reflect earlier events in the chalazal region of the embryo sac. Finally, the growing pollen tube could potentially mark this region of the kernel.

### 6.6.4 What is the genetic program that confers aleurone identity? Is aleurone homologous to epidermis?

While quite a number of mutants that disrupt normal aleurone development are known in maize and other cereals, none are truly aleurone-specific. All maize aleurone mutants described thus far show pleiotropic effects and many also affect the epidermis of the shoot (Becraft and Yi, 2011). The aleurone is often referred to as the epidermis of the endosperm and the shared genetic regulation suggests there might be evolutionary homology. However, even within the context of the endosperm, most mutants that have been studied in depth show pleiotropic effects on starchy endosperm tissue (Jin *et al.*, 2000; Young and Gallie, 2000; Becraft *et al.*, 2002; Yi *et al.*, 2011, 2015; Qi *et al.*, 2017).

Interestingly, whole genome coexpression network analysis (WGCNA) in 8 DAP kernels showed the aleurone transcriptome is more similar to the embryo than to other endosperm cell types (Zhan *et al.*, 2015). Nonetheless, an aleurone-specific gene expression module was identified containing just over 800 genes that are presumably important for aleurone-specific functions. Among these were 43 transcription factors that likely hold the key to understanding the "aleurone program." Analysis of these transcription factors will provide information about the genes and biological functions they regulate. What would be particularly exciting is identification of a "master regulator," but whether a single gene controls aleurone identity is questionable. Given the intensive mutant screens that have been conducted in pigmented aleurone genotypes and the obvious nature of an aleurone-deficient phenotype, it seems likely such a factor would have been discovered. Hence, a combinatorial mechanism appears more likely, which will only become clear with continued study of the gene regulatory networks.

### 6.6.5 Is aleurone patterning related to the radial patterning mechanism in Arabidopsis roots?

Several lines of evidence suggest the possibility that aleurone could be patterned by mechanisms related to the well-known root radial patterning mediated by SCARECROW (SCR) and SHORT-ROOT (SHR), members of the GRAS family of transcription factors (Nakajima *et al.*, 2001). SHR is expressed in the stele of the root and the protein is trafficked to the cortical/endodermal initial where it activates expression of SCR. This system is required to regulate an asymmetric cell division that gives rise to the cortex and endodermal layers of the root. In addition, four IDD proteins, JACKDAW (JKD), BALDIBIS (BIB), NUTCRACKER (NUC), and MAGPIE, have overlapping functions in the specification of the cortical cell layer of the root. JKD, NUC, and MAGPIE form a transcription factor complex with SCR and SHR (Ogasawara *et al.*, 2011; Long *et al.*, 2015). This complex is essential for specification of endodermal cell fate in the Arabidopsis root and for nuclear retention of SHR, leading to just a single layer of endodermal cells. In *jkd bib* double mutant roots, the single endodermal layer is replaced with multiple layers of cells with indistinct identity (Ogasawara *et al.*, 2011; Long *et al.*, 2015).

The *jkd bib* mutant phenotype is reminiscent of *nkd1 nkd2* mutants, where multiple layers of cells with indistinct identities replace the single aleurone layer (Becraft and Asuncion-Crabb, 2000; Yi *et al.*, 2015). Furthermore, a *scarecrow-like1* (*scl1*) gene encoding a maize member of the GRAS family is differentially expressed between wild-type and *nkd1 nkd2* mutant endosperm and is predicted to be a direct target of NKD1 and NKD2 transcriptional activation. This suggests the intriguing possibility that NKD1 and NKD2 might similarly interact with GRAS family members to

### 6.6.6 How are hormones involved in aleurone development?

Phytohormones seem to be involved in most, if not all plant processes, and endosperm development is no exception. A detailed account of all the hormone studies conducted in endosperm is beyond the scope of this chapter, but they are described throughout the book. Suffice to say, many studies examined global hormone levels and biological activities. For example, a global increase in auxin levels is linked to increased rates of cell division and endoreduplication (Lur and Setter, 1993), while ethylene promotes PCD (Young *et al.*, 1997). However, hormones are now known to have highly localized functions and to be intimately involved in many examples of cell and tissue-level patterning, such as organ primordia formation on the shoot apical meristem (Bar and Ori, 2014). Such fine-scale analyses have yet to be undertaken in the endosperm. Nonetheless, there have been a number of reports that provide intriguing hints at the importance of various hormones for regulating aleurone differentiation and patterning.

ABA and GA are well-known regulators of late aleurone maturation and germination, and aleurone cells served as a classic system in early studies on hormone-mediated gene regulation (Skriver *et al.*, 1991). While currently there is no strong evidence for these hormones functioning in early patterning events, there have been enticing hints as to the involvement of auxin and cytokinin in regulating aleurone development. Plants watered with a solution containing the auxin transport inhibitor NPA produced kernels with multiple aleurone layers (Forestan *et al.*, 2010). There was a commensurate expansion in the expression of the auxin transporter ZmPIN1 to multiple cell layers as well as an expanded region of auxin accumulation. These observations

regulate aleurone patterning and cell fate. Future experiments are needed to test this hypothesis.

support the hypothesis that an auxin maximum directs aleurone differentiation.

Cytokinins have been implicated in aleurone development through transgenic expression of an IPT gene under the regulation of the senescence inducible SAG12 promoter. IPT encodes a rate-limiting enzyme in cytokinin biosynthesis, and the endosperm surfaces of transgenic kernels showed mosaics with interspersed aleurone and starchy endosperm cells, suggesting that cytokinins can inhibit aleurone differentiation (Geisler-Lee and Gallie, 2005). Despite these reports, there is no strong evidence directly implicating endogenous auxin or cytokinin in directing normal aleurone differentiation. The tools are available to test these hypotheses, but such studies have not yet been conducted.

### 6.6.7 Is there a functional link between aleurone development and carotenoid biosynthesis?

Many aleurone defective mutants, including *dek1*, *cr4*, and *nkd*, are carotenoid deficient (Jin *et al.*, 2000; Becraft *et al.*, 2002; Yi *et al.*, 2015). Carotenoids themselves do not appear to be required for aleurone differentiation, because carotenoid deficient mutants appear able to produce a normal aleurone (unpublished observations). In the case of the NKD1 and NKD2 transcription factors, it appears the link could be transcriptional regulation of the *y1* and *vp5* genes, which encode key carotenoid biosynthetic enzymes (Gontarek *et al.*, 2016). It is currently unknown whether this represents a link in common among other defective aleurone mutants, or if it is coincidental that these particular mutants happen to share aspects of their pleiotropic phenotypes. Interestingly, the *thk1* mutant, which is epistatic to *dek1* for producing aleurone cells, is also able to rescue the carotenoid deficiency of *dek1* endosperm (Yi *et al.*, 2011). This would be consistent with a shared mechanism regulating both aleurone differentiation and carotenoid biosynthesis, which is not an obvious expectation. In-depth transcriptomic analysis of all these mutants would likely reveal whether there is a link.

### 6.6.8 How can we manipulate metabolic pathways in the aleurone to improve biological functions, grain quality traits, and dietary health benefits?

As described above, aleurone contains many compounds responsible for the health benefits associated with dietary cereal bran. What are the regulatory mechanisms that control the levels of these various healthful compounds? How much natural variation is available for the accumulation of these compounds? Are the traits favorable for breeding? Are the metabolic pathways amenable to engineering? Will there be adverse side effects, compensatory loss of other desirable compounds, or yield penalties for increasing particular desirable compounds? Addressing these questions will provide fertile research topics for many years.

### Acknowledgment

Research in the Becraft lab is currently funded by the U.S. National Science Foundation Award 1444568.

## References

Bar, M. and Ori, N. (2014) Leaf development and morphogenesis. *Development* 141, 4219–4230.

Becraft, P.W. and Asuncion-Crabb, Y.T. (2000) Positional cues specify and maintain aleurone cell fate in maize endosperm development. *Development* 127, 4039–4048.

Becraft, P.W. and Gutiérrez-Marcos, J.F. (2012) Endosperm development: dynamic processes and cellular innovations underlying sibling altruism. *Wiley Interdisciplinary Review of Developmental Biology* 1, 579–593.

Becraft, P.W. and Yi, G. (2011) Regulation of aleurone development in cereal grains. *Journal of Experimental Botany* 62, 1669–1675.

Becraft, P.W., Stinard, P.S. and McCarty, D.R. (1996) CRINKLY4: a TNFR-like receptor kinase involved in maize epidermal differentiation. *Science* 273, 1406–1409.

Becraft, P.W., Li, K., Dey, N. and Asuncion-Crabb, Y.T. (2002) The maize *dek1* gene functions in embryonic pattern formation and in cell fate specification. *Development* 129, 5217–5225.

Bethke, P.C. and Jones, R.L. (2001) Cell death of barley aleurone protoplasts is mediated by reactive oxygen species. *The Plant Journal* 25, 19–29.

Brink, R.A. (1956) A genetic change associated with the R locus in maize which is directed and potentially reversible. *Genetics* 41, 872–889.

Brouns, F., Hemery, Y., Price, R. and Anson, N.M. (2012) Wheat aleurone: separation, composition, health aspects, and potential food use. *Critical Reviews in Food Science and Nutrition* 52, 553–568.

Cao, X., Costa, L.M., Biderre-Petit, C., Kbhaya, B., Dey, N., *et al.* (2007) Abscisic acid and stress signals induce *viviparous1* (*vp1*) expression in seed and vegetative tissues of maize. *Plant Physiology* 143, 720–731.

Casacuberta, J.M., Puigdomenech, P. and San Segundo, B. (1991) A gene coding for a basic pathogenesis-related (PR-like) protein from *Zea mays*: molecular cloning and induction by a fungus (*Fusarium moniliforme*) in germinating maize seeds. *Plant Molecular Biology* 16, 527–536.

Costa, L.M., Gutiérrez-Marcos, J.F., Brutnell, T.P., Greenland, A.J. and Dickinson, H.G. (2003) The *globby1-1* (*glo1-1*) mutation disrupts nuclear and cell division in the developing maize seed causing alterations in endosperm cell fate and tissue differentiation. *Development* 130, 5009–5017.

Forestan, C., Meda, S. and Varotto, S. (2010) ZmPIN1-mediated auxin transport is related to cellular differentiation during maize embryogenesis and endosperm development. *Plant Physiology* 152, 1373–1390.

Galletti, R. and Ingram, G.C. (2015) Communication is key: reducing DEK1 activity reveals a link between cell–cell contacts and epidermal cell differentiation status. *Communicative & Integrative Biology* 8, e1059979.

Geisler-Lee, J. and Gallie, D.R. (2005) Aleurone cell identity is suppressed following connation in maize kernels. *Plant Physiology* 139, 204–212.

Goff, S.A., Cone, K.C. and Chandler, V.L. (1992) Functional analysis of the transcriptional activator encoded by the maize B gene: evidence for a direct functional interaction between two classes of regulatory proteins. *Genes & Development* 6, 864–875.

Gontarek, B.C., Neelakandan, A.K., Wu, H. and Becraft, P.W. (2016) NKD transcription factors are central regulators of maize endosperm development. *Plant Cell* 28, 2916–2936.

Goyal, K., Walton, L.J. and Tunnacliffe, A. (2005) LEA proteins prevent protein aggregation due to water stress. *Biochemical Journal* 388, 151–157.

Gruis, D., Guo, H., Selinger, D., Tian, Q. and Olsen, O.-A. (2006) Surface position, not signaling from surrounding maternal tissues, specifies aleurone epidermal cell fate in maize. *Plant Physiology* 141, 898–909.

Hattori, T., Vasil, V., Rosenkrans, L., Hannah, L.C., McCarty, D.R. and Vasil, I.K. (1992) The *Viviparous-1* gene and abscisic acid activate the C1 regulatory gene for anthocyanin biosynthesis during seed maturation in maize. *Genes & Development* 6, 609–618.

Henderson, A.J., Ollila, C.A., Kumar, A., Borresen, E.C., Raina, K., Agarwal, R. and Ryan, E.P. (2012) Chemopreventive properties of dietary rice bran: current status and future prospects. *Advances in Nutrition: An International Review Journal* 3, 643–653.

Islam, M.S., Nagasaka, R., Ohara, K., Hosoya, T., Ozaki, H., Ushio, H. and Hori, M. (2011) Biological abilities of rice bran-derived antioxidant phytochemicals for medical therapy. *Current Topics in Medicinal Chemistry* 11, 1847–1853.

Iwatsuki, K., Akihisa, T., Tokuda, H., Ukiya, M., Higashihara, H., *et al.* (2003) Sterol ferulates, sterols, and 5-alk(en)ylresorcinols from wheat, rye, and corn bran oils and their inhibitory effects on Epstein-Barr virus activation. *Journal of Agricultural and Food Chemistry* 51, 6683–6688.

Jain, D., Ebine, N., Jia, X., Kassis, A., Marinangeli, C., *et al.* (2008) Corn fiber oil and sitostanol decrease cholesterol absorption independently of intestinal sterol transporters in hamsters. *Journal of Nutritional Biochemistry* 19, 229–236.

Jakobsen, J.V., Knox, R.B. and Pyliotis, N.A. (1971) The structure and composition of aleurone grains in the barley aleurone layer. *Planta* 101, 189–209.

Jerkovic, A., Kriegel, A.M., Bradner, J.R., Atwell, B.J., Roberts, T.H. and Willows, R.D. (2010) Strategic distribution of protective proteins within bran layers of wheat protects the nutrient-rich endosperm. *Plant Physiology* 152, 1459–1470.

Jin, P., Guo, T. and Becraft, P.W. (2000) The maize CR4 receptor-like kinase mediates a growth factor-like differentiation response. *Genesis: The Journal of Genetics and Development* 27, 104–116.

Johnson, K.L., Faulkner, C., Jeffree, C.E. and Ingram, G.C. (2008) The phytocalpain DEFECTIVE KERNEL 1 is a novel *Arabidopsis* growth regulator whose activity is regulated by proteolytic processing. *Plant Cell* 20, 2619–2630.

Kermicle, J.L. (1970) Dependence of the R-mottled aleurone phenotype in maize on mode of sexual transmission. *Genetics* 66, 69–85.

Kiesselbach, T.A. (1949) *The Structure and Reproduction of Corn.* University of Nebraska Press, Lincoln, Nebraska.

Kowles, R.V. and Phillips, R.L. (1988) Endosperm development in maize. *International Review of Cytology* 112, 97–136.

Kumar, S.B., Venkateswaran, K. and Kundu, S. (2010) Alternative conformational model of a seed protein DeK1 for better understanding of structure–function relationship. *Journal of Proteins and Proteomics* 1, 77–90.

Lafiandra, D., Riccardi, G. and Shewry, P.R. (2014) Improving cereal grain carbohydrates for diet and health. *Journal of Cereal Science* 59, 312–326.

Leroux, B.M., Goodyke, A.J., Schumacher, K.I., Abbott, C.P., Clore, A.M., *et al.* (2014) Maize early endosperm growth and development: from fertilization through cell type differentiation. *American Journal of Botany* 101, 1259–1274.

Liang, Z., Brown, R.C., Fletcher, J.C. and Opsahl-Sorteberg, H.-G. (2015) Calpain-mediated positional information directs cell wall orientation to sustain plant stem cell activity, growth and development. *Plant & Cell Physiology* 56, 1855–1866.

Lid, S.E., Gruis, D., Jung, R., Lorentzen, J.A., Ananiev, E., *et al.* (2002) The *defective kernel 1 (dek1)* gene required for aleurone cell development in the endosperm of maize grains encodes a membrane protein of the calpain gene superfamily. *Proceedings of the National Academy of Sciences of the United States of America* 99, 5460–5465.

Lillioja, S., Neal, A.L., Tapsell, L. and Jacobs, D.R., Jr. (2013) Whole grains, type 2 diabetes, coronary heart disease, and hypertension: links to the aleurone preferred over indigestible fiber. *Biofactors* 39, 242–258.

Lin, X.X., Shi, Y., Cao, Y.L. and Liu, W. (2016) Recent progress in stem cell differentiation directed by material and mechanical cues. *Biomedical Materials* 11, 014109.

Long, Y., Smet, W., Cruz-Ramírez, A., Castelijns, B., De Jonge, W., *et al.* (2015). Arabidopsis BIRD zinc finger proteins jointly stabilize tissue boundaries by confining the cell fate regulator SHORT-ROOT and contributing to fate specification. *Plant Cell* 27, 1185–1199.

Louveaux, M., Julien, J.-D., Mirabet, V., Boudaoud, A. and Hamant, O. (2016) Cell division plane orientation based on tensile stress in *Arabidopsis thaliana. Proceedings of the National Academy of Sciences of the United States of America* 113, E4294–E4303.

Lur, H.S. and Setter, T.L. (1993) Role of auxin in maize endosperm development (timing of nuclear DNA endoreduplication, zein expression, and cytokinin). *Plant Physiology* 103, 273–280.

McClintock, B. (1950) The origin and behavior of mutable loci in maize. *Proceedings of the National Academy of Sciences of the United States of America* 36, 344–355.

Nakajima, K., Sena, G., Nawy, T. and Benfey, P.N. (2001) Intercellular movement of the putative transcription factor SHR in root patterning. *Nature* 413, 307–311.

Ogasawara, H., Kaimi, R., Colasanti, J. and Kozaki, A. (2011) Activity of transcription factor JACKDAW is essential for SHR/SCR-dependent activation of SCARECROW and MAGPIE and is modulated by reciprocal interactions with MAGPIE, SCARECROW and SHORT ROOT. *Plant Molecular Biology* 77, 489–499.

Olsen, O.-A., Linnestad, C. and Nichols, S.E. (1999) Developmental biology of the cereal endosperm. *Trends in Plant Science* 4, 253–257.

Qi, X., Li, S., Zhu, Y., Zhao, Q., Zhu, D. and Yu, J. (2017) *ZmDof3*, a maize endosperm-specific Dof protein gene, regulates starch accumulation and aleurone development in maize endosperm. *Plant Molecular Biology* 93, 7–20.

Randolph, L.F. (1936) Developmental morphology of the caryopsis in maize. *Journal of Agricultural Research* 53, 881–916.

Regvar, M., Eichert, D., Kaulich, B., Gianoncelli, A., Pongrac, P., Vogel-Mikuš, K. and Kreft, I. (2011) New insights into globoids of protein storage vacuoles in wheat aleurone using synchrotron soft X-ray microscopy. *Journal of Experimental Botany* 62, 3929–3939.

Reyes, F.C., Chung, T., Holding, D., Jung, R., Vierstra, R. and Otegui, M.S. (2011) Delivery of prolamins to the protein storage vacuole in maize aleurone cells. *Plant Cell* 23, 769–784.

Royo, J., Gómez, E., Sellam, O., Gerentes, D., Paul, W. and Hueros, G. (2014) Two maize *END-1* orthologs, *BETL9* and *BETL9like*, are transcribed in a non-overlapping spatial pattern on the outer surface of the developing endosperm. *Frontiers in Plant Science* 5, 180.

Shen, B., Li, C., Min, Z., Meeley, R.B., Tarczynski, M.C. and Olsen, O.A. (2003) *sal1* determines the number of aleurone cell layers in maize endosperm and encodes a class E vacuolar sorting protein. *Proceedings of the National Academy of Sciences of the United States of America* 100, 6552–6557.

Skriver, K., Olsen, F.L., Rogers, J.C. and Mundy, J. (1991) Cis-acting DNA elements responsive to gibberellin and its antagonist abscisic acid. *Proceedings of the National Academy of Sciences of the United States of America* 88, 7266–7270.

Stuckey, R.E., Niblack, T.L., Nyvall, R.F., Krausz, J.P. and Horne, C.W. (1985) Corn disease management. *National Corn Handbook*. Iowa State University, Cooperative Extension Service.

Suzuki, M., Latshaw, S., Sato, Y., Settles, A.M., Koch, K.E., *et al.* (2008) The maize *viviparous8* locus, encoding a putative ALTERED MERISTEM PROGRAM1-like peptidase, regulates abscisic acid accumulation and coordinates embryo and endosperm development. *Plant Physiology* 146, 1193–1206.

Tian, Q., Olsen, L., Sun, B., Lid, S.E., Brown, R.C., *et al.* (2007) Subcellular localization and functional domain studies of DEFECTIVE KERNEL1 in maize and *Arabidopsis* suggest a model for aleurone cell fate specification involving CRINKLY4 and SUPERNUMERARY ALEURONE LAYER1. *Plant Cell* 19, 3127–3145.

Vicente-Carbajosa, J. and Carbonero, P. (2005) Seed maturation: developing an intrusive phase to accomplish a quiescent state. *The International Journal of Developmental Biology* 49, 645–651.

Woo, Y.M., Hu, D.W., Larkins, B.A. and Jung, R. (2001) Genomics analysis of genes expressed in maize endosperm identifies novel seed proteins and clarifies patterns of zein gene expression. *Plant Cell* 13, 2297–2317.

Yi, G., Lauter, A.M., Scott, M.P. and Becraft, P.W. (2011) The *thick aleurone1* mutant defines a negative regulation of maize aleurone cell fate that functions downstream of *dek1*. *Plant Physiology* 156, 1826–1836.

Yi, G., Neelakandan, A.K., Gontarek, B.C., Vollbrecht, E. and Becraft, P.W. (2015) The *naked endosperm* genes encode duplicate INDETERMINATE domain transcription factors required for maize endosperm cell patterning and differentiation. *Plant Physiology* 167, 443–456.

Young, T.E. and Gallie, D.R. (2000) Regulation of programmed cell death in maize endosperm by abscisic acid. *Plant Molecular Biology* 42, 397–414.

Young, T.E., Gallie, D.R. and Demason, D.A. (1997) Ethylene-mediated programmed cell death during maize endosperm development of wild-type and *shrunken2* genotypes. *Plant Physiology* 115, 737–751.

Yuan, L., Liu, Z., Song, X., Johnson, C., Yu, X. and Sundaresan, V. (2016) The CKI1 histidine kinase specifies the female gametic precursor of the endosperm. *Developmental Cell* 37, 34–46.

Zentella, R., Yamauchi, D. and Ho, T.H. (2002) Molecular dissection of the gibberellin/abscisic acid signaling pathways by transiently expressed RNA interference in barley aleurone cells. *Plant Cell* 14, 2289–2301.

Zhan, J., Thakare, D., Ma, C., Lloyd, A., Nixon, N.M., *et al.* (2015) RNA sequencing of laser-capture microdissected compartments of the maize kernel identifies regulatory modules associated with endosperm cell differentiation. *Plant Cell* 27, 513–531.

# 7   Embryo Development


**William F. Sheridan* and Janice K. Clark**
*Department of Biology, University of North Dakota, North Dakota, USA*


## 7.1   Introduction

The maize embryo develops over a 40–50-day period from a single-celled zygote into a miniature plant consisting of five or six leaf primordia and a single primary root. The first detailed description of the development of the maize embryo and caryopsis, wherein it is formed, was by Randolph (1936). This was followed by an extensive report on the structure and reproduction of corn by Kiesselbach (1949). In both publications the authors utilized ink drawings and photographic images to illustrate embryo morphogenesis throughout its development. Genetic analysis of this process began early in the 20th century with the reports of Jones (1920), Demerec (1923), Mangelsdorf (1923, 1926), and Wentz (1930). Following the iconic publications of Randolph (1936) and Kiesselbach (1949), a third descriptive paper was published by Abbe and Stein (1954). These authors introduced the terminology currently used to describe the stages of embryo development. In this chapter we describe the process of maize embryo morphogenesis and mutations that have been shown to disrupt this process.

The goals of this research, which involved mutant screens that evolved over several decades, address two different but related questions. The project related to the first question began in the spring of 1973 with a collaboration involving Dr. Gerry Neuffer at the University of Missouri. The goal was to determine if it was possible to identify and characterize maize auxotrophic mutants. The strategy was to screen a large collection of EMS-induced *defective kernel* mutants that affect endosperm and embryo development, and rescue the lethal (*dek*) mutants by culturing the embryos on supplemented media. The concept tested was whether single gene mutations can identify "housekeeping" genes essential for growth and cell maintenance. Several years of research led us to recognize that many of these mutants fail to undergo sufficient morphogenesis for the immature embryo to be viable (see Section 7.3 below). This led us to ask a second conceptual question: Is it possible to identify genes responsible for maize developmental mutations that regulate embryo morphogenesis? Our goal was to identify mutations amenable to gene cloning and sequencing in order to dissect developmental pathways. To answer this question we used transposable element mutagenesis, because it creates the possibility of molecularly tagging genes of interest and cloning them.

---


*Corresponding author e-mail: william.sheridan@email.und.edu

In November 1985, Dr. Donald S. Robertson invited us to visit his laboratory at Iowa State University to screen a large collection of self-pollinated ears of active *Mutator* stocks. We identified a large collection of *embryo-specific* (*emb*) mutants with normal or nearly normal endosperm development, but manifesting defective embryos. Over the next several years, these mutants were characterized genetically and their phenotypes described. We were not successful in cloning any of the mutant genes, but this was accomplished by other investigators (see Section 7.4 below).

In January 2010 we were encouraged by Dr. Thomas Brutnell to screen for EMS-induced *embryo-specific* mutations, because recent advances in DNA sequencing set the stage for whole genome sequencing and mapping. We initiated this approach the following summer and produced a large collection of *emb* mutants. We believe their analysis will identify the molecular mechanisms underlying maize embryo morphogenesis.

From the start of our search for *embryo-specific* mutations, we anticipated that many of these mutants could identify genes encoding transcriptional regulators and plant hormone-responsive genes that interact in controlling embryo morphogenesis. Recent reports show that some of these *dek* and *emb* mutations involve alterations in nuclear-coded mitochondrial and plastid proteins, respectively (see Sections 7.5 and 7.6 below).

## 7.2 A Description of Maize Embryo Morphogenesis

Prior to Randolph's report in 1936, little attention was paid to maize embryo development at the cellular level. Randolph stated: "The present account (of maize embryogeny) is limited to a consideration of the ontogenetic development of the embryo" (Randolph 1936). His pen and ink-drawn images (figures 4–7 in the same article) beautifully depict cellular events occurring in the embryo sac before and after fertilization and through the nine stages of embryo morphogenesis (Abbe and Stein, 1954). These exceptionally detailed drawings reveal details of three crucial morphogenetic events that must be tightly regulated during passage of the zygote (and its cellular progeny) from the first cell division through formation of the first leaf primordium. We know of no drawings or photographic images that equal his exquisitely drawn figures.

The first important ontogenetic event of embryo development is asymmetric division of the zygote, resulting in a small lens-shaped apical cell with dense protoplasm above a much larger basal cell with less dense cytoplasm and numerous vacuoles. This cell division is followed by a series of additional mitoses in varying planes that produce a proembryo comprising an upper region of small dense cells subtended by a few much longer cells filled mostly with large vacuoles. Randolph noted: "especially noteworthy in connection with the early history of the maize embryo is the fact that the proembryo develops very irregularly with no very definite or orderly arrangement of the cells or sequences of cell divisions." The fact that: "the proembryo of maize does not regularly conform to any definite growth pattern" distinguishes it from the regular cell division patterns characteristic of most dicot proembryos and those of some grasses. The apical cells, which are much richer in plastids and mitochondria than the basal cells, form the embryo proper, while the basal cells form the suspensor. This asymmetry in organelle distribution provisions the embryo proper for its future morphogenesis into the fully formed embryo. It is consistent with the hypothesis that these organelle-rich cells must meet the energy and nutrient demands for passage of the proembryo into the transition stage embryo.

The second important ontogenetic event is changing morphogenesis from radial symmetry in the proembryo to bilateral symmetry in the transition stage embryo. In materials studied by Randolph, this began about 8 days after pollination (DAP). At first, an epidermis appears over the embryo apical region and then gradually spreads down to the suspensor. Up to that point, cell divisions in the upper region occur randomly, but within 2 days the radial symmetry is lost. This happens as the plane of cell divisions in the upper region becomes progressively more anticlinal and less periclinal. By 9 DAP, the pattern of cell division in the subapical region creates cell walls at right angles to the axis of the embryo, resulting in a longitudinal increase in meristematic tissue between the suspensor and the tip of the embryo. By 9 or 10 DAP, the axis of embryo is evident. This occurs internally and can be recognized in radial longitudinal sections or in cross sections as a: "group of densely protoplasmic and actively dividing cells at the front (anterior face) of the embryo, slightly below the tip." This is a roughly triangular group of cells dividing more rapidly than the other cells in the embryo, so as to produce a slight protuberance of the anterior face of the embryo. The internal asymmetric location of the wedge of cells of the external anterior protuberance, resulting from their rapid division, defines passage of the proembryo into the transition stage embryo (Fig. 7.1). It is noteworthy that the most significant morphogenetic feature of the earlier developmental sequence, as the zygote develops into the proembryo, is the sequestering of organelles in the small, dense cells of the upper region of the proembryo during the period of radial symmetry. The most significant feature of the onset of the transition stage is that it begins establishment of the axis of the mature embryo: "an axis which does not coincide with the axis of the proembryo." Whereas provision of the upper cells of the proembryo might be considered as occurring through activity of cytoskeletal elements, it appears that formation and orientation of the wedge of dividing cells, producing the protuberance on the anterior face of the transition stage embryo, results from regulated signaling by transcription factors or phytohormones (see below).

The third ontogenetic event is a more prolonged process whereby the single group of meristematic cells increases in size and by 13 DAP differentiates into two distinct groups. The upper group: "is the primordium of the stem portion (SAM) of the new axis; the lower group is the primordium of the corresponding root portion." Randolph noted that by this point the forerunner of the vascular supply to the scutellum is evident as a strand of elongated cells with dense cytoplasm that leads into the scutellum from the region between the two meristems. This strand of elongated cells appears coincidently with differentiation of the two meristems from the single precursor meristematic region. As a result of the appearance of the shoot–root axis as a lateral structure, there is a shift in the polarity of the embryo from a vertical to an oblique orientation, with the ensuing stages of embryogenesis occurring along the axis of the shoot–root meristems. As the SAM (shoot apical meristem) enlarges, the coleoptile primordium appears as a ridge of tissue, first protruding from the enlarging scutellum above the SAM and extending down and around the SAM to nearly encircle it. The scutellum proceeds through a period of broadening, lengthening and thickening over the following 30 or more days. Following formation of the coleoptile primordium at 16 DAP, the first leaf primordium appears as a ridge of tissue located along the lower edge of the SAM and opposite the coleoptile primordium. This ridge of tissue expands upward to cover the SAM and is subsequently covered by the leaf primordia. The suspensor ceases to enlarge by 20 DAP.

A fourth ontogenetic event is the iterative formation of four or five leaf primordia and the continued enlargement of the shoot–root axis and scutellum. During this period and until about 45 DAP, the scutellum enlarges to cover the coleoptile and the leaf primordia enclosed within it to create a narrow vertical cleft when viewed from the anterior (frontal) face of the mature embryo (Abbe and Stein, 1954, see Fig. 7.2).

## 7.3 Genetic Analysis of *Defective Kernel* (*dek*) Mutants

A major advance facilitating genetic analysis of maize embryo development came from mutagenesis of maize pollen by treatment with ethyl methane sulfanate (EMS) in paraffin oil (Neuffer and Coe, 1978; Neuffer, 1978, 1994). This innovation led to production and analysis of over 200 kernel mutants

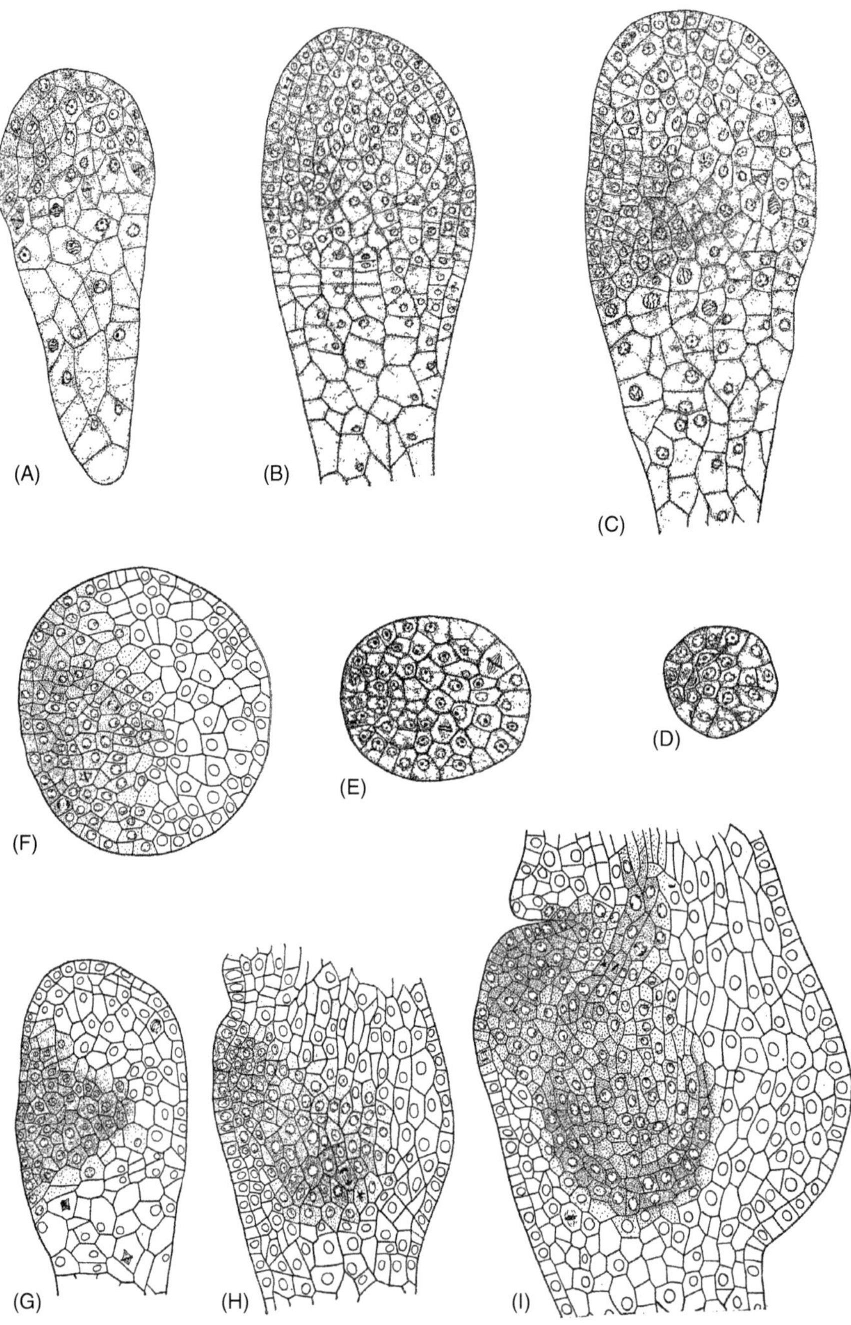

**Fig. 7.1.** Embryos 8–13 days after pollination. (A)–(C) Epidermis differentiating over the tip of the embryo. (C) At 10 days after pollination the initial stage of differentiation of the shoot–root axis is evident as a wedge-shaped group of cells in the anterior portion of the embryo. (D)–(F) Transverse sections in the subapical region of the embryo at 6, 7, and 10 days after pollination. (G)–(I) Longitudinal sections of the subapical portion of the embryos at 10, 12, and 13 days after pollination. Note in (H) the presence of two meristematic regions that become the forerunners of the forming shoot apical meristem and root apical meristem evident in (I). Reproduced from Randolph (1936).

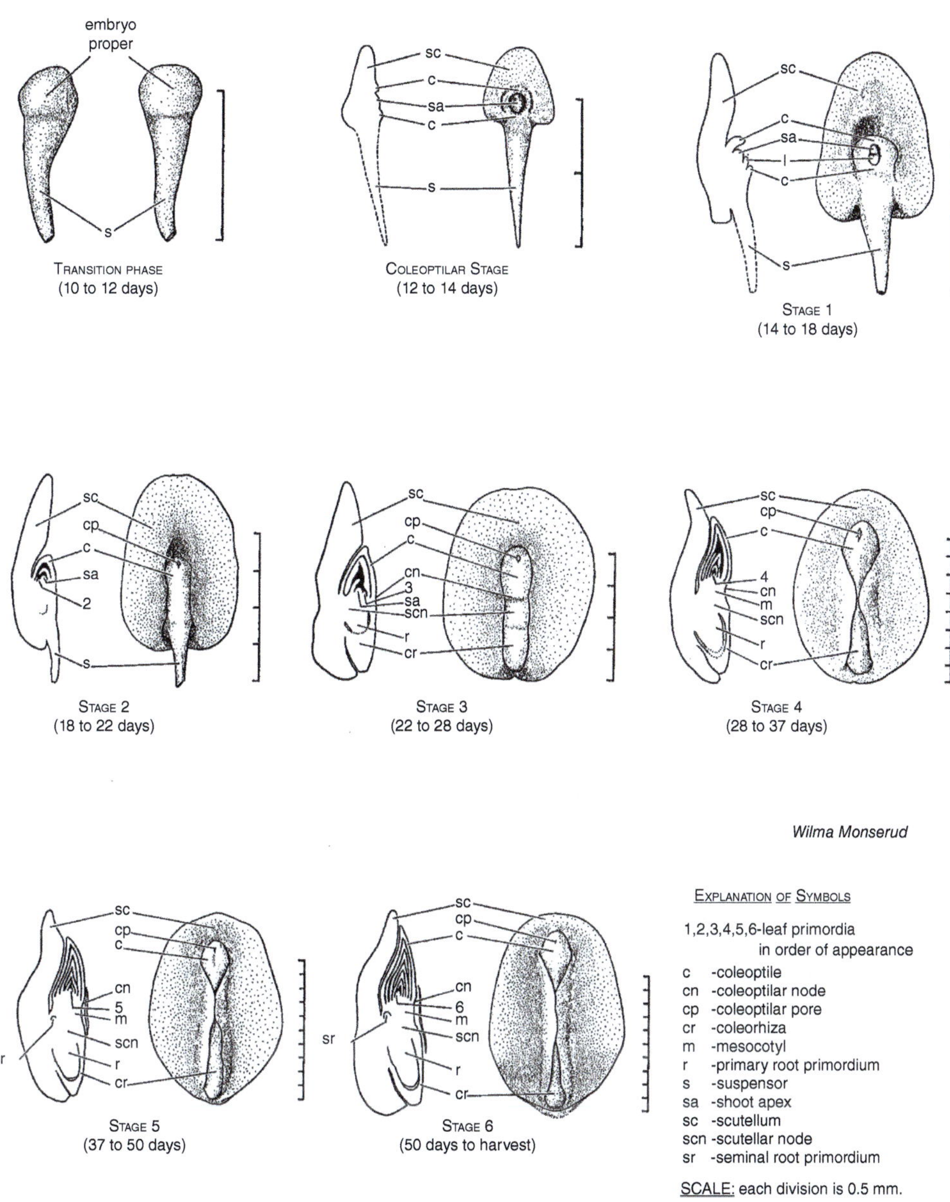



**Fig. 7.2.** The proembryo stage is not shown, but the other eight stages of embryogenesis are shown in side and face view. These present the standard nomenclature used for describing maize embryo morphogenesis. Note that by stage 2 (the presence of the second leaf primordium) the coleoptilar pore is mostly closed so that the second and subsequent leaf primordia are not visible in a surface view. Note the progressive enclosure of the embryonic axis by the scutellum. Reproduced from Abbe and Stein (1954) with permission of the Botanical Society of America.

and introduction of the term *defective kernel (dek)* mutation (Neuffer and Sheridan 1980), a generalization of the defective seed terminology of Jones (1920) and Mangelsdorf (1923). Following initial reports (Neuffer and Sheridan, 1980; Sheridan and Neuffer, 1980), there were many additional papers describing this type of kernel phenotype, including *dek* mutants, wherein embryos are blocked early in development (Scanlon *et al.*, 1994). These studies led to a search for mutations affecting only embryo development and resulted in identification of a large group of putative *Mutator*-induced mutants wherein embryo morphogenesis is profoundly disturbed, whereas that of the endosperm is essentially normal or only slightly affected. This resulted in the introduction of the term *"embryo-specific" (emb)* mutations (Clark and Sheridan, 1991) and was followed by analysis of these and other mutants with the *emb* phenotype. Recently, new insight emerged whereby it is apparent that many mutations resulting in defective kernel and embryo-specific phenotypes occur in nuclear genes encoding mitochondrial and plastid proteins.

A large study of EMS-induced mutations that affect maize kernel development was reported by Neuffer and Sheridan (1980). These were all found to be single gene, recessive mutations that mapped throughout the genome, with mutants defective in both endosperm and embryo development and for the most part not viable. Forty mutants were found to have defects early in kernel development and were either blocked in embryogenesis prior to initiation of leaf primordia, or if primordia formed, the embryos were unable to germinate when tissue cultured or germinated at kernel maturity. A major goal of the study was to screen for auxotrophic mutants. A total of 102 defective kernel mutants were examined by culturing mutant and normal embryos on basal or enriched culture media. The embryos of 21 mutants simply enlarged or completely failed to grow on any of the media tested; 81 produced shoots and roots on at least one medium. Among the 10 mutants with responses on enriched medium indicating they may be auxotrophs, a proline-requiring mutant was identified that is allelic to maize *pro-1* (Gavazzi *et al.*, 1975).

Extensive screening of immature mutant embryos by tissue culturing identified 17 blocked at the proembryo, transition, or coleoptilar stage, and therefore prior to initiation of the first leaf primordium. The stage of blockage was confirmed by examination of embryos in mature mutant kernels. Fourteen of these were examined in greater detail, confirming the notion that embryonic development in plants must involve regulated expression of a genetic program, just as in animals (Sheridan and Neuffer, 1981, 1982, 1986); Clark and Sheridan, 1986, 1988; Sheridan and Thorstenson, 1986; Neuffer *et al.*, 1986; and Clark, 1996).

## 7.4    Genetic Analysis of *Embryo-specific* (*emb*) Mutants

In the Neuffer *et al.* (1986) review, we estimated there are about 250 loci that, upon mutating, alter embryo and endosperm development. However, we also noted this did not include an estimate of the frequency of loci affecting only embryo development, and although we had no basis for estimating that frequency, we noted: "at least in theory they could be numerous." Subsequently, we noted: "some loci are known that only affect endosperm development, and it is likely that a careful search would uncover loci controlling only embryo development, although without the aid of an accompanying endosperm alteration such a search would be difficult" (Sheridan and Clark, 1987).

The results with *dek* mutants described in the previous section led to a search for mutations that specifically affect maize embryo development. This resulted in identification of 51 *embryo-specific (emb)* mutations representing 45 independent events; these were isolated by examining self-pollinated ears of Mutator stocks from Dr. Donald S. Robertson (Clark and Sheridan, 1991; Sheridan and Clark, 1993). In these reports we noted morphogenesis of the maize embryo occurs in three phases. The first is a period in which basal–apical asymmetry is established: the zygote develops into the proembryo, with the early transition stage accompanied by differentiation of the lower region as the suspensor and the upper region as the "embryo proper." Among the 51 mutants analyzed, 12 were blocked during the first phase of morphogenesis. During the second phase, radial symmetry converts to bilateral symmetry as the adaxial (front) face of the embryo proper flattens and expands to form the scutellum, with the protruding shoot apical meristem nearly surrounded by the coleoptilar ring. This phase extends from the late transition stage until appearance of the first leaf primordium. There were 29 mutants blocked during this second phase. The third phase of morphogenesis is the period when vegetative structures are elaborated, culminating in formation of the sixth leaf primordium. There were ten mutants blocked in the third phase. All 51 of these mutants were retarded in development and morphologically abnormal. Among 42 mutants tested, germination capacity was low for most of them. For nine of the ten mutants blocked at the third phase of development, only three had a germination rate above 3%. Some *emb* mutants were shared with other investigators; this resulted in analysis (Heckel *et al.*, 1999; Elster *et al.*, 2000) and cloning and sequencing of *emb-8516* (Magnard *et al.*, 2004) and *emb-8522* (Sosso *et al.*, 2012).

Another collection of EMS-induced *emb* mutants provided 57 new embryo-specific mutations (Brunelle *et al.*, unpublished results). Among the 44 mutants tested, nearly all are lethal. The embryo phenotypes of 34 mutants are developmentally and morphologically abnormal; half of them are blocked at the proembryo and transition stages. This group includes several with phenotypes blocked in the transition stage, with necrotic embryo proper regions that resemble phenotypes of mutations in nuclear genes encoding plastid proteins (see below). The other 17 mutants are mainly blocked, at a low frequency, in the coleoptilar or later stages. This group appears to include mutations in genes regulating completion of shoot apical meristem development and the accompanying morphogenetic processes.

and embryo development (Chapter 4). When a nuclear gene-encoded protein is targeted for mitochondria, both the embryo and endosperm are affected, resulting in a dek phenotype. Five reports describe mutations in nuclear genes encoding a pentatricopeptide repeat (PPR) protein that is required for mitochondrial RNA editing (Liu *et al.*, 2013; Li *et al.*, 2014; Chen *et al.*, 2016; Sun *et al.*, 2015; Xiu *et al.*, 2016). All of these mutations result in a dek phenotype, with embryos blocked early in development and showing necrosis.

When a nuclear gene mutation corresponds to a plastid protein, the result is quite different: the embryo is severely affected, while the endosperm is not affected or only slightly reduced in size relative to kernels with normal embryos (Chapter 4). However, a variety of nuclear genes are involved in the six mutants affecting plastid functions. They include a gene encoding a chloroplast-targeted ribosomal protein (Ma and Dooner, 2004), a gene encoding a ribosomal subunit protein (Magnard *et al.*, 2004), a gene encoding a PPR protein (Sosso *et al.*, 2012), a gene encoding a plastid translation initiation factor (Shen *et al.*, 2013), a gene required for ribosome formation in plastids (Zhang *et al.*, 2013), and a gene encoding a GTPase proposed to function in assembly of the 20S subunit of the chloroplast ribosome (Li *et al.*, 2015). Among 34 new mutants with similar phenotypes (Brunelle *et al.*, unpublished results), we consider nine mutants that exhibit necrosis in embryo proper regions at the transition stage of embryo development (Clark and Sheridan, 1991; Sheridan and Clark, 1993) and 16 mutants blocked at the transition stage to be good candidates for mutations affecting nuclear genes encoding proteins required for plastid function, based on their mutant embryo phenotypes.

## 7.5 Effects of Mutations in Nuclear Genes Encoding Proteins of Plastids and Mitochondria

Recent advances in the analysis of *dek* and *emb* mutants provided insight into the effects of nuclear gene mutations on endosperm

## 7.6 What Causes the Important Ontogenetic Events?

The first ontogenetic event, preferential segregation of organelles into the apical region of the zygote and proembryo, most likely involves the zygote cytoskeleton and those of subsequently formed cells. We found that microtubules play a crucial role for nuclei distribution in the normal developing embryo sac (Huang and Sheridan,1994) and those of the *indeterminate gametophyte1* (Huang and Sheridan, 1996) and *lethal ovule2* mutants (Sheridan and Huang, 1997). Prior to mitosis of the zygote, mitochondria and plastids are mostly located in the apical region of the cell. As apical cells produced by the zygote and its progeny cells continue to divide, organelles multiply and maintain a high density. As these processes continue between 5 and 8 DAP, there is a rapid decrease in cell size and a change in cell shape, so: "the relatively large, irregularly shaped cells of the 5-day embryo are replaced subsequently by smaller isodiametric cells" (Randolph, 1936).

Causes of the first ontogentic event must involve action of cytoskeletal elements, presumably microtubes and actin filaments, moving organelles into the apical cell of the early proembryo and positioning the preprophase band and mitotic spindle of dividing cells in the apical region. Additionally, some mechanism must titer the ratio of nuclear to cytoplasmic volume and influence cell divisions that progressively reduce cell size in the apical region during the 5–8-day duration of this event. With formation of the epidermis, differentiated cells could influence subsequent events.

The second ontogenetic event, the shift from radial symmetry to bilateral symmetry, occurs as the SAM forms; initially, it is a wedge-shaped group of cytoplasmically-dense cells interior to the anterior surface of the early transition stage embryo. This process likely involves auxin and cytokinin and their interaction. Auxin transport is mediated by the PINFORMED1 (PIN1) transport system. By *in situ* hybridization of RNA transcripts and immunolocalization of proteins, Forestan *et al.* (2010) showed auxin and *Zm*PIN1 proteins co-localize in developing maize embryos. The authors proposed a model based on the distribution of ZmPIN1 proteins in the embryo plasma membrane (observed) and polar auxin fluxes (deduced) (Fig. 7.3A). A more recent study utilized reporter lines expressing DR5-RFP and PIN1-YFP to examine polar auxin transport mediated by *ZmPIN1a* (Chen *et al.*, 2014). The earliest auxin response in the embryo was observed at the late transition stage, at about 8 DAP (Fig. 7.3B). The patterns of DR5 and ZmPIN1a localization during the late transition stage were different from those based on *in situ* hybridization and immunolocalization (Forestan *et al.*, 2010). One major difference between these two studies is that Forestan *et al.* (2010) found earlier evidence of auxin than Chen *et al.* (2014) during embryo development. Secondly, whereas the earlier report (Forestan *et al.*, 2010) indicates *in situ* origin of auxin (within the early transition stage embryo) at the site where the SAM initiates, the more recent report by Chen *et al.* (2014) proposes a possible auxin flux from the endosperm to the site of SAM initiation on the adaxial side of the transition stage embryo.

The protoderm (epidermis) could be essential in the switch from radial to bilateral symmetry. Elster *et al.* (2000) analyzed four *embryo-specific* mutants blocked early in embryogenesis and found that incomplete radial organization of the proembryo interferes with development. This could reflect a possible lack of cytokinin production in such mutants. Cytokinin in the protoderm of the early transition stage embryo could play an important role by inducing *ZmPIN1* in the SAM initiation site in the early (8 DAP) transition stage embryo, thereby enabling an auxin flux into that region (Lee *et al.*, 2009).

The third ontogenetic event, development of the shoot–root embryonic axis, which is composed of the SAM and RAM (root apical meristem), likely involves continued spread and functioning of the polar PIN1-auxin flux in the transition and coleoptilar stage embryo. This would likely involve continued cytokinin interaction, resulting in enlargement of the SAM and scutellum. When the SAM has developed sufficiently, a primordium, the coleoptilar ring, initiates in the scutellum just above the upper edge of the SAM. The ring expands and eventually nearly encircles the SAM. This process could be regulated by cytokinin controlling both the shape and formation of the tube that becomes the coleoptile.

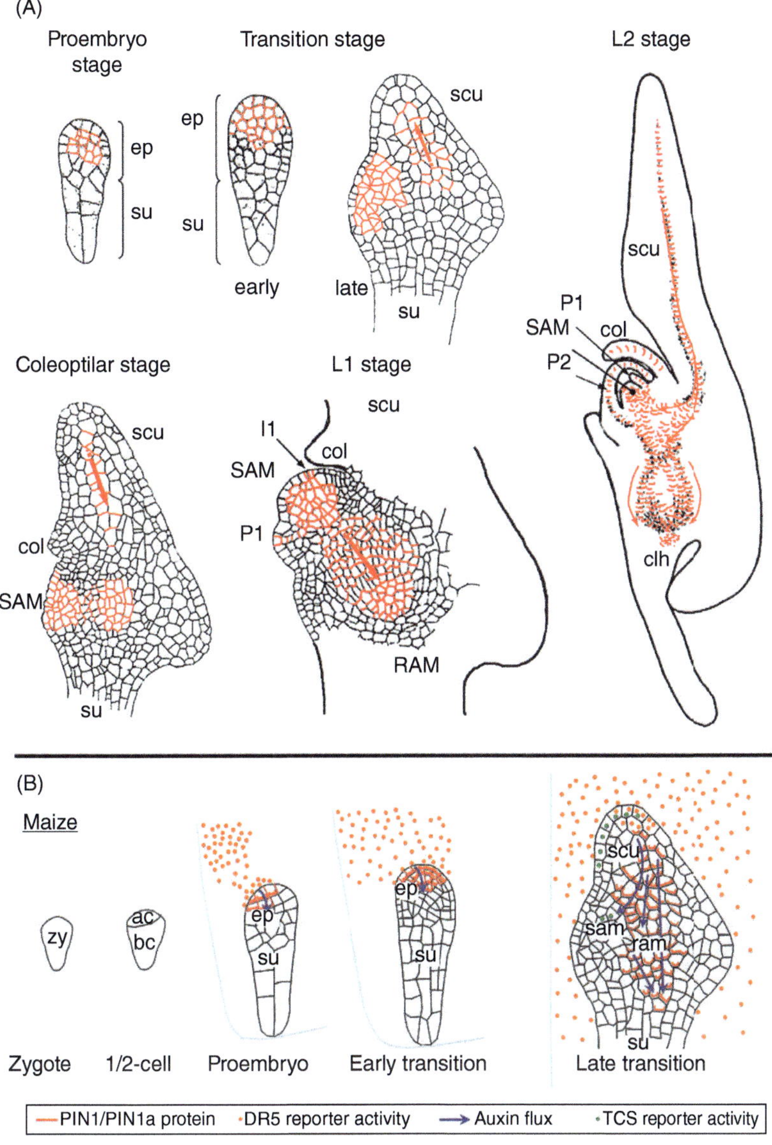



**Fig. 7.3.** Distribution of PIN1 auxin transporter protein, and observed or deduced distribution of auxin and cytokinin in the developing maize embryo. (A) The localization of ZmPIN1 proteins by immunostaining in embryo plasma proteins is reported in red and first observed in the apical region of the proembryo; arrows indicate the polar auxin fluxes deduced from the PIN1 localized pattern. Modified from Forestan *et al.* (2010). (B) Model showing ZmPIN1a protein localization and auxin and cytokinin responses detected by fluorescent protein reporters. Modified from Chen *et al.* (2014). Major differences in the results between A and B are the timing of appearance and distribution of PIN1 in the early embryo and direction of the auxin fluxes in the late embryo transition stages. Forestan *et al.* (2010) reported auxin moving in the apical direction (A), while Chen *et al.* (2014) reported auxin moving downward from the apical region (B). This figure is composed of Figure 3 from Chen *et al.* (2014) and of Figure 8 from Forestan *et al.* (2010) (adapted from Bommert and Werr, 2001), reprinted with permission from the American Society of Plant Biologists.

Opposite the origin of the coleoptile, on the lower face of the SAM, the first leaf primordium appears; its position and enlargement could also depend on the auxin–cytokinin system that regulates the initiation of SAM formation.

## 7.7 Questions Remaining to Be Answered

Clearly, there are many maize genes that when mutated affect embryo nutrition, growth, and development (Chapter 4); however, *how many specifically regulate morphogenesis?* Answering this question can best be pursued by conducting screens of EMS-induced mutants to obtain a large number of independent events that disrupt embryogenesis. This sets the stage for addressing a second important question: *What is the spectrum of embryo-specific phenotypes resulting from these mutations?* Answering this question requires examination of the physical characteristics of mutant embryos during development, and it can best be pursued by dissecting embryos of mature kernels to discover the impact on embryo morphogenesis, followed by examination of earlier stages of development.

Investigation of mutant embryo phenotypes provides data that support genetic complementation tests of the mutants. These studies address the first question regarding the frequency of mutations affecting the same gene. Phenotypic characterization of a large number of *embryo-specific* mutants provides insight regarding independently occurring or co-occurring embryo morphology abnormalities that can aid in answering the next question to be addressed.

The most fundamental question is: *What is the nature of gene products of embryo-specific genes, are they proteins or regulatory RNAs, and what are the roles and interactions among them and their relationships with other molecules acting in the embryo morphogenetic system?* This question can be most directly answered by conducting whole genome sequencing and mapping, wherein genomes of plants that are heterozygous for *emb* mutant alleles are compared with those of plants homozygous for their normal allelic counterparts. This approach has become feasible because of whole genome sequencing and annotation of many maize inbred lines, enabling the use of single nucleotide polymorphisms and advances in bioinformatics to precisely map and sequence mutant alleles.

## 7.8 Research on *emb* Mutants

For those wishing to investigate the genetic regulation of maize embryo morphogenesis, there are two key considerations: the time required and the work and field space available. It is important to have experience and technical expertise in four research activities: (i) production of *emb* mutants; (ii) their phenotypic characterization; (iii) determination of their genetic relationships; and (iv) mapping and sequencing of the mutated genes. The number of mutants studied will influence the length of the project and the space required.

### 7.8.1 Production of new *emb* mutants

In the first (summer) nursery, one can grow 200 plants of an inbred, such as B73 or Mo17, for which the entire genome has been sequenced and annotated; one of these will be used as the female parent. In addition, 100 plants can be grown of a different inbred with a sequenced and annotated genome, such as W22 or another inbred of a different heterotic group than the female parent. Both inbreds should have yellow (colorless) kernels. Kernels of the female parent should be planted on a single date, but male parent kernels should be divided into three staggered plantings to assure nicking (flowering simultaneously) when female ears are silking. At that time, pollen collected from male parent plants is suspended in mineral oil containing EMS and, after intermittent shaking for 30 to 70 minutes, applied to silks of the female ears (Neuffer, 1994; Brunelle *et al.*, unpublished).

After harvesting, drying, and tagging ears, 10- to 15-kernel samples should be removed from 40 to 50 partially filled ears and planted in a second nursery to produce approximately 500 to 600 robust plants. These M1 plants should be self-pollinated, and the resulting (400–500) ears screened for normal kernels containing mutant embryos. It is helpful to keep the ears intact and not shell all the kernels, because intact ears reveal kernels with a defective endosperm, helping avoid misclassifying *dek* mutants as *emb* mutants. Ample storage space for the ears is required, as well as a large surface for sorting, tagging, and sample preparation. Screening *emb* mutants can be performed by examining 100 kernels from the self-pollinated M1 ears. This is aided by using a large 2x magnifier with supplemented lighting, with the 100-kernel sample viewed embryo side up. When mutant embryos are observed, their number is counted and recorded on the envelope as the percentage showing phenotypic segregation. The mutant kernels are placed in a coin envelope and included with the normal kernels in a larger envelope. With a mutation frequency of approximately 12–13%, 400–500 ears from the M1 nursery should yield about 50 to 60 ears segregating for candidate *emb* mutations. Screening the large number of ears from the M1 nursery to identify these 50 to 60 ears is labor-intensive, requiring several weeks if performed by a single individual. For the second (summer) nursery, a 15- to 20-kernel sample with normal embryos from ears segregating for the putative *emb* mutation, should be grown and tested for heritability. In anticipation of future genetic analyses, an equal number of plants of a recurring parent with a colored (purple) kernel genotype can be planted. For each of the mutant families, 10 or more plants should be self-pollinated and crossed onto the colored recurrent parent inbred.

### 7.8.2   Phenotyping

Examination of the mature embryo phenotype is performed by placing 15 kernels with mutant embryos on moist filter paper in glass Petri dishes. The dishes are sealed with Parafilm and left at room temperature for 2 days. Dissection under magnification with photography of at least 10 embryos is performed for each mutant in order to assess phenotype variability. This normally requires about 2½ hours of focused effort at a dissecting microscope, and best results are obtained with one mutant sample per day.

### 7.8.3   Genetic analysis

Genetic complementation tests can be performed by planting the yellow kernel source of one mutant (to be self-pollinated and cross-pollinated) and a purple kernel source of a second mutant. The yellow kernel (female) parent is self-pollinated on half its silks and cross-pollinated the following day by the purple kernel (male) parent on the remaining silks; the purple kernel parent is then self-pollinated. This can also be done in the reverse order, or all pollinations can be done on the same day. Whichever sequence is used, the procedure requires careful attention and focused concentration. An unambiguous test is obtained when the self-pollinated ear of the male plant segregates for its *emb*, the self-pollinated side of the female plant ear segregates for its *emb*, and the cross-pollinated side of the female plant ear sets sufficient kernels for phenotypic scoring. If the cross-pollinated kernels on the yellow kernel ear are all normal, then the two *emb* mutations complement each other. If the kernels segregate for an emb phenotype, the two *emb* mutants failed to complement and are therefore allelic, i.e. occur in the same gene.

### 7.8.4   Sequencing and mapping

Mapping and sequencing of a mutation is most rewarding when it results in identification of the affected gene, and its nucleotide sequence reveals the nature of the gene product. This requires two or more different mutant alleles to confirm the mutated gene. When a pair of allelic mutants is obtained, one can proceed with mapping and sequencing. For each of the alleles to be analyzed,

100 to 120 kernels with normal embryos should be removed from a self-pollinated segregating M1 founder ear, or an advanced generation self-pollinated ear. Seeds should be planted to produce self-pollinated ears. When silks are trimmed ("cut back") the day prior to self-pollination, ear husk tissue can be collected and saved in a Zip-loc plastic bag, noting the family and plant number on the bag. These tissue samples are used to prepare a DNA sample of each plant. Following harvest, the self-pollinated ears are phenotypically scored and the genotype assessed for the corresponding DNA sample. The DNA samples are used to prepare a sequencing library, and each DNA sample is "bar-coded" to mark its identity. Samples are combined into pools of heterozygous mutant and homozygous normal alleles and subjected to DNA sequencing using a high-throughput system. The resulting sequences are aligned and analyzed with the goal of identifying the DNA sequences of the normal and *emb* mutant alleles. These data are used to search sequence databases and obtain information about the gene and its product.

## Acknowledgment

Research in our laboratory reported here was supported by grants from the National Science Foundation Plant Genome Research Program to WFS. We thank Victoria Swift for assistance with the graphics.

## References

Abbe, E.C. and Stein, O.L. (1954) The growth of the shoot apex in maize: embryogeny. *American Journal of Botany* 41, 285–293.

Chen, J., Lausser, A. and Dresselhaus, T. (2014) Hormonal responses during early embryogenesis in maize. *Biochemical Society Transactions* 42, 325–331.

Chen, X., Feng, W.Q.F., Yao, D. Wang, Q. and Song, R. (2016) Dek35 encodes a PPR protein that affects cis-splicing of mitochondrial nad4 intron 1 and seed development in maize. *Molecular Plant* 10, 427–441. DOI:10.1016/j.molp.2016.08.008

Clark, J.K. (1996) Maize embryogenesis mutants. In: Wang, T. and Cuming, A.C. (eds.) *Embryogenesis: The Generation of a Plant.* Bios Scientific Publishers, Oxford, UK, pp. 89–112.

Clark, J.K. and Sheridan, W.F. (1986) Developmental profiles of the maize embryo-lethal mutants dek 22 and dek 23. *Journal of Heredity* 77, 83–92.

Clark, J.K. and Sheridan, W.F. (1988) Characterization of two maize embryo-lethal defective kernel mutants rgh*-1210 and fl*-1253B: effects on embryo and gametophyte development. *Genetics* 120, 279–290.

Clark, J.K. and Sheridan, W.F. (1991) Isolation and characterization of 51 embryo-specific mutations of maize. *Plant Cell* 3, 935–951.

Bommert, P. and Werr, W. (2001) Gene expression patterns in the maize caryopsis: clues to decisions in embryo and endosperm development. *Gene* 271, 131–142

Demerec, M. (1923) Heritable characters of maize. XV. Germless seeds. *Journal of Heredity* 14, 297–300.

Elster, R., Bommert, P., Sheridan, W.F. and Werr, W. (2000) Analysis of four embryo-specific mutants in *Zea mays* reveals that incomplete radial organization of the proembryo interferes with subsequent development. *Development Genes and Evolution* 210, 300–310.

Forestan, C., Meda, S. and Varotto, S. (2010) ZmPIN1-mediated auxin transport is related to cellular differentiation during maize embryogenesis and endosperm development. *Plant Physiology* 152, 1373–1390.

Gavazzi, G., Nava-Rachi, M. and Tonelli, C. (1975) A mutation causing proline requirement in maize. *Journal of Theoretical and Applied Genetics* 46, 339–346.

Heckel, T., Werner, K., Sheridan, W.F., Dumas, C. and Rogowsky, P.M. (1999) Novel phenotypes and developmental arrest in early embryo specific mutants of maize. *Planta* 210, 1–8.

Huang, B.-Q. and Sheridan, W.F. (1994) Female gametophyte development in maize: microtubular organization and embryo sac polarity. *Plant Cell* 6, 845–861.

Huang, B.-Q. and Sheridan, W.F. (1996) Embryo sac development in the maize *indeterminate gametophyte1* mutant: abnormal nuclear behavior and defective microtubule organization. *Plant Cell* 8, 1391–1407.

Jones, D.G. (1920) Heritable characters of maize. IV. A lethal factor – defective seeds. *Journal of Heredity* 11, 161–167.

Kiesselbach, T.A. (1949) *The Structure and Reproduction of Corn.* University of Nebraska Press, Lincoln, Nebraska.

Lee, B.-H., Johnston, R., Yang, Y., Gallavotti, A., Kojima, M., *et al.* (2009) Studies of aberrant phyllotaxy1 mutants of maize indicates complex interactions between auxin and cytokinin signaling in the shoot apical meristem. *Plant Physiology* 150, 205–216.

Li, X.-J., Zhang, Y.-F., Hou, M., Sun, F., Shen, Y., *et al.* (2014) *Small kernel 1* encodes a pentatricopeptide repeat protein required for mitochondrial *nad7* transcript editing and seed development in maize (*Zea mays*) and rice (*Oryza sativa*). *Plant Journal* 79, 797–809.

Li, C., Shen, Y., Meely, R., McCarty, D.R. and Tan, B.C. (2015) Embryo defective 14 encodes a plastid-targeted GTPase essential for embryogenesis in maize. *Plant Journal* 84, 785–799.

Liu, Y.-J., Xui, Z.-H., Meeley, R. and Tan, B.C. (2013) *Empty Pericarp5* encodes a pentatricopeptide repeat protein that is required for mitochondrial RNA editing and seed development in maize. *Plant Cell* 25, 868–883.

Ma, Z. and Dooner, H.K. (2004) A mutation in the nuclear-encoded plastid ribosomal protein S9 leads to early embryo lethality in maize. *Plant Journal* 37, 92–103.

Magnard, J.L., Hecle, T., Massonneau, A., Wisniewski, J.P., Codelier, S., *et al.* (2004) Morphogenesis of maize embryos requires *ZmPRPL35-1* encoding a plastid ribosomal protein. *Plant Physiology* 134, 649–663.

Mangelsdorf, P.C. (1923) The inheritance of defective seeds in maize. *Journal of Heredity* 14, 119–125.

Mangelsdorf, P.C. (1926) The genetics and morphology of some endosperm characters in maize. *Connecticut Agricultural Experiment Station Bulletin* 279, 509–614.

Neuffer, M.G. (1978) Induction of genetic variability. In: Walden, D.B. (ed.) *Genetics and Breeding of Maize.* Wiley-Interscience, New York, pp. 579–600.

Neuffer, M.G. (1994) Mutagenesis. In: Freeling, M. and Walbot, V. (eds.) *The Maize Handbook.* Springer, New York, pp. 212–218.

Neuffer, M.G. and Coe, E.H. (1978) Paraffin oil technique for treating mature corn pollen with chemical mutagens. *Maydica* 23, 21–28.

Neuffer, M.G. and Sheridan, W.F. (1980) Defective kernel mutants of maize. I. Genetic and lethality studies. *Genetics* 95, 929–944.

Neuffer, M.J., Chang, M.T., Clark, J.K. and Sheridan, W.F. (1986) The genetic control of maize kernel development. In: Shannon, J.C., Knievel, D.P. and Boyer, C.D. (eds.) *Regulation of Carbon and Nitrogen Reduction and Utilization in Maize.* American Society of Plant Physiologists, Rockville, Maryland, pp. 35–50.

Randolph, L.F. (1936) Developmental morphology of the caryopsis in maize. *Journal of Agricultural Research* 53, 881–916.

Scanlon, M.J., Stinard, P.S., James, M.G., Myers, A.M. and Robertson, D.S. (1994) Genetic analysis of 63 mutations affecting maize kernel development isolated from Mutator stocks. *Genetics* 136, 281–294.

Shen, Y., Li, C., McCarty, D.R., Meeley, R. and Tan, B.C. (2013) Embryo defective12 encodes the plastid initiation factor 3 and is essential for embryogenesis in maize. *Plant Journal* 74, 792–804.

Sheridan, W.F. and Clark, J.K. (1987) Allelism testing by double pollination of lethal maize dek mutants. *Journal of Heredity* 78, 49–50.

Sheridan, W.F. and Clark, J.K. (1993) Mutational analysis of morphogenesis of the maize embryo. *Plant Journal* 3, 347–358.

Sheridan, W.F. and Huang, B.-Q. (1997) Nuclear behavior is defective in the maize (*Zea mays* L.) lethal *ovule2* female gametophyte. *Plant Journal* 11, 1029–1041.

Sheridan, W.F. and Neuffer, M.G. (1980) Defective kernel mutants of maize. II. Morphological and embryo culture studies. *Genetics* 95, 945–960.

Sheridan, W.F. and Neuffer, M.G. (1981) Maize mutants altered in embryo development. In: Subtelney, S. and Abbott, U. (eds.) *Levels of Genetic Control and Development. The 39th Annual Symposium of the Society for Developmental Biology.* Alan Liss, Inc., New York, pp. 137–156.

Sheridan, W.F. and Neuffer, M.G. (1982) Maize developmental mutants: embryos unable to form leaf primordia. *Journal of Heredity* 73, 318–329.

Sheridan, W.F. and Neuffer, M.G. (1986) Genetic control of embryo and endosperm development in maize. In: Reddy, G.M. and Coe, E.H., Jr. (eds.) *Gene Structure and Function in Higher Plants.* Oxford & IBH Publishing Co., New Delhi, India, pp. 105–122.

Sheridan, W.F. and Thorstenson, Y.R. (1986) Developmental profiles of three maize embryo-lethal mutants lacking leaf primordia: bn*-747B, ptd*-1130, and cp*-1418. *Developmental Genetics* 7, 35–49.

Sosso, D., Canut, M., Gendrot, G., Dedieu, A., Chambrier, P., *et al.* (2012) *PPR8522* encodes a chloroplast-targeted pentatricopeptide repeat protein necessary for maize embryogenesis and vegetative development. *Journal of Experimental Botany* 63, 5843–5857.

Sun, F., Wang, X., Bonnard, G., Shen, Y., Xiu, Z., *et al.* (2015) *Empty pericarp7* encodes a mitochondrial E-subgroup pentatricopeptide repeat protein that is required for $ccmF_N$ editing, mitochondrial function and seed development in maize. *Plant Journal* 84, 283–295.

Wentz, J.B. (1930) The inheritance of germless seeds in maize. *Iowa Agricultural Experiment Station Research Bulletin* 121, 347–379.

Xiu, Z., Sun, F., Shen, Y., Zhang, X., Jiang, R., *et al.* (2016) EMPTY PERICARP16 is required for mitochondrial *nad2* intron 4 *cis*-splicing, complex I assembly and seed development in maize. *Plant Journal* 85, 507–519.

Zhang, Y.F., Hou, M.M. and Tan, B.C. (2013) The requirement of WHIRLY1 for embryogenesis is dependent on genetic background in maize. *PLOS ONE* 8, e67369.

# 8 Embryo–Endosperm–Sporophyte Interactions in Maize Seeds


Thomas Widiez[1,*], Gwyneth C. Ingram[1] and José F. Gutiérrez-Marcos[2]

[1]*Laboratoire Reproduction et Développement des Plantes, Université de Lyon, ENS de Lyon, France;* [2]*School of Life Sciences, University of Warwick, Coventry, UK*


## 8.1 Introduction

Maize seeds, like those of all other angiosperms, are highly complex biological systems. This complexity is a consequence of the fact that the angiosperm seed is composed of tissues that evolved from three genetically distinct organisms: the mother plant (maternal sporophyte—specifically the nucellus, integuments, and in the case of maize and other cereals, other floral organs that fuse with the integuments to form the pericarp); the developing embryo (zygotic sporophyte); and the endosperm (arising through fertilization-dependent proliferation of a second fertilization competent cell of the female gametophyte). These tissues are organized one inside the other like Russian dolls. Nutrients are transported through the endosperm from the maternal ovule-derived seed coat tissues to drive embryo growth. The three different seed components must therefore interact both physically and chemically to allow successful development.

In evolutionary terms, the two main interactions that will be addressed in this chapter initially arose well before emergence of the angiosperms. The first and probably most ancient association is "retention and nourishment" of the sporophytic

embryo by the female gametophyte. Moss gametophytes support the whole of sporophyte development, and in the lycophyte *Selaginella,* sporophytic embryos grow invasively into a specialized gametophyte-derived nutritive tissue from which they emerge in a process that looks surprisingly like seed germination (Webster, 1967). Invasive growth of the developing embryo into a female gametophyte-derived nutrient storing tissue (the endosperm in angiosperms) is also observed during the development of both gymnosperm and angiosperm seeds (Cairney and Pullman, 2007; Yang *et al.,* 2008). The second association is derived from the "retention of the megaspore" by the maternal sporophyte, and the subsequent extended interaction of maternal tissues (the nucellus and integuments) with the developing female gametophyte (and subsequently, in angiosperms, the endosperm). This conversion of the female gametophyte, from an organism with an autonomous autotrophic existence to one whose existence depended completely upon the maternal sporophyte for nutrition and protection, must have arisen rather early during seed plant evolution. These key evolutionary innovations during the rise of seed plants presumably required recruitment of novel communication pathways


*Corresponding author e-mail: thomas.widiez@ens-lyon.fr

necessary for both coordinated development of the three seed compartments and regulated nutrient provision to offspring. Much later in evolutionary terms, and specifically in angiosperms, the acquisition of fertilization competence by the female gametophyte central cell (the endosperm precursor) meant development of the embryo and surrounding nutritive tissue (endosperm) became temporally synchronized. While the introduction of a copy of the paternal genome into the nutrient-storing endosperm is thought to have enhanced parental conflict regarding the translocation of nutrients to offspring (see Chapter 9), the temporal synchronization of zygotic compartment development may have imposed other constraints, potentially generating a need for refinement to existing interactions within the developing seed.

The profound and tightly coordinated changes that occur within the three compartments of the maize seed during its development highlight the need for constant communication. Nonetheless, how communication pathways/interactions are established and regulated, and the identity of the molecular effectors responsible remain largely elusive. Understanding these interactions is a key challenge in the drive to clarify mechanisms responsible for resource partitioning during cereal seed filling. At a more fundamental level, it will also provide a critical complement to the work carried out in model dicotyledonous species (such as Arabidopsis) in picking apart the evolutionary history of signaling pathways and how their modification in different angiosperm lineages could have led to the very different seed developmental strategies we see today. With these questions in mind, we address interactions that occur at compartment interfaces. After a short description of their structures, we focus mainly on three aspects: (i) the potential role of signaling peptides (and other mobile substances) and their receptors as signaling components in cell fate decisions and intercompartmental communication; (ii) the role of sugars both as nutrients and as signaling molecules; and (iii) the role and control of cell death processes as a consequence of physical interactions and as a potential source of signaling molecules for intercompartmental communication.

Besides "classic" signaling events (involving hormones/peptides then receptor recognition followed by signal transduction), the three seed compartments have to import and export (with the exception of the embryo) metabolites to support growth and development. Although many metabolites (sugars, amino acids and ions) transit between these compartments, we focus mainly on sugars, as they are best documented and represent the main carbon source for storage compounds. Sugars have a well-established dual functionality, acting not only as metabolites but also as signaling molecules with multiple regulatory roles (Rolland *et al.*, 2006). We specifically highlight potential roles for sugar signaling in intercompartmental communication in the seed. Other metabolic considerations, including the control of kernel sink strength, are described in Chapter 15.

The results of studies in several species, based on ultrastructural examination and monitoring of labeled molecules (Oparka and Gates, 1981a,b; Stadler *et al.*, 2005), have converged to show the mature female gametophyte and zygote (or at least the very young embryo) become effectively symplastically isolated from surrounding tissues at or soon after fertilization. This isolation appears to be an active process, the regulation and functional significance of which has not been investigated in any depth. However, for the purposes of this review these results signify that all interorganismal communication in the developing seed must occur across the apoplast. The intercompartmental apoplastic interfaces and the membranes of the cells on each side of these interfaces thus represent critical zones for understanding exchanges of nutritional and signaling cues. In this context, the presence of apoplastic modifications that could regulate or gate movement of key molecules could play an important role in regulating the growth and development of neighboring seed compartments, and must therefore be taken into account when studying embryo, endosperm, and sporophyte interactions.

## 8.2 Interactions between Endosperm and Sporophyte

### 8.2.1 The basal endosperm transfer cell layer–placenta–chalazal interface

One of the first endosperm cell types specified in the maize kernel is the basal endosperm transfer cell layer (BETL). At the BETL, the endosperm is separated from the maternal phloem-unloading region by the placenta–chalazal region, a compound tissue formed both from nucellus and integument-derived cells (see Chapter 5). The factors implicated in early BETL differentiation are poorly understood; however, there is increasing evidence supporting the view that critical determinants of transfer cell fate are already laid down prior to fertilization. The genetic dissection of maize seed mutants displaying parent-of-origin effects revealed several maternal factors that are critical for the fate and differentiation of BETL cells (Gutiérrez-Marcos *et al.*, 2004; Bai *et al.*, 2016). Notably, most maize maternal effect seed mutants that display abnormal BETL development also manifest female gametophyte defects, ranging from abnormal central and egg cell morphology to irregular numbers of accessory antipodal cells (Gutiérrez-Marcos *et al.*, 2006; Chettoor *et al.*, 2015, 2016). This maternal regulation is likely further supported by the phytohoromone auxin, which accumulates in a polar fashion in the integuments surrounding the female gametophyte (Lituiev *et al.*, 2013). The critical role of the maternal integument in auxin signaling for maize gametophytic cell fate determination and seed development was revealed through analysis of leaf polarity *Lax midrib1* (*lxm1*) mutants (Schichnes *et al.*, 1997). Maize *lmx1* mutants interfere with auxin signaling in integument cells; this in turn is associated with abnormal development of antipodal cells, causing maternal defects in seed size presumably by interfering with BETL development (Chettoor *et al.*, 2015). Further analysis of maize auxin signaling mutants will clarify the interplay between the maternal sporophyte and endosperm development.

After fertilization and during the early stages of endosperm development, sugars—in particular glucose—accumulate significantly at the maternal/filial interface. Accumulation of glucose in the placenta–chalazal region of the integuments coincides with the transition from coenocytic to cellular endosperm, a stage critical for BETL development. The asymmetric distribution of glucose has hitherto been proposed as the major regulatory signal for transfer cell differentiation (Yuan *et al.*, 2016). The recent discovery of the maize hexose transporter SWEET4c revealed the importance of polar transport of sugars from maternal tissues to endosperm (Sosso *et al.*, 2015), while active transport of hexoses from maternal integuments, which coincides with the transcriptional activation of a critical transcription factor (MYB RELATED PROTEIN 1, MRP1) in endosperm transfer cells, has been long known (see Chapter 5). *MRP1* is highly expressed in plant tissues that actively transport sugars, suggesting that a sugar signaling pathway contributes to temporal and spatial regulation of this transcriptional regulator in transfer cells. However, this model implies that an alternative pathway must operate in the endosperm before sugars are actively translocated from maternal tissues. The best candidate for a sugar-independent signaling component in maize endosperm is MATERNALLY EXPRESSED GENE1 (MEG1), which accumulates in the free nuclear endosperm soon after fertilization and activates the expression of *MRP1* and other BETL-specific transcripts in naïve endosperm cells (Chapter 5). Future studies should be directed at dissecting the roles of these signaling pathways and their interactions with maternally-deposited phytohormones.

During BETL formation, cells from the placenta–chalazal region undergo a form of cell death, leading to the production of a nutrient transferring structure (Chapter 5). Interestingly, the initiation of the cell death process is dependent on fertilization, which implies a role for an undefined endosperm-derived signal. In addition to transport functions, the placenta–chalazal region has been proposed to have a defensive role based on expression of "defensin-like" cysteine-rich peptides on the BETL side of the interface and because phenolic compounds accumulate in this structure (Chapter 5). However, one should bear in mind that secreted peptides could well play important signaling roles that have yet to be uncovered. A recent study identified a subtilisin protease-encoding gene (*ZmSBT2*) specifically expressed in this filial–maternal interface (López *et al.*, 2016). It is tempting to speculate this putative cell wall-located protease takes part in either cell fate determination or defense functions.

### 8.2.2 Aleurone–nucellus–integument interface

Together with the BETL, the aleurone forms the outermost layer of endosperm. Whereas the BETL is limited to the placenta–chalazal region, the aleurone surrounds the rest of the endosperm and thus represents the largest surface interface between sporophytic tissues and endosperm. In maize, the aleurone, comprising cells with specific cytological characteristics and biochemical composition, forms a single cell layer with an epidermis-like structure (Chapter 6). The most well-known function of aleurone occurs during seed germination, where it provides hydrolytic enzymes that catalyze breakdown of metabolic reserves to nourish the embryo.

Two important genes have been shown to maintain aleurone cell identity: *DEK1 (DEFECTIVE KERNEL1)* and *CR4 (CRIN-KLY4)*. In their mutants, aleurone cell layer specification is impaired (completely disappearing in null *dek1* mutants) and peripheral cells adopt a starchy endosperm fate (Chapter 6). CR4 encodes a receptor-like kinase in the plasma membrane; DEK1 is also a membrane-localized protein with a cytosolic calpain-like protease domain (Chapter 6). Several genetic and molecular studies led to the hypothesis that DEK1 and CR4 are involved in signaling pathways involving perception of extracellular signals/ligands (Chapter 6). A recent study proposed DEK1 activity may be necessary to maintain epidermal cell–cell contact zones,

permitting the maintenance of intercellular signaling necessary for epidermal cell fate maintenance (Galletti *et al.*, 2015). A similar scenario could be true in the aleurone. It is also possible that hormone signaling, and especially auxin signaling, play a role in aleurone cell identity, as high auxin concentration is correlated with aleurone cell fate (Forestan *et al.*, 2010). Immunological indoleacetic acid detection shows that aleurone cells have high levels of auxin compared to surrounding cell layers. Interestingly, treatment with NPA (a non-degradable auxin) leads to the production of a multilayered aleurone that has an ectopic expression of the auxin transporter-encoding gene *ZmPIN1*. Thus, auxin seems to positively trigger aleurone cell fate and could take part in a positive feedback loop involved in the aleurone cell fate decision. The origin (endosperm or sporophyte) of auxin accumulating in the maize aleurone layer remains an open question (Forestan *et al.*, 2010; Chen *et al.*, 2014). However, recent studies in Arabidopsis suggest that the endosperm is a key site for the production of the auxin necessary both for endosperm development and, non-autonomously, seed coat differentiation during post-fertilization seed development (Figueiredo *et al.*, 2015, 2016). Intriguingly, altered activity of endosperm-specific ZmYUCCA1, a putative auxin-biosynthetic enzyme, has been implicated in endosperm defects in the *defective endosperm18 mutant* (Bernardi *et al.*, 2012). Whether zygotically-derived auxin drives pericarp development in maize is unclear. It should be noted that the presence of nucellus tissues between the endosperm and much of the pericarp until relatively late in kernel development could affect such an interaction.

Several studies showed that the surface (peripheral) position of endosperm cells, rather than proximity to maternal tissues, is critical for aleurone cell fate specification (Chapter 6). Thus, in connated maize kernels aleurone fate is lost at fused endosperm surfaces (Geisler-Lee and Gallie, 2005), whilst it has been shown that aleurone cell fate can be specified *in vitro* in the absence of neighboring sporophyte tissues

(Gruis *et al.*, 2006). However, it should be noted that in the latter study isolated endosperms were cultured in a medium containing extremely high levels of sucrose (150g/L). It is possible sucrose released from phloem can enter the nucellus from surrounding maternal cell layers and perfuse the entire zone surrounding the developing endosperm. Since sucrose released into the wheat seed cavity from maternal tissues has been found to traverse the nucellus (Wang and Fisher, 1995), it is possible that the surface of the young cereal endosperm (excepting the BETL) is bathed in a sucrose-rich solution with developmental as well as nutritional roles. In this scenario, hexoses could promote BETL cell differentiation (see Section 8.2.1), whereas sucrose would promote aleurone cell fate. Interestingly, epidermal cell fates in explanted legume cotyledons are altered when cotyledons are incubated with hexose versus sucrose (Weber *et al.*, 1996; Wobus and Weber, 1999), further supporting a potential role for sugars in regulating cell fate decisions. How specific sugars are sensed in the apoplast is not clear, although the recent discovery of a sucrose-sensing receptor kinase in Arabidopsis may provide an important indicator (Wu *et al.*, 2013). In this context, the *globby1-1* mutant, which causes aberrant globular embryo and endosperm morphology, could provide interesting material. This mutant does not follow the "surface rule," because some internal endosperm cells show aleurone identity (Costa *et al.*, 2003). It would be interesting to revisit the *globby1-1* mutant with sugar-signaling in mind, as both BETL and aleurone cell fate acquisition are impaired in this mutant.

During early maize seed development, the aleurone is surrounded by nucellus tissues and then becomes juxtaposed to the maternal integuments as the nucellus disappears. The process underlying nucellus disintegration involves cytoplasmic degeneration and cell collapse, and has been described to some extent for maize and other cereals (Domínguez *et al.*, 2001; Radchuk *et al.*, 2011; Lammeren *et al.*, 2005), but it remains remarkably poorly studied. In maize, most of the nucellus is lost by around 12 days after pollination (DAP) (Leroux *et al.*, 2014; Rousseau *et al.*, 2015), although the nucellar epidermis survives longer (Lammeren *et al.*, 2005). Whether this cell death process is chemically triggered by the growing endosperm, as has been suggested in the cucurbit *Sechium edule* (Lombardi *et al.*, 2012), or is simply due to compression between the endosperm and surrounding maternal tissues, as has been suggested for barley (Radchuk *et al.*, 2011), or is caused by a mixture of chemical and physical cues, remains to be investigated. In addition, to what extent nucellus cell death plays a nutritive or signaling role for the developing endosperm is also an open question. Interestingly, the nucellar epidermis is covered by a cuticle, which separates it from the overlying integuments (Lammeren *et al.*, 2005). This cuticle survives even after the nucellar epidermis has degenerated. The presence of an apparent apoplastic barrier might be expected to directly affect chemical communication between maternal tissues and the endosperm after nucellar epidermis degradation (Moussu *et al.*, 2013); however, the question of whether this is indeed the case has not been directly addressed.

## 8.3 Interactions between Embryo and Surrounding Tissues

### 8.3.1 Regulation of embryo development by pre-zygotic factors

The development of the embryo in higher plants has traditionally been considered to follow a predictable set of intrinsically established developmental rules. However, recent work has brought to light the increasingly important role played by non-cell autonomous factors, namely those derived from the surrounding tissues before and after fertilization. The first zygotic division is critical for embryo development as it establishes the apical–basal axis that later develops into the accessory suspensor cells and the embryo proper. The apical–basal axis in Arabidopsis is determined after zygote division by the asymmetric expression of Wuschel Related Homebox (*WOX*) genes (Jeong *et al.*, 2016). Although related *WOX* genes have been identified in maize, and their products are also asymmetrically localized in early embryos, their precise functions remain unknown (Nardmann *et al.*, 2007). Formation of the apical–basal axis is influenced by pre-fertilization factors. For instance, the sperm-derived cytoplasmic receptor-like kinase, SSP, and central cell-derived ESF1 cysteine-rich peptides are implicated in asymmetric division of the zygote in arabidopsis (Costa *et al.*, 2014; Bayer *et al.*, 2009). SSP evolved recently in the Brassicaceae and is related to brassinosteroid signaling kinases (BSK), which are conserved in maize and implicated in growth (Baute *et al.*, 2015). ESF1 peptides are also conserved in maize, where they are specifically expressed in central cells and implicated in early embryo development (unpublished data). Further studies in maize will undoubtedly focus on identifying the pre-zygotic and zygotic components that regulate the early stages of embryo development.

### 8.3.2 The embryo surrounding region (ESR): an important tissue for embryo/endosperm interactions?

Development of the interface between the embryo and its surrounding tissues is complex. The embryo is surrounded by endosperm cells. In maize, these can be classified into two types: those of the embryo surrounding region (ESR) proper (Cosségal *et al.*, 2007), which surround the embryo at its earliest stages of development and are made up of a very particular population of cytoplasm-rich small cells; and those of the embryo surrounding endosperm, with which the embryo comes into contact upon elongation out of the ESR. This section deals with the ESR proper, whereas the following one (8.3.3) deals with the endosperm/embryo interface once the ESR has vanished (after 14–18 DAP).

In addition to particular cytological characteristics, specific gene expression patterns characterize the ESR proper (Doll *et al.*, 2017). Interestingly, the proteins encoded by the three first genes (*ESR1*, *ESR2*, and *ESR3*) found to be ESR-specific encode the founding members, together with the Arabidopsis CLAVATA3 protein, of the CLE (CLavata3/Endosperm surrounding region) peptide family. The CLE peptide family represents the largest such family in plants, and numerous CLE peptides have been implicated in cell-to-cell communication (Katsir *et al.*, 2011). Unfortunately, the absence of loss-of-function mutants for CLE-encoding genes only allows speculation regarding their functions. For example, ESR peptides could promote ESR cell fate, similar to the function of MEG1 peptides in BETL cells (see 8.2.1). Alternatively, they could be involved in endosperm–embryo crosstalk, as has been suggested for the peptide CLE8, with the important difference that ESR-encoding genes are specifically expressed in the ESR and are not also expressed in embryo (Fiume and Fletcher, 2012). The identification and characterization of receptors involved in ESR perception and downstream signaling actors would be an important breakthrough in understanding maize seed development. Another unrelated gene, *ESR6*, encoding a cysteine-rich peptide (CRP) is also specifically expressed in ESR cells. ESR6 shares structural homology with plant defensins and the antimicrobial activity of this protein has been demonstrated *in vitro* by growth inhibition assays of bacterial and fungal plant pathogens (Balandín *et al.*, 2005). The expression of a second CRP-encoding gene, *ZmAE1*, was also found to be ESR specific (Magnard *et al.*, 2000), but only a functional analysis will clarify its potential role in seed development.

A recent study of hormonal responses in the maize kernel highlighted an interesting potential role for the ESR in preventing auxin fluxes from the endosperm to the embryo at very early developmental stages (Chen *et al.*, 2014). Using a hormone responsive promoter (DR5) and a ZmPIN1a reporter, Chen *et al.* (2014) showed that ESR cells do not respond to auxin, and the first embryo hormonal responses, which are observed in the apical regions, coincide with its emergence from the ESR (Chen *et al.*, 2014).

The localization of auxin efflux carriers (ZmPINs) in the endomembrane compartment rather than the ESR cell plasma membrane reinforce this hypothesis (Forestan *et al.*, 2010). High auxin levels in the ESR are detected by immunolocalization (Rijavec *et al.*, 2011). It is therefore tempting to speculate that these cells could physically "trap" endosperm-derived auxin, preventing it from reaching the early embryo and, subsequently, the suspensor. Interestingly, Chen *et al.* (2014) observed elevated auxin signaling activity in the embryo-surrounding endosperm above the ESR, and correlated with this the activation of basally directed auxin fluxes and auxin responses in the apical regions of the embryo once it emerged from the ESR. Based on these observations, they propose that in maize, endosperm-derived auxin could be critically important in early embryo patterning (Chen *et al.*, 2014). This is contrary to the situation in Arabidopsis, where auxin fluxes are initially directed to the embryo proper from the suspensor, and then reorient to direct embryo-proper-derived auxin to the root pole and subsequently to the tips of the cotyledons (Wabnik *et al.*, 2013). Interestingly, auxin responses in maize endosperm are stronger in adaxial regions, leading to the tempting idea that this exogenous asymmetry could be required to orient the morphogenesis of the asymmetric maize embryo. The continuing presence of the ESR "buffer" around the basal regions of the embryo could be critical in maintaining a basally-oriented auxin flux within the embryo until patterning is well established (Doll *et al.*, 2017).

Since the ESR is an interface tissue and completely surrounds the embryo during early development (3–5 DAP), it is also likely to be implicated in transport of nutrients towards the developing embryo (Cosségal *et al.*, 2007). Compared to another interface tissue, the BETL, in which a crosstalk between sugar transport and sugar signaling has been shown to be important for both BETL development and endosperm nutrition (see 8.2.1), less is known about the nutritional function of ESR cells. Interestingly, however, an invertase inhibitor, ZM-INVINH1, was found to be secreted into the maize ESR apoplast (Bate *et al.*, 2004). This enzyme could potentially play an important role in post-translational regulation of cell wall invertase activity in the ESR and thus control sucrose cleavage in a spatially- and temporally-specific manner to regulate metabolic partitioning between endosperm and embryonic tissues. In Arabidopsis, a role for the embryo-surrounding endosperm in transporting sucrose to the embryo was suggested (Baud *et al.*, 2005). This study showed that mutants in the sucrose transporter-encoding gene, *SUC5*, are defective in accumulation of embryonic storage lipids during early seed filling. A more recent study in Arabidopsis revealed a complex situation in which transport of sugars from maternal tissues and their uptake by filial tissues is mediated by a suite of sugar transporters, including some expressed in the embryo-surrounding endosperm (Chen *et al.*, 2015).

Disappearance of maize ESR cells occurs at around 14–18 DAP, but their elimination remains a poorly characterized process that appears to be controlled, in part, by the transcription factor ZmZOU (ZHOUPI) (Grimault *et al.*, 2015). The Arabidopsis ZOU protein is required for the more extensive endosperm cell-elimination processes (Yang *et al.*, 2008). However, it is still unclear whether endosperm cell elimination occurs in the embryo surrounding region in maize or in other cereals without a well defined ESR. The Arabidopsis ZOU protein is also involved in regulating an endosperm–embryo signaling pathway that affects embryo surface formation (Xing *et al.*, 2013; Yang *et al.*, 2008; Moussu *et al.*, 2013). Although this pathway appears to act at least partially independently of endosperm breakdown, it is possible degenerating endosperm cells in Arabidopsis contribute to mechanisms involved in embryo surface reinforcement (Xing *et al.*, 2013). To what extent cell death in the ESR plays a role in limiting inter-tissue cross talk, or, indeed, affects nutrient movement between the embryo and endosperm, has not been investigated.

The embryo suspensor has been shown in several species to play important roles in nutrient uptake from the endosperm during seed development (Kawashima and Goldberg, 2010) and thus represents a potential site of inter-compartmental communication in maize, particularly in light of its strategic position next to the developing ESR. As in other plants, the maize suspensor undergoes programmed cell death during seed development (Giuliani *et al.*, 2002; Domínguez and Cejudo, 2014). This process is slow in maize kernels with reduced expression of the endosperm specific *ZmZOU* gene (Grimault *et al.*, 2015). The mechanisms underlying the correlation between delayed ESR degradation and retarded suspensor loss is unclear, but it could indicate that communication, either at the level of nutrient transfer or another signaling mechanism, links the fate of these two tissues.

### 8.3.3    Embryo–endosperm interface, when ESR is not there

Here we consider the embryo/endosperm interface that is generated once the embryo has emerged from the ESR (around 16-18 DAP). The tissue at this interface is referred to as the "embryo surrounding endosperm." As described above, auxin derived from endosperm in this zone has been proposed to play key roles in embryo patterning (Doll *et al.*, 2017). However, this may only be part of the story. Thanks to an elegant screen, expression of *Rough Endosperm3* (*Rgh3*) was shown to be required in the endosperm for embryo development, illustrating the potential importance of endosperm/embryo interactions in maize (Fouquet *et al.*, 2011). However, target(s) of this RNA splicing factor, as well as the molecule(s) directly involved in endosperm-to-embryo signaling, remain to be identified. In Arabidopsis, as highlighted above, the endosperm-specific transcription factor ZOU/RGE1 and the ZOU-regulated subtilisin protease ABNORMAL LEAF SHAPE1 (ALE1) are implicated in the endosperm-to-embryo signaling required for normal embryo cuticle development (Xing *et al.*, 2013; Tanaka *et al.*, 2001). To date, however, there is no evidence for a role of the endosperm in embryonic surface formation in maize. Maize embryos have very different surface characteristics to those in Arabidopsis, with a hydrophobic cuticle

present on the developing embryonic axis (Grimault *et al.*, 2015; Rocca *et al.*, 2015). In contrast, the scutellum, which some consider to be the unique cotyledon in the maize embryo (Chandler, 2008), is in close contact with the endosperm and does not have an obvious hydrophobic cuticle.

Molecular players involved in controlling the size balance between embryo and endosperm have recently been identified in maize and rice thanks to cloning of the *GIANT EMBRYO* (*GE*) gene, loss of function of which leads to production of a large embryo at expense of the endosperm (Nagasawa *et al.*, 2013; Zhang *et al.*, 2012). The underlying *GE/CYP78A13* (rice) and *ZmGE2* (maize) genes were found to encode enzymes of the cytochrome P450 protein (CYP) superfamily and could therefore be involved in production of an unknown signal molecule that could regulate embryo and endosperm growth. The closest Arabidopsis orthologs of rice and maize GE were shown to work as fatty acid hydroxylases *in vitro* and in plant cells (Kai *et al.*, 2009), but this finding has not provided significant insight into the substrates of these enzymes in the seed. Interestingly, it has been postulated that one Arabidopsis GE ortholog, CYP78A5/ KLU, could be involved in generating a mobile growth signal that acts non-cell autonomously (Anastasiou *et al.*, 2007; Adamski *et al.*, 2009). *CYP78A5/KLU* is expressed in the inner integument of developing Arabidopsis ovules and acts as a maternal regulator of seed size. However, this is not the case for maize and rice *GE* genes, which are expressed in the embryo and endosperm. *In situ* hybridization in rice kernels clearly revealed *GE/CYP78A13* expression in juxtaposed tissues at the embryo–endosperm interface, strengthening the idea of crosstalk occurring between these filial tissues to co-ordinate their growth. Mutations in a MATE (Multidrug And-Toxin-Extrusion)-type transporter lead to a Big embryo 1 (Bige1) phenotype in which embryos produce an abnormally large scutellum compared to wild-type (Suzuki *et al.*, 2015) and thus phenocopy *ge* mutants. The fact that *ZmGE1* is upregulated in *bige1* mutant embryos has led to speculation about a potential role for

BIGE1 in transporting a metabolite associated with the CYP78A signaling pathway (Suzuki *et al.*, 2015).

## 8.4 Summary and Perspectives

Based on the research described in this chapter, we highlight several important principles. The first and possibly the most important is that even where there are parallels between maize and other developmentally divergent species, such as Arabidopsis, there appear to be appreciable alterations at the mechanistic level in pathways involved in intercompartmental communication. This is best illustrated by auxin-mediated communication. In maize there is evidence suggesting that spatially-regulated provision of exogenous auxin from the endosperm could be critical for embryo patterning (Chen *et al.*, 2014), while in Arabidopsis, endosperm-derived auxin has thus far been proposed to only affect development of the testa in a non-cell autonomous manner (Figueiredo *et al.*, 2016). It is clear that, in angiosperms, creation of interorganismal auxin fluxes could have been critical during seed evolution for promoting interactions between the gametophyte and zygotic sporophyte (embryo) and establishing communication between the gametophyte and maternal sporophyte. However, questions remain. Do the apparently profound differences in auxin-mediated communication in maize and Arabidopsis reflect a fundamental divergence in the role of auxin during seed development, or are they merely a reflection of the complementary strengths of the two species as systems for addressing specific developmental questions? Neither possibility can currently be excluded, and only in-depth studies in maize and other species will allow distinction between "basal" developmental mechanisms common to all angiosperms and those responsible for the unique features (embryo asymmetry for example) of maize seed development.

A second theme throughout this chapter is the potential role of sugars as signaling molecules. Perhaps unsurprisingly, given the size of the maize seed and its agronomic importance, this aspect of intercompartmental communication arises frequently in developmental studies. However, frustratingly, despite much correlative evidence, pinpointing signaling roles of sugars is incredibly difficult due to the challenge of separating nutritive roles and our relatively poor understanding of how sugar signals are perceived and transduced. Future research in this area, to which maize is particularly well suited, is of critical importance for understanding this enigmatic aspect of seed development.

Thirdly, it is clear that small secreted peptides are likely involved in many of the signaling processes highlighted in this review. The tissue-specific expression of these peptides at tissue interfaces in maize seeds is particularly tantalizing and hints at potential regulatory roles. However, the role of these molecules is opaque because of the absence of tools for their analysis. Due to their small size, peptide-encoding genes are notoriously under-represented in insertional mutant collections. To complicate matters, their receptors are often encoded by redundantly-acting multigene families. The recent successful adaptation of CRISPR-CAS9-mediated mutagenesis to maize (Char *et al.*, 2016) is likely to provide an opening to this research area, and as a result we predict that our understanding of the importance of peptide signaling during maize seed development will explode over the next decade.

Last, but not least, we touched on the structure of intercompartmental interfaces in maize seeds. It is clear that these interfaces, which have compound origins and are impacted by apoplastic modifications and cell death throughout seed development, are likely to have specific and dynamic physiological properties that could profoundly influence intercompartmental signaling. Describing the biophysical characteristics of these interfaces—such as permeability to diffusible molecules (hormones, sugars, and peptides) and mechanisms of membrane-localized hormone and sugars transporters—is a challenging but indispensable step for clarifying how intercompartmental signaling is controlled, and could lead to identification of key players in signaling regulation.

# References

Adamski, N.M., Anastasiou, E., Eriksson, S., O'Neill, C.M. and Lenhard, M. (2009) Local maternal control of seed size by *KLUH/CYP78A5*-dependent growth signaling. *Proceedings of the National Academy of Sciences of the United States of America* 106, 20115–20120.

Anastasiou, E., Kenz, S., Gerstung, M., MacLean, D., Timmer, J., Fleck, C. and Lenhard, M. (2007) Control of plant organ size by *KLUH/CYP78A5*-dependent intercellular signaling. *Developmental Cell* 13, 843–856.

Bai, F., Daliberti, M., Bagadion, A., Xi, N., Li, Y., *et al.* (2016) Parent-of-origin effect rough endosperm mutants in maize. *Genetics* 204, 221–231. Available at: https://doi.org/10.1534/genetics.116.191775 (accessed May 3, 2017).

Balandín, M., Royo, J., Gómez, E., Muniz, L.M., Molina, A. and Hueros, G. (2005) A protective role for the embryo surrounding region of the maize endosperm, as evidenced by the characterisation of *ZmESR-6*, a defensin gene specifically expressed in this region. *Plant Molecular Biology* 58, 269–282.

Bate, N.J., Niu, X., Wang, Y., Reimann, K.S. and Helentjaris, T.G. (2004) An invertase inhibitor from maize localizes to the embryo surrounding region during early kernel development. *Plant Physiology* 134, 246–254.

Baud, S., Wuillème, S., Lemoine, R., Kronenberger, J., Caboche, M., Lepiniec, L. and Rochat, C. (2005) The AtSUC5 sucrose transporter specifically expressed in the endosperm is involved in early seed development in Arabidopsis. *Plant Journal* 43, 824–836.

Baute, J., Herman, D., Coppens, F., De Block, J., Slabbinck, B., *et al.* (2015) Correlation analysis of the transcriptome of growing leaves with mature leaf parameters in a maize RIL population. *Genome Biology* 16, 168.

Bayer, M., Nawy, T., Giglione, C., Galli, M., Meinnel, T. and Lukowitz, W. (2009) Paternal control of embryonic patterning in *Arabidopsis thaliana*. *Science* 323, 1485–1488.

Bernardi, J., Lanubile, A., Li, Q.-B., Kumar, D., Kladnik, A., *et al.* (2012) Impaired auxin biosynthesis in the *defective endosperm18* mutant is due to mutational loss of expression in the *ZmYuc1* gene encoding endosperm-specific YUCCA1 protein in maize. *Plant Physiology* 160, 1318–1328.

Cairney, J. and Pullman, G.S. (2007) The cellular and molecular biology of conifer embryogenesis. *New Phytologist* 176, 511–536.

Chandler, J.W. (2008) Cotyledon organogenesis. *Journal of Experimental Botany* 59, 2917–2931.

Char, S.N., Neelakandan, A.K., Nahampun, H., Frame, B., Main, M., *et al.* (2016) An *Agrobacterium*-delivered CRISPR/Cas9 system for high-frequency targeted mutagenesis in maize. *Plant Biotechnology Journal* 15, 257–268.

Chen, J., Lausser, A. and Dresselhaus, T. (2014) Hormonal responses during early embryogenesis in maize. *Biochemical Society Transactions* 42, 325–331.

Chen, L.-Q., Lin, I.W., Qu, X.-Q., Sosso, D., McFarlane, H.E., *et al.* (2015) A cascade of sequentially expressed sucrose transporters in the seed coat and endosperm provides nutrition for the Arabidopsis embryo. *Plant Cell* 27, 607–619.

Chettoor, A.M., Yi, G., Gomez, E., Hueros, G., Meeley, R.B. and Becraft, P.W. (2015) A putative plant organelle RNA recognition protein gene is essential for maize kernel development. *Journal of Integrative Plant Biology* 57, 236–246.

Chettoor, A.M., Phillips, A.R., Coker, C.T., Dilkes, B. and Evans, M.M.S. (2016) Maternal gametophyte effects on seed development in maize. *Genetics* 204, 233–248.

Cosségal, M., Vernoud, V., Depège, N. and Rogowsky, P.M. (2007) The embryo surrounding region. In: Olsen, O.-A. (ed.) *Endosperm*. Springer, Berlin, Heidelberg, Germany, pp. 57–71.

Costa, L.M., Gutiérrez-Marcos, J.F., Brutnell, T.P., Greenland, A.J. and Dickinson, H.G. (2003) The *globby1-1* (*glo1-1*) mutation disrupts nuclear and cell division in the developing maize seed causing alterations in endosperm cell fate and tissue differentiation. *Development* 130, 5009–5017.

Costa, L.M., Marshall, E., Tesfaye, M., Silverstein, K.A.T., Mori, M., *et al.* (2014) Central cell-derived peptides regulate early embryo patterning in flowering plants. *Science* 344, 168–172.

Doll, N.M., Depège-Fargeix, N., Rogowsky, P.M. and Widiez, T. (2017) Signaling in early maize kernel development. *Molecular Plant* 10, 375–388.

Domínguez, F. and Cejudo, F.J. (2014) Programmed cell death (PCD): an essential process of cereal seed development and germination. *Plant Evolution and Development* 5, 366.

Domínguez, F., Moreno, J. and Cejudo, F.J. (2001) The nucellus degenerates by a process of programmed cell death during the early stages of wheat grain development. *Planta* 213, 352–360.

Figueiredo, D.D., Batista, R.A., Roszak, P.J. and Köhler, C. (2015) Auxin production couples endosperm development to fertilization. *Nature Plants* 1, 15184.

Figueiredo, D.D., Batista, R.A., Roszak, P.J., Hennig, L. and Köhler, C. (2016) Auxin production in the endosperm drives seed coat development in *Arabidopsis*. *eLife* 5, e20542.

Fiume, E. and Fletcher, J.C. (2012) Regulation of *Arabidopsis* embryo and endosperm development by the polypeptide signaling molecule CLE8. *Plant Cell* 24, 1000–1012.

Forestan, C., Meda, S. and Varotto, S. (2010) ZmPIN1-mediated auxin transport is related to cellular differentiation during maize embryogenesis and endosperm development. *Plant Physiology* 152, 1373–1390.

Fouquet, R., Martin, F., Fajardo, D.S., Gault, C.M., Gómez, E., *et al.* (2011) Maize rough endosperm3 encodes an RNA splicing factor required for endosperm cell differentiation and has a nonautonomous effect on embryo development. *Plant Cell* 23, 4280–4297.

Galletti, R., Johnson, K.L., Scofield, S., San-Bento, R., Watt, A.M., Murray, J.A.H. and Ingram, G.C. (2015) DEFECTIVE KERNEL 1 promotes and maintains plant epidermal differentiation. *Development* 142, 1978–1983.

Geisler-Lee, J. and Gallie, D.R. (2005) Aleurone cell identity is suppressed following connation in maize kernels. *Plant Physiology* 139, 204–212.

Giuliani, C., Consonni, G., Gavazzi, G., Colombo, M. and Dolfini, S. (2002) Programmed cell death during embryogenesis in maize. *Annals of Botany* 90, 287–292.

Grimault, A., Gendrot, G., Chamot, S., Widiez, T., Rabillé, H., *et al.* (2015) ZmZHOUPI, an endosperm-specific basic helix–loop–helix transcription factor involved in maize seed development. *Plant Journal* 84, 574–586.

Gruis, D.F., Guo, H., Selinger, D., Tian, Q. and Olsen, O.-A. (2006) Surface position, not signaling from surrounding maternal tissues, specifies aleurone epidermal cell fate in maize. *Plant Physiology* 141, 898–909.

Gutiérrez-Marcos, J.F., Costa, L.M., Biderre-Petit, C., Khbaya, B., O'Sullivan, D.M., *et al.* (2004) *Maternally expressed gene1* is a novel maize endosperm transfer cell–specific gene with a maternal parent-of-origin pattern of expression. *Plant Cell* 16, 1288–1301.

Gutiérrez-Marcos, J.F., Costa, L.M. and Evans, M.M.S. (2006) Maternal gametophytic *baseless1* is required for development of the central cell and early endosperm patterning in maize (*Zea mays*). *Genetics* 174, 317–329.

Jeong, S., Eilbert, E., Bolbol, A. and Lukowitz, W. (2016) Going mainstream: How is the body axis of plants first initiated in the embryo? *Developmental Biology* 419, 78–84.

Kai, K., Hashidzume, H., Yoshimura, K., Suzuki, H., Sakurai, N., Shibata, D. and Ohta, D. (2009) Metabolomics for the characterization of cytochromes P450-dependent fatty acid hydroxylation reactions in *Arabidopsis*. *Plant Biotechnology* 26, 175–182.

Katsir, L., Davies, K.A., Bergmann, D.C. and Laux, T. (2011) Peptide signaling in plant development. *Current Biology* 21, R356–R364.

Kawashima, T. and Goldberg, R.B. (2010) The suspensor: not just suspending the embryo. *Trends in Plant Science* 15, 23–30.

Lammeren, A.A.M. van, Kieft, H., and Schel, J.H.N. (2005) Cell differentiation in the pericarp and endosperm of developing maize kernels (*Zea mays* L.). In: Batygina, T.B. (ed.) *Embryology of Flowering Plants: Terminology and Concepts. Volume 2: The Seed.* Science Publishers, Enfield, New Hampshire, pp. 131–139.

Leroux, B.M., Goodyke, A.J., Schumacher, K.I., Abbott, C.P., Clore, A.M., *et al.* (2014) Maize early endosperm growth and development: from fertilization through cell type differentiation. *American Journal of Botany* 101, 1259–1274.

Lituiev, D.S., Krohn, N.G., Müller, B., Jackson, D., Hellriegel, B., Dresselhaus, T. and Grossniklaus, U. (2013) Theoretical and experimental evidence indicates that there is no detectable auxin gradient in the angiosperm female gametophyte. *Development* 140, 4544–4553.

Lombardi, L., Mariotti, L., Picciarelli, P., Ceccarelli, N. and Lorenzi, R. (2012) Ethylene produced by the endosperm is involved in the regulation of nucellus programmed cell death in *Sechium edule* Sw. *Plant Science* 187, 31–38.

López, M., Gómez, E., Faye, C., Gerentes, D., Paul, W., *et al.* (2016) *Zmsbt1* and *zmsbt2*, two new subtilisin-like serine proteases genes expressed in early maize kernel development. *Planta* 245, 409–424.

Magnard, J.-L., Deunff, E.L., Domenech, J., Rogowsky, P.M., Testillano, P.S., *et al.* (2000) Genes normally expressed in the endosperm are expressed at early stages of microspore embryogenesis in maize. *Plant Molecular Biology* 44, 559–574.

Moussu, S., San-Bento, R., Galletti, R., Creff, A., Farcot, E. and Ingram, G. (2013) Embryonic cuticle establishment: the great (apoplastic) divide. *Plant Signaling & Behavior* 8, e27491.

Nagasawa, N., Hibara, K., Heppard, E.P., Vander Velden, K.A., Luck, S., *et al.* (2013) *GIANT EMBRYO* encodes CYP78A13, required for proper size balance between embryo and endosperm in rice. *Plant Journal* 75, 592–560.

Nardmann, J., Zimmermann, R., Durantini, D., Kranz, E. and Werr, W. (2007) *WOX* gene phylogeny in *Poaceae*: a comparative approach addressing leaf and embryo development. *Molecular Biology and Evolution* 24, 2474–2484.

Oparka, K.J. and Gates, P. (1981a) Transport of assimilates in the developing caryopsis of rice (*Oryza sativa* L.): the pathways of water and assimilated carbon. *Planta* 152, 388–396.

Oparka, K.J. and Gates, P. (1981b) Transport of assimilates in the developing caryopsis of rice (*Oryza sativa* L.): ultrastructure of the pericarp vascular bundle and its connections with the aleurone layer. *Planta* 151, 561–573.

Radchuk, V., Weier, D., Radchuk, R., Weschke, W. and Weber, H. (2011) Development of maternal seed tissue in barley is mediated by regulated cell expansion and cell disintegration and coordinated with endosperm growth. *Journal of Experimental Botany* 62, 1217–1227.

Rijavec, T., Jain, M., Dermastia, M. and Chourey, P.S. (2011) Spatial and temporal profiles of cytokinin biosynthesis and accumulation in developing caryopses of maize. *Annals of Botany* 107, 1235–1245.

Rocca, N.L., Manzotti, P.S., Cavaiuolo, M., Barbante, A., Vecchia, F.D., *et al.* (2015) The maize *fused leaves1* (*fdl1*) gene controls organ separation in the embryo and seedling shoot and promotes coleoptile opening. *Journal of Experimental Botany* 66, 5753–5767.

Rolland, F., Baena-Gonzalez, E. and Sheen, J. (2006) Sugar sensing and signaling in plants: conserved and novel mechanisms. *Annual Review of Plant Biology* 57, 675–709.

Rousseau, D., Widiez, T., Tommaso, S., Rositi, H., Adrien, J., *et al.* (2015) Fast virtual histology using X-ray in-line phase tomography: application to the 3D anatomy of maize developing seeds. *Plant Methods* 11, 55.

Schichnes, D., Schneeberger, R. and Freeling, M. (1997) Induction of leaves directly from leaves in the maize mutant *Lax midrib1-O. Developmental Biology* 186, 36–45.

Sosso, D., Luo, D., Li, Q.-B., Sasse, J., Yang, J., *et al.* (2015) Seed filling in domesticated maize and rice depends on SWEET-mediated hexose transport. *Nature Genetics* 47, 1489–1493.

Stadler, R., Lauterbach, C. and Sauer, N. (2005) Cell-to-cell movement of green fluorescent protein reveals post-phloem transport in the outer integument and identifies symplastic domains in arabidopsis seeds and embryos. *Plant Physiology* 139, 701–712.

Suzuki, M., Sato, Y., Wu, S., Kang, B.-H. and McCarty, D.R. (2015) Conserved functions of the MATE transporter BIG EMBRYO1 in regulation of lateral organ size and initiation rate. *Plant Cell* 27, 2288–2300.

Tanaka, H., Onouchi, H., Kondo, M., Hara-Nishimura, I., Nishimura, M., Machida, C. and Machida, Y. (2001) A subtilisin-like serine protease is required for epidermal surface formation in *Arabidopsis* embryos and juvenile plants. *Development* 128, 4681–4689.

Wabnik, K., Robert, H.S., Smith, R.S. and Friml, J. (2013) Modeling framework for the establishment of the apical-basal embryonic axis in plants. *Current Biology* 23, 2513–2518.

Wang, N. and Fisher, D.B. (1995) Sucrose release into the endosperm cavity of wheat grains apparently occurs by facilitated diffusion across the nucellar cell membranes. *Plant Physiology* 109, 579–585.

Weber, H., Borisjuk, L. and Wobus, U. (1996) Controlling seed development and seed size in *Vicia faba*: a role for seed coat-associated invertases and carbohydrate state. *Plant Journal* 10, 823–834.

Webster, T.R. (1967) Induction of selaginella sporelings under greenhouse and field conditions. *American Fern Journal* 57, 161–166.

Wobus, U. and Weber, H. (1999) Sugars as signal molecules in plant seed development. *Biological Chemistry* 380, 937–944.

Wu, X.N., Sanchez Rodriguez, C., Pertl-Obermeyer, H., Obermeyer, G. and Schulze, W.X. (2013) Sucrose-induced receptor kinase SIRK1 regulates a plasma membrane aquaporin in Arabidopsis. *Molecular & Cellular Proteomics* 12, 2856–2873.

Xing, Q., Creff, A., Waters, A., Tanaka, H., Goodrich, J. and Ingram, G.C. (2013) ZHOUPI controls embryonic cuticle formation via a signalling pathway involving the subtilisin protease ABNORMAL LEAF-SHAPE1 and the receptor kinases GASSHO1 and GASSHO2. *Development* 140, 770–779.

Yang, S., Johnston, N., Talideh, E., Mitchell, S., Jeffree, C., Goodrich, J. and Ingram, G. (2008) The endosperm-specific *ZHOUPI* gene of *Arabidopsis thaliana* regulates endosperm breakdown and embryonic epidermal development. *Development* 135, 3501–3509.

Yuan, J., Bateman, P. and Gutiérrez-Marcos, J. (2016) Genetic and epigenetic control of transfer cell development in plants. *Journal of Genetics and Genomics* 43, 533–539.

Zhang, P., Allen, W.B., Nagasawa, N., Ching, A.S., Heppard, E.P., *et al.* (2012) A transposable element insertion within *ZmGE2* gene is associated with increase in embryo to endosperm ratio in maize. *Theoretical and Applied Genetics* 125, 1463–1471.

# 9 Aneuploidy and Ploidy in the Endosperm: Dosage, Imprinting, and Maternal Effects on Development

James A. Birchler* and Adam F. Johnson

*Division of Biological Sciences, University of Missouri, Columbia, Missouri, USA*

## 9.1 Introduction

For nearly 100 years it has been recognized that changes in the dosage of parts of plant genomes (aneuploidy) impact their development, stature, and vigor, even more than changes in copies of the entire genome (polyploidy). The first studies of aneuploidy and polyploidy, using *Datura stramonium*, were conducted by Blakeslee and colleagues (Blakeslee *et al.*, 1920; Blakeslee, 1934), who made an extensive set of changes in each chromosome as well as a dosage series of various ploidies. The effects of aneuploidy were manifested in all aspects of the life cycle. The question we address in this chapter is how these effects, which appear to be manifested somewhat differently, apply to endosperm development. We propose that the stoichiometry of regulatory complexes for carrying out the maternally contributed program for endosperm development and the primary endosperm nucleus dosage contributions affect endosperm development at a critical early stage.

We begin by reviewing evidence for the importance of "genomic balance," i.e. the impact of relative changes within the genome. Starting with the aforementioned work of Blakeslee and extending to many other plant species, including maize (Lee *et al.*, 1996; Sheridan and Auger, 2008; Birchler and Veitia, 2012; Brunelle and Sheridan, 2014; Henry *et al.*, 2015), it is a common observation that adding a chromosome to a genotype has a more detrimental effect than increasing the ploidy to triploid, tetraploid, and greater. The impact of aneuploids can be mimicked by expression of single genes that exhibit dosage effects when varied alone (Rabinow *et al.*, 1991; Birchler *et al.*, 2001). The types of genes that exhibit these dosage effects tend to be those that are involved with macromolecular complexes, such as transcription factors and components of signal transduction complexes (Birchler *et al.*, 2001, 2005, 2007; Veitia *et al.*, 2008; Birchler and Veitia, 2012), and they are associated with quantitative trait loci that exhibit some level of additive behavior in hybrids between parents of extreme phenotypes (Frary *et al.*, 2000; Cong *et al.*, 2002, 2008; Liu *et al.*, 2002). This relationship also seems to be reflected in evolutionary genomics in terms of the gradual loss of genes after whole genome duplication (WGD) events (Maere *et al.*, 2005; Freeling and Thomas, 2006; Thomas *et al.*, 2006; Freeling *et al.*, 2008; Freeling, 2009). Following tetraploid formation, one of the

---

*Corresponding author e-mail: BirchlerJ@missouri.edu

 

duplicate pair of genes is lost over evolutionary time but the spectrum is not random; genes involved with macromolecular complexes tend to be retained for longer periods of time (Freeling, 2009). In population studies, these gene classes are underrepresented in copy number variants (Freeling, 2009). In each case, they potentially mimic a dosage change that is detrimental and would be selected against. In the case of WGD, loss would be detrimental; in the case of copy number variation, extra copies would be detrimental. These results illustrate that aneuploidy/ploidy balance effects play out over evolutionary time.

The rationale for why genes involved with multi-subunit complexes behave in this manner is based on several considerations. In a yeast study, nulls of essential genes were tested for haplo-insufficiency (Papp *et al.*, 2003). As the involvement in multi-subunit complexes increased, there was a negative impact on culture fitness.

Secondly, the kinetics of assembly of multi-subunit complexes in general indicate that for bridge molecules that interact with multiple, different subunits, varying the quantity of the bridge component while holding the others constant results in a positive effect on assembly of the holoenzyme at low bridge subunit concentration; the amount of the assembly eventually peaks and at higher concentrations leads to the formation of smaller amounts of completed assembly due to subunits being tied up in non-functional partial complexes (Bray and Lay, 1997; Veitia, 2002). Figure 9.1 illustrates how changing the ratio of a bridge molecule to other interacting subunits impacts the kinetics of assembly of the whole complex. In a cell, aneuploidy is likely to produce such an outcome because only some components of a complex will change in dosage, while polyploidy is less likely to do so because the whole genome is varied in concert. These considerations provide an experimental and theoretical framework with which to explain why relative changes in parts of the genome are considerably more detrimental than those involving the whole genome (Birchler and Veitia, 2012).

## 9.2 Aneuploidy in the Endosperm

Aneuploidy can be generated in maize endosperm as in any other tissue, but its

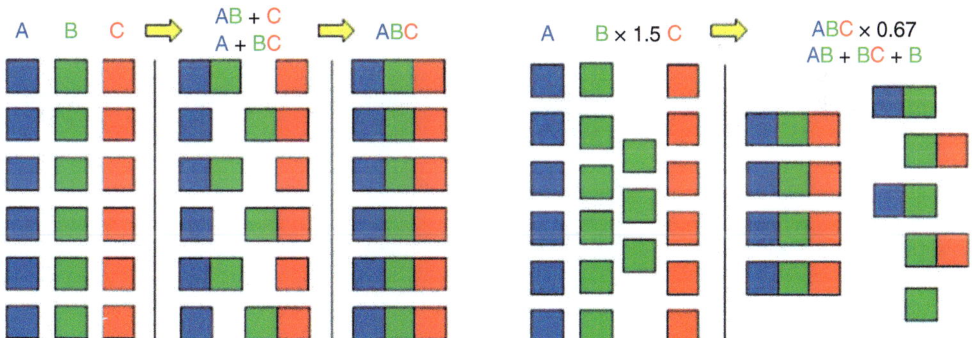



**Fig. 9.1.** A theoretical framework for genomic balance. A, B, and C represent protein subunits of a larger complex, which must be assembled into ABC to function. As shown on the left, when the three subunits are present in the cell in equal amounts, the number of ABC complexes formed matches the number of each type of subunit. B can bind with A or C in any order and still find its other binding partner. As shown on the right, when the bridge subunit B is present in excess of the others, A and C are consumed in incomplete complexes, and there is a decrease in the number of complete ABC complexes. In this model, the excess of B is inversely proportional to the shortage of ABC. If ABC is a regulatory complex that acts upon target genes, then varying one subunit will alter the expression of the target genes because of the effect of a stoichiometric shift on the whole. When aneuploidy occurs, the opportunity for stoichiometric shifts is present because part of the genome is varied relative to the rest. However, when ploidy changes occur, all of the genome is varied together so stoichiometric changes are less likely.

effects seem to be manifested somewhat differently—a phenomenon we attempt to resolve in this chapter. Trisomic plants will segregate chromosomes in meiosis such that some spores will carry an extra chromosome. Thus, the polar nuclei in female gametophytes borne on trisomics can possess 20 chromosomes or 22 (aneuploid). Through the male parent, aneuploidy can be produced in the endosperm with the use of B-A translocations. These translocations append a portion of a normal maize (A) chromosome to the supernumerary (B) chromosome. The B chromosome is of unknown origin but is essentially inert, being neither necessary nor detrimental at low copy number. Being unnecessary, the B chromosome has evolved properties that maintain itself. It frequently nondisjoins at the second pollen mitosis that produces the two sperm, and the sperm with the B chromosomes preferentially fertilizes the egg as opposed to the polar nuclei in double fertilization (Roman, 1947, 1948). Thus, when portions of normal A chromosomes are translocated to the B chromosome, they too undergo nondisjunction. This produces different gene doses (0,1,2) in a sperm and thus in the endosperm after fertilization. B-A translocation genotypes that have extra chromosomes can also transmit them to the endosperm through the female side in a similar manner as described above for trisomics.

In what ways are aneuploidy effects manifested in the endosperm? Compared to individuals with a normal genome, trisomics will segregate for progeny with an extra chromosome as noted above. The extra chromosome can be transmitted through the female parent, but pollen with a duplicated chromosome is seldom successful in competing with normal haploid male gametophytes. When the additional chromosome is transmitted through the female, kernels carrying the extra copy of some but not all chromosomes show a slight reduction in endosperm size.

The most dramatic aneuploidy effects, however, arise when a paternal genome segment is missing in the sperm. Absence of the paternal copy of chromosome arms 1S, 1L, 4S, 5S, 7L, and 10L typically results in a small endosperm, but this effect is dependent to some degree upon the genetic background (Birchler and Hart, 1987). The sibling kernels with an extra copy of these chromosome arms do not change in size. In a set of experiments involving the 10L region, Lin (1982) found that extra copies of 10L introduced through the female parent could not rescue the phenotype conditioned by the absence of the paternal copy. The results were interpreted to imply that the paternal source of a gene responsible for an endosperm size factor is active, but not the maternal one, i.e. parental gene imprinting, as elaborated upon below. This could explain why a missing genetic element from the sperm is detrimental and addition of this region through the female is ineffective at rescuing the small endosperm phenotype.

However, examination of self-pollination events of B-A translocations from various regions of the genome revealed that for all regions that produced a paternal absence small kernel phenotype, introduction through the female parent actually made the small kernel phenotype more severe (Birchler and Hart, 1987). Furthermore, by crossing translocations of one arm with those of another, it was found that extra copies of some chromosomal regions worsened the phenotype associated with the missing paternal region. By comparing those regions that caused an enhancement, it was found that the regions producing the maternal enhancement were identical to those that created small kernels when missing from the male parent (Birchler and Hart, 1987) (Fig. 9.2).

The question arises: Are these effects on endosperm size the equivalent of the aneuploidy and stoichiometric effects on the whole plant phenotype? If so, then it appears that a stoichiometry of polar nuclei to primary endosperm nucleus, typically a 2:3 ratio, is critically important for proper endosperm development. Clearly, the egg and polar nuclei, while genetically identical and differing only in ploidy, are differentially programmed for the subsequent embryo and endosperm development following fertilization, because the two sperm are randomly involved in double fertilization in maize (Carlson, 1969) and cannot contribute to this difference. Below, we explain how the polar nuclei to endosperm

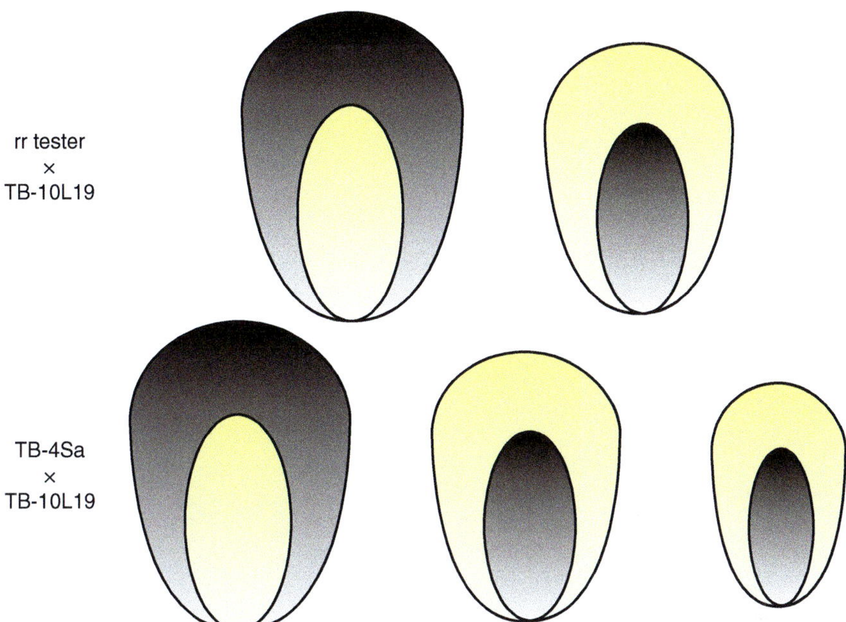



**Fig. 9.2.** Enhancement of the small kernel effect of paternal deficiencies. With TB-10L19, the long arm of chromosome 10 carries a dominant allele of the *r* gene, which produces kernel color. 10L has been translocated onto a B chromosome centromere. A hyperploid male parent carries two copies of the B-A chromosome. When crossed to a female recessive tester, the dosage of B-A chromosomes is indicated by the color, or lack of it, in both the endosperm and embryo. For example, if the embryo has color and the endosperm does not, nondisjunction has occurred and the embryo has received both copies. These kernels are missing the paternal contribution of that chromosome arm to the endosperm. In the lower panel, hyperploid heterozygotes of TB-4Sa are used as females for hyperploid heterozygotes of TB-10L19. When used as a male parent, both of these chromosome arms, when missing as the paternal contribution, produce the small kernel effect. However, in this cross, the extra copies of 4S introduced through the female parent make the 10L small kernel effect more severe. This relationship—of a correlation of an impact on kernel size when missing a chromosome in the paternal parent, to increasing the effect when it is introduced as extra copies through the female parent—extends to all chromosome arms tested. (Adapted from Birchler and Hart, 1987.)

chromosome ratio for the whole genome appears to be important. In such interploidy crosses, an early developmental time window is particularly critical for proper development. This timeframe also appears to be an integral aspect for the segmental aneuploidy effects. Many B-A translocations carry dominant anthocyanin pigment markers that express in the aleurone layer of the endosperm. When there are somatic losses of the B-A chromosome in early endosperm development, these kernels are readily apparent because of their chimeric appearance. When regions of the genome that produce a small kernel effect when absent from the sperm succeed in fertilization, but then are lost during the early cell divisions of the endosperm, the

mosaic kernels produced are normal in size (Birchler, 1980). In other words, the only time that the small kernel phenotype is observed is within an early developmental timeframe. And the timing of this effect is coincident with the timing for sensitivity to whole genome dosage changes, as described in more detail below.

Is imprinting, in which there is preferential parental allelic expression, involved with the small kernel effect? It does not appear to be so, because the chromosomal regions critical for the small kernel effect in the paternal gamete are identical to those that produce an enhancement when introduced through the female. Thus, it seems unlikely that there is absence of expression through one parent. Indeed, it might be the case that the maternal effect results from expression in the female gametophyte and that this expression is what interacts with the primary endosperm nucleus gene dosage. Nevertheless, parental imprinting of genes involved in endosperm development might in fact mimic the dosage effect, because parental imprinting effectively represents allelic dosage effects (Beaudet and Jiang, 2002; Dilkes and Comai, 2004). Indeed, there is no known region of the maize genome that is lethal to the endosperm when missing in the sperm. Thus, there can be no *vital* gene in the assayed regions that is expressed solely from the paternal genome. Further, there are scores of qualitative mutations that affect endosperm characteristics, but none of them are apparent in endosperms lacking paternal contributions. These considerations add to the argument that the small kernel phenotype is a form of aneuploidy dosage response.

There have been, however, mutations recovered that impact various aspects of endosperm development when transmitted solely maternally or paternally (Gutiérrez-Marcos *et al.*, 2004, 2006; Bai and Settles, 2015; Bai *et al.*, 2016; Chettoor *et al.*, 2016). Most of these act maternally, and defects in the female gametophyte can be detected as the potential trigger for the endosperm phenotype (Chettoor *et al.*, 2016). Some mutations produce an endosperm phenotype when inherited paternally (Bai and Settles, 2015). These might be candidates for imprinted genes with an effect on the endosperm. However, Bai and Settles (2016) note that these mutations might also reflect an altered interaction of the polar nuclei and the primary endosperm nucleus.

## 9.3 The Ploidy Hybridization Barrier

In angiosperms, endosperm development provides a hybridization barrier between individuals of different ploidy. Endosperm development occurs normally for diploid and tetraploid plants with triploid and hexaploid tissues, respectively, but in most species crosses between the two fail in both directions (Cooper, 1951; Johnston *et al.*,

1980; Johnston and Hanneman, 1982; Birchler, 1993). Chromosome doubling occurs spontaneously, and when it does essentially reproductive isolation ensues. Alternatively, crosses can occur between related species, but meiosis fails because of divergent homologs. This can produce an allotetraploid that is blocked from crossing with diploid progenitors. In this section, we use maize as an example to discuss some of the issues involved with this process, and we suggest future directions of research to understand further the phenomenon in greater detail.

There are many barriers to hybridization in the plant and animal kingdoms that involve pre- or post-mating. These barriers maintain species integrity, de facto, whether or not there is any evolutionary pressure for them to evolve. Divergent genomes must retain functionality within the context of the respective related species involved, but when brought back together in hybrids, incompatibilities can result (Dobzhansky, 1937; Muller, 1942). The incompatibilities might occur through changes within a species over time, but diverging patterns of gene expression or developmental programs will not be compatible in interspecies hybrids. This is a regularly occurring situation that was recognized by Dobzhansky (1937) and Muller (1942), and thus has been dubbed Muller–Dobzhansky incompatibilities.

The hybridization barrier involving interploidy crosses is an instantaneous incompatibility (Birchler and Veitia, 2010). Diploid maize crossed by tetraploid maize results in endosperm collapse, as does the reciprocal tetraploid by diploid cross (Cooper, 1951; Birchler, 1993; Leblanc *et al.*, 2002). Various ideas have been put forth to explain the basis of this occurrence, but a long-standing hypothesis is that a ratio of two maternal (m) to one paternal (p) genomes in the endosperm is requisite for normal development (Lin, 1984). In a diploid, the two polar nuclei of the female gametophyte join with one of the two sperm involved in double fertilization to form the triploid nucleus of the primary endosperm cell, which has a 2m:1p genome ratio. In a tetraploid, the same process occurs, but each nucleus has double the number of chromosomes. Therefore, the polar nuclei contribute four copies of the genome and each sperm contributes two. Thus, the ratio of maternal to paternal genomes is also 2:1 in the hexaploid endosperm.

However, in a diploid by tetraploid cross, the female gametophyte contributes two sets of chromosomes and the sperm two sets, creating a primary endosperm nucleus with a maternal to paternal ratio of 2:2. In a tetraploid by diploid cross, the polar nuclei contribute four copies of the chromosome set and the paternal parent donates a single set; this creates a ratio of 4m:1p. Other contribution ratios have been studied by utilizing the maize *indeterminate gametophyte* (*ig*) mutation that produces various numbers of chromosome sets in the central cell (Lin, 1984). However, the only combinations that produce normal endosperm development involve a 2m:1p ratio. Early explanations for the ploidy hybridization barrier included the proposal that the genomic relationship of the embryo and endosperm is important, but experiments with the *ig* mutant indicate that the critical parameters involve the endosperm irrespective of the embryo ploidy.

In other maize tissues, changes in ploidy as complete genomes are viable (Rhoades and Dempsey, 1966; Yao *et al.*, 2011), thus raising the issue of why balanced genotypes resulting from interploidy crosses in the endosperm are not. One explanation for this phenomenon was provided by Haig and Westoby (1989, 1991) and comes from the theoretical concept of parental conflict affecting allocation of resources to progeny as the driving force for "gene imprinting," which would effectively alter the parental contributions. They hypothesized that mothers distribute resources equally to all progeny to foster success of her genotype in future generations. In contrast, in species in which there can be multiple fathers, each male tries to partition resources to his future progeny; this creates potential for competition among different fathers. This concept is the basis for suggesting an evolutionary rationale to the emergence of parental gene imprinting in the endosperm.

Parental gene imprinting refers to the situation in which alleles of genes in the progeny are only expressed from the genome of one or the other parent based on their history. In other words, for some genes only the maternally derived allele is expressed, while for others only the paternally derived allele is expressed. The first example of parental gene imprinting in any organism was described in maize for the *R* anthocyanin gene by Kermicle (1970). Its expression in the aleurone is uniform when transmitted through the female parent, but mottled when transmitted through the pollen, regardless of the dosage. Imprinting of genes involved with metabolite allocation has the potential to be utilized as a vehicle for differential resource distribution and thus was postulated to be the basis of interploidy hybridization failure.

While there is ample evidence that parental imprinting of genes occurs in the endosperm (Chaudhuri and Messing, 1994; Gutiérrez-Marcos *et al.*, 2004, 2006; Zhang *et al.*, 2011, 2014; Waters *et al.*, 2011; Xin *et al.*, 2013), the involvement with differential parental resource allocation is questioned by the fact that endosperm failure occurs in both directions of interploidy crosses and there is no advantage to either parent (Birchler, 1993, 2014). Moreover, endosperm size in maize is largely determined by the maternal parent, presumably by the developmental program laid down in the female gametophyte, more specifically embodied in the central cell with the polar nuclei, with little paternal influence. Paternal parents with large or small endosperms have little to no impact on endosperm size (Birchler, 1980). Thus, different male parents show little potential for differential advantage.

In the course of generating new tetraploid derivatives of inbred lines, Kato and Birchler (2006) treated newly pollinated diploid plants in a chamber containing nitrous oxide. Nitrous oxide will arrest metaphase cells, causing chromosome doubling. While the object of the experiment was to double the genomes of zygotic embryonic nuclei, the central cell was also affected during the treatment. It was obvious at maturity that many of the endosperms on treated ears showed defective development, similar to interploidy crosses (see Fig. 9.3).

**Fig. 9.3.** A comparison of ears borne on hybrid tetraploid plants with and without nitrous oxide treatment at the time of fertilization. The treated ear (top) shows many kernels with defective endosperms reminiscent of interploidy cross endosperm failure. A normal hybrid tetraploid ear is shown below. In the treated ear, the defective endosperms have no change of the maternal to paternal ratio within the endosperm itself but rather there is a change of the target copy number of genomes in the primary endosperm nucleus, without changing the megagametophyte central cell regulatory machinery contributions to the endosperm developmental program. Interploidy crosses have two variables: one is the change of dosage of genomes in the endosperm itself and the other is the change of female gametophyte developmental regulatory factors to the endosperm targets. The chromosome doubling experiment separates these variables and illustrates that the components of the regulatory program contributed by the dosage of the polar nuclei relative to the genomic copy number in the endosperm will certainly cause endosperm failure without changing the maternal to paternal ratio of genomes in the endosperm itself.

Bauer (2006) examined this effect on endosperm development in greater detail (see Birchler, 2014). A time course of treatment was conducted and chromosome counts were performed on individual endosperms. In this case, doubling of the triploid endosperm genome produced a hexaploid that was defective, in contrast to the normal pattern of endosperm development that occurs in tetraploid plants. Moreover, the maternal to paternal ratio of parental contributions was not changed and remained at 2m:1p within the endosperm. Chromosome doubling created by treatment at slightly later times during endosperm development indeed occurred but did not produce defective endosperms. Thus, there appears to be a narrow window of developmental time that is particularly responsive to the maternal–paternal genome interaction. It is noteworthy that this also appears to be the case for expression of varying segmental parts of the genome, as noted above.

These chromosome doubling experiments illustrate that interploidy crosses have multiple variables. One variable is the maternal to paternal ratio of genomes in the primary endosperm nucleus, but also varied are the amounts of cellular components contributed from the female gametophyte (central cell) to the endosperm developmental program, relative to those from the primary endosperm nucleus. In diploids, this relationship is 2:3 but with doubling of the primary endosperm nucleus, it becomes 2:6. The chromosome doubling experiments separate these variables. They keep the endosperm maternal to paternal genomic ratio the same, but change the relationship between the gametophytically supplied regulatory machinery for endosperm development and the dosage of the early endosperm genome itself. The collective evidence suggests the latter parameter is the critical one responsible for the interploidy hybridization barrier in maize.

What about parental imprinting? In Arabidopsis, crosses of diploid by tetraploid plants produce progeny with larger seeds than normally occur on diploids (Scott *et al.*, 1998; Adams *et al.*, 2000). This is apparently consistent with the resource allocation concept, in that paternal excess generates a larger endosperm containing protein and carbohydrate reserves. Mutations in genes that are parentally imprinted and that affect development of the endosperm will impact this response (Kinoshita *et al.*, 1999, 2004; Kiyosue *et al.*, 1999; Ohad *et al.*, 1999).

The finding of larger endosperms resulting from paternal excess appears to be ecotype specific to some degree (Dilkes and Comai, 2004) and, as noted above, there is endosperm collapse in both directions of maize interploidy crosses. Nevertheless, the impact of imprinting should not be dissociated from the hybridization barrier, because imprinting will de facto change the dosage contributions of the affected genes. Indeed, some evidence does in fact demonstrate a dosage impact of imprinted genes in Arabidopsis (Kradolfer *et al.*, 2013). However, as noted above, a large fraction of the maize genome has been examined for an effect of the absence of a paternal contribution on endosperm development. In no case is there evidence for sole paternal expression of any vital gene, because all regions tested produced viable seeds when missing paternal contributions. Furthermore, these kernels are without any qualitative effect on the endosperm in spite of the fact that many genes produce such effects when mutant. Collectively, these observations and considerations would seem to argue that gene imprinting is not a major contributor to the endosperm effects described here.

## 9.4  Future Directions

In this era of transcriptomics and single cell genomics, the potential now exists to identify the genes responsible for the small kernel effects in maize. The small kernel effect loci and their interactions have the potential to influence kernel size, and understanding the function of these loci could guide experiments to improve endosperm size and composition, or at least prevent defective endosperm formation. Yet, if as postulated, these effects are the endosperm manifestation of a "genomic imbalance" phenomenon, then the relative expression of interacting regulatory genes will need to be considered, and this presents a challenge to genetic engineering.

This volume is devoted to identifying areas of research that will advance our understanding of endosperm biology. With regard to the interploidy hybridization barrier, creative approaches to isolate and interrogate the effect of gene dosage variables during early endosperm development seem to be critical. Fertilization of the polar nuclei in the central cell of the megagametophyte to form the primary endosperm nucleus occurs in tissues encased deep in the maternal tissue, and thus is difficult to examine. However, laser-dissection methods to isolate single cells and follow their molecular features at the RNA and protein levels have been developed and can be applied to this question. Genetic and gene editing approaches to identify the contributing gene products could also be utilized. Transgenic studies that temporally express critical proteins and RNAs might also help define the developmental timing aspects that seem to be affected by genetic imbalances.

The endosperm of cereal grains contributes substantially to human nutrition, both directly and indirectly. Creative approaches to understand how dosage effects of critical genes in the endosperm are manifested, how their products interact with other gene products, and how and when stoichiometric relationships influence the outcome could move the field forward and help ensure world food security.

## Acknowledgment

Funding for work on this topic was provided by NSF grant IOS-1545780.

# References


Adams, S., Vinkenoog, R., Spielman, M., Dickinson, H.G. and Scott, R.J. (2000) Parent-of-origin effects on seed development in *Arabidopsis thaliana* require DNA methylation. *Development* 127, 2493–2502.

Bai, F. and Settles, A.M. (2015) Imprinting in plants as a mechanism to generate seed phenotypic diversity. *Frontiers in Plant Science* 5, 780.

Bai, F., Daliberti, M., Bagadion, A., Xu, M., Li, Y., *et al.* (2016) Parent-of-origin-effect rough endosperm mutants in maize. *Genetics* 204, 221–231.

Bauer, M.J. (2006) The interploidy hybridization barrier in *Zea mays* L. Ph.D. dissertation, University of Missouri-Columbia.

Beaudet, A.L. and Jiang, Y.-H. (2002) A rheostat model for a rapid and reversible form of imprinting-dependent evolution. *American Journal of Human Genetics* 70, 1389–1397.

Birchler, J.A. (1980) On the nonautonomy of the small kernel phenotype produced by B-A translocations in maize. *Genetical Research* 36, 111–116.

Birchler, J.A. (1993) Dosage analysis of maize endosperm development. *Annual Review of Genetics* 27, 181–204.

Birchler, J.A. (2014) Interploidy hybridization barrier of endosperm as a dosage interaction. *Frontiers in Plant Science* 5, 281.

Birchler, J.A. and Hart, J.R. (1987) Interaction of endosperm size factors in maize. *Genetics* 117, 309–317.

Birchler, J.A. and Veitia, R.A. (2010) The gene balance hypothesis: implications for gene regulation, quantitative traits and evolution. *The New Phytologist* 186, 54–62.

Birchler, J.A. and Veitia, R.A. (2012) Gene balance hypothesis: connecting issues of dosage sensitivity across biological disciplines. *Proceedings of the National Academy of Sciences of the United States of America* 109, 14746–14753.

Birchler, J.A., Bhadra, U., Bhadra, M.P. and Auger, D.L. (2001) Dosage-dependent gene regulation in multicellular eukaryotes: implications for dosage compensation, aneuploid syndromes, and quantitative traits. *Developmental Biology* 234, 275–288.

Birchler, J.A., Riddle, N.C., Auger, D.L. and Veitia, R.A. (2005) Dosage balance in gene regulation: biological implications. *Trends in Genetics* 21, 219–226.

Birchler, J.A., Yao, H., and Chudalayandi, S. (2007) Biological consequences of dosage dependent gene regulatory systems. *Biochimica et Biophysica Acta* 1769, 422–428.

Blakeslee, A.F. (1934) New Jimson weeds from old chromosomes. *Journal of Heredity* 25, 81–108.

Blakeslee, A.F., Belling, J. and Farnham, M.E. (1920) Chromosomal duplication and Mendelian phenomena in *Datura* mutants. *Science* 52, 388–390.

Bray, D. and Lay, S. (1997) Computer-based analysis of the binding steps in protein complex formation. *Proceedings of the National Academy of Sciences of the United States of America* 94, 13493–13498.

Brunelle, D.C. and Sheridan, W.F. (2014) The effects of varying chromosome arm dosage on maize plant morphogenesis. *Genetics* 198, 171–180.

Carlson, W.R. (1969) Factors affecting preferential fertilization in maize. *Genetics* 62, 543–554.

Chaudhuri, S. and Messing, J. (1994) Allele-specific parental imprinting of *dzr1*, a posttranscriptional regulator of zein accumulation. *Proceedings of the National Academy of Sciences of the United States of America* 91, 4867–4871.

Chettoor, A.M., Phillips, A.R., Coker, C.T., Dilkes, B. and Evans, M.M.S. (2016) Maternal gametophytic effects on seed development in maize. *Genetics* 204, 233–248.

Cong, B., Liu, J. and Tanksley, S.D. (2002) Natural alleles at a tomato fruit size quantitative trait locus differ by heterochronic regulatory mutations. *Proceedings of the National Academy of Sciences of the United States of America* 99, 13606–13611.

Cong, B., Barrero, L.S. and Tanksley S.D. (2008) Regulatory change in YABBY-like transcription factor led to evolution of extreme fruit size during tomato domestication. *Nature Genetics* 40, 800–804.

Cooper, D.C. (1951) Caryopsis development following matings between diploid and tetraploid strains of *Zea mays*. *American Journal of Botany* 38, 702–708.

Dilkes, B.P. and Comai, L. (2004) A differential dosage hypothesis for parental effects in seed development. *Plant Cell* 16, 3174–3180.

Dobzhansky, T. (1937) *Genetics and the Origin of Species*. Columbia University Press, New York.

Frary, A., Nesbitt, T.C., Grandillo, S., Knaap, E., Cong, B., *et al.* (2000) *fw2.2*: a quantitative trait locus key to the evolution of tomato fruit size. *Science* 289, 85–88.

Freeling, M. (2009) Bias in plant gene content following different sorts of duplication: tandem, whole-genome, segmental, or by transposition. *Annual Review of Plant Biology* 60, 433–453.

Freeling, M. and Thomas, B.C. (2006) Gene-balanced duplications, like tetraploidy, provide predictable drive to increase morphological complexity. *Genome Research* 16, 805–814.

Freeling, M., Lyons, E., Pedersen, B., Alam, M., Ming, R. and Lisch, D. (2008) Many or most genes in *Arabidopsis* transposed after the origin of the order Brassicales. *Genome Research* 18, 1924–1937.

Gutiérrez-Marcos, J.F., Costa, L.M., Biderre-Petit, C., Khbaya, B., O'Sullivan, D.M., *et al.* (2004) *maternally expressed gene1* is a novel maize endosperm transfer cell-specific gene with a maternal parent-of-origin pattern of expression. *The Plant Cell* 16, 1288–1301.

Gutiérrez-Marcos, J.F., Costa, L.M. and Evans, M.M.S. (2006) Maternal gametophytic *baseless1* is required for development of the central cell and early endosperm patterning in maize (*Zea mays*). *Genetics* 174, 317–329.

Haig, D. and Westoby, M. (1989) Parent specific gene expression and the triploid endosperm. *The American Naturalist* 134, 147–155.

Haig, D. and Westoby, M. (1991) Genomics imprinting in endosperm: its effect on seed development in crosses between species, and between different ploidies of the same species, and its implications for the evolution of apomixes. *Philosophical Transactions of the Royal Society of London B* 333, 1–13.

Henry, I.M., Zinkgraf, M.S., Groover, A.T. and Comai, L. (2015) A system for dosage-based functional genomics in poplar. *Plant Cell* 27, 2370–2383.

Johnston, S.A. and Hanneman, R.E., Jr. (1982) Manipulations of endosperm balance number overcome crossing barriers between diploid *Solanum* species. *Science* 217, 446–448.

Johnston, S.A., den Nijs, T.P.M., Peloquin, S.J. and Hanneman, R.E., Jr. (1980) The significance of genic balance to endosperm development in interspecific crosses. *Theoretical and Applied Genetics* 57, 5–9.

Kato, A. and Birchler, J.A. (2006) Induction of tetraploid derivatives of maize inbred lines by nitrous oxide gas treatment. *Journal of Heredity* 97, 39–44.

Kermicle, J.L. (1970) Dependence of the R-mottled phenotype in maize on mode of sexual transmission. *Genetics* 66, 69–85.

Kinoshita, T., Yadegari, R., Harada, J.J., Goldberg, R.B. and Fischer, R.L. (1999) Imprinting of the *MEDEA* polycomb gene in the Arabidopsis endosperm. *Plant Cell* 11, 1945–1952.

Kinoshita, T., Miura, A., Choi, Y., Kinoshita, Y., Cao, X., *et al.* (2004) One-way control of FWA imprinting in *Arabidopsis* endosperm by DNA methylation. *Science* 303, 521–523.

Kiyosue, T., Ohad, N., Yadegari, R., Hannon, M., Dinneny, J., *et al.* (1999) Control of fertilization-independent endosperm development by the *MEDEA* polycomb gene in *Arabidopsis*. *Proceedings of the National Academy of Sciences of the United States of America* 96, 4186-4191.

Kradolfer, D., Hennig, L. and Kohler, C. (2013) Increased maternal genome dosage bypasses the requirement of the FIS Polycomb repressive complex 2 in Arabidopsis seed development. *PLOS Genetics* 9, e1003163.

Leblanc, O., Pointe, C. and Hernandez, M. (2002) Cell cycle progression during endosperm development in *Zea mays* depends on parental dosage effects. *Plant Journal* 32, 1057–1066.

Lee, E.A., Darrah, L.L. and Coe, E.H. (1996) Dosage effects on morphological and quantitative traits in maize aneuploids. *Genome* 39, 898–908.

Lin, B.-Y. (1982) Association of endosperm reduction with parental imprinting in maize. *Genetics* 100, 475–486.

Lin, B.-Y. (1984) Ploidy barrier to endosperm development in maize. *Genetics* 107, 103–115.

Liu, J., Van Eck, J ., Cong, B. and Tanksley, S.D. (2002) A new class of regulatory genes underlying the cause of pear-shaped tomato fruit. *Proceedings of the National Academy of Sciences of the United States of America* 99, 13302–13306.

Maere, S., De Bodt, S., Raes, J., Casneuf, T., Van Montagu, M., Kuiper, M. and Van de Peer, Y. (2005) Modeling gene and genome duplications in eukaryotes. *Proceedings of the National Academy of Sciences of the United States of America* 102, 5454–5459.

Muller, H.J. (1942) Isolating mechanisms, evolution and temperature. *Biology Symposium* 6, 71–125.

Ohad, N., Yadegari, R., Margossian, L., Hannon, M., Michaeli, D., *et al.* (1999) Mutations in FIE, a WD polycomb group gene, allow endosperm development without fertilization. *Plant Cell* 11, 407–416.

Papp, B., Pal, C. and Hurst, L.D. (2003) Dosage sensitivity and the evolution of gene families in yeast. *Nature* 424, 194–197.

Rabinow, L., Nguyen-Huynh, A.T. and Birchler, J.A. (1991) A trans-acting regulatory gene that inversely affects the expression of the *white*, *brown* and *scarlet* loci in Drosophila. *Genetics* 129, 463–480.

Rhoades, M.M. and Dempsey, E. (1966) Induction of chromosome doubling at meiosis by the elongate gene in maize. *Genetics* 54, 505–522.

Roman, H. (1947) Mitotic nondisjunction in the case of interchanges involving the B-type chromosome in maize. *Genetics* 32, 391–409.

Roman, H. (1948) Directed fertilization in maize. *Proceedings of the National Academy of Sciences of the United States of America* 34, 36–42.

Scott, R.J., Spielman, M., Bailey, J. and Dickinson, H.G. (1998) Parent-of-origin effects on seed development in *Arabidopsis thaliana. Development* 125, 3329–3341.

Sheridan, W.F. and Auger, D.L. (2008) Chromosome segmental dosage analysis of maize morphogenesis using B-A-A translocations. *Genetics* 180, 755–769.

Thomas, B.C., Pedersen, B. and Freeling, M. (2006) Following tetraploidy in an *Arabidopsis* ancestor, genes were removed preferentially from one homeolog leaving clusters enriched in dose-sensitive genes. *Genome Research* 16, 934–946.

Veitia, R.A. (2002) Exploring the etiology of haploinsufficiency. *Bioessays* 24, 175–184.

Veitia, R.A., Bottani, S. and Birchler, J.A. (2008) Cellular reactions to gene dosage imbalance: genomic, transcriptomic and proteomic effects. *Trends in Genetics* 24, 390–397.

Waters, A.J., Makarevitch, I., Eichten, S.R., Swanson-Wagner, R.A., Yeh, C. T., *et al.* (2011) Parent-of-origin effects on gene expression and DNA methylation in the maize endosperm. *Plant Cell* 23, 4221–4233.

Xin, M., Yang, R., Li, G., Chen, H., Laurie, J., *et al.* (2013) Dynamic expression of imprinted genes associates with maternally controlled nutrient allocation during maize endosperm development. *Plant Cell* 25, 3212–3227.

Yao, H., Kato, A., Mooney, B. and Birchler, J.A. (2011) Phenotypic and gene expression analyses of a ploidy series of maize inbred Oh43. *Plant Molecular Biology* 75, 237–251.

Zhang, M., Zhao, H., Xie, S., Chen, J., Xu, Y., *et al.* (2011) Extensive, clustered parental imprinting of protein-coding and noncoding RNAs in developing maize endosperm. *Proceedings of the National Academy of Sciences of the United States of America* 108, 20042–20047.

Zhang, M., Xie, S., Dong, X., Zhao, X., Zeng, B., *et al.* (2014) Genome-wide high resolution parental-specific DNA and histone methylation maps uncover patterns of imprinting regulation in maize. *Genome Research* 24, 167–176.

# 10 Cell Cycle and Cell Size Regulation during Maize Seed Development: Current Understanding and Challenging Questions

Paolo A. Sabelli*

*School of Plant Sciences, University of Arizona, Tucson, Arizona, USA*

## 10.1 Introduction

Formation of the maize seed and that of related cereals occurs through coordination of different biological processes, including cell proliferation, cell fate specification, endoreduplication, cell differentiation, accumulation of storage metabolites, and programmed cell death (PCD). Development of the three genetically distinct seed compartments, the sporophyte (i.e. the embryo), the triploid endosperm, and the maternal pericarp, involves extensive crosstalk and tight regulation between and within maternal and filial structures, with genetic, epigenetic, and environmental factors playing important roles. The objective of this chapter is to provide a perspective on the roles of cell cycle and cell size regulation during maize seed development, with an emphasis on what is not yet understood about these processes. Only a cursory overview of the state of our knowledge of cell cycle control and cell size regulation will be provided, given that they have been extensively reviewed elsewhere (Sabelli and Larkins, 2009b; Dante *et al.*, 2014a; Sabelli, 2014). While considerable information has been obtained about the molecular control of the cell cycle in maize endosperm and its interplay with cell size regulation, much less is known regarding the embryo and pericarp. Cell cycle regulatory mechanisms in the latter two tissues are very much an open question. Consequently, this chapter focuses primarily on the endosperm, where a relatively coherent picture is emerging.

## 10.2 An Overview of Cell Cycle and Cell Size Patterns during Maize Seed Development

Seed development typically begins with a double fertilization within the central cell of the female gametophyte, where one of the two sperm nuclei fuses with the egg cell nucleus and the other fuses with two polar nuclei (Bleckmann *et al.*, 2014). As a result, the embryo (originating from the former fertilization) is diploid (2n, 2C, with n indicating the number of chromosomes and C the DNA content of the haploid nucleus), whereas the endosperm (originating from the latter fertilization) is triploid (3n, 3C) (Sabelli and Larkins, 2009c; Dante *et al.*, 2014a). Following fertilization, the endosperm primary nucleus is the first filial component to resume cell cycle activity, and it does so before the zygote begins cell division. The DNA in the primary

---

*Corresponding author e-mail: psabelli1@gmail.com



"

endosperm nucleus replicates and divides mitotically, and synchronously, a number of times, producing many nuclei (up to 512 have been reported) (Leroux *et al.*, 2014). The mitotic cell cycles of the early endosperm nuclei are uncoupled from cytokinesis (until approximately 2–4 days after pollination (DAP)), resulting in a single multinucleate cell (i.e. a syncytium) (Lopes and Larkins, 1993; Olsen, 2004).

Meanwhile, after a cell cycle hiatus that apparently involves polarization of the cell contents and extensive cytoskeletal rearrangements, the zygote begins proliferating through mitotic cell divisions, following a highly conserved pattern among angiosperms (Olsen, 2004; Sabelli, 2012b). The initial zygotic cell division is asymmetric and results in a cytoplasm-dense apical cell at the chalazal end of the embryo sac, and a vacuolated basal cell toward the micropyle. The fates of these two cells are clearly distinct, as the apical cell gives rise to a cluster of small, dense cells comprising the proembryo and later the embryo proper, whereas the basal cell generates a series of large, highly vacuolated cells, the ephemeral suspensor, involved in nourishing the growing embryo. Past the proembryo stage, though, the pattern of embryonic cell division becomes rather unpredictable, but it typically results in rapid growth of the single cotyledon, the scutellum, at the expense of the embryo proper, which becomes laterally displaced off the embryo axis (Randolph, 1936) (Chapter 7, this volume). Throughout seed development, embryo cells proliferate through conventional mitotic cell cycles and remain small with a dense cytoplasm.

As in other members of the Poaceae family, the fruit coat of the maize caryopsis (the pericarp) is fused to the testa. Initially, this structure grows through a phase of cell division that is followed, particularly in the placento-chalazal cell layers, by cell elongation and PCD. Cell expansion and PCD in the pericarp are roughly concomitant with cellularization of nuclear domains in the endosperm (Olsen, 2004). These processes in the pericarp, similar to those described in wheat and barley (Domínguez and Cejudo, 2014; Radchuk *et al.*, 2011), depend on fertilization, which presumably underscores the presence of signals linking the development of maternal seed tissue to those of filial tissue (Kladnik *et al.*, 2004). However, PCD patterns differ between maize and members of the Triticeae, because of anatomical differences influencing the distribution of cell populations charged with ensuring nutrient flux from maternal to filial tissues. In any case, it appears that PCD in these cells is completed before onset of the storage metabolite deposition phase, and the ensuing redistribution of their metabolites could in fact support the accumulation of storage compounds in the endosperm.

In most cases, at around 3–4 DAP the endosperm syncytium becomes cellularized through an inward alveolation process that involves repeated rounds of anticlinal cell wall formation and expansion, nuclear division, and periclinal cell wall deposition, followed by a less organized partitioning of the residual central vacuole, until the central cavity is completely filled with cells (Monjardino *et al.*, 2007; Leroux *et al.*, 2014). Interestingly, cytoskeletal rearrangements during alveolar cell divisions occur in the absence of the pre-prophase band (PPB) that typically marks the site of cell wall deposition early in the M-phase of the somatic cell cycle (Olsen, 2001, 2004; Sabelli and Larkins, 2009b). Complete cellularization of the endosperm takes place relatively fast, in about one day. From approximately 4 DAP, the endosperm grows through mitotic cell divisions, which generate the vast majority of endosperm cells eventually found in mature seeds (Kiesselbach, 1949; Kowles and Phillips, 1985; Lur and Setter, 1993). By the end of the mitotic phase of endosperm development (~10–12 DAP), the maternally-derived nucellus is completely degraded (through PCD) and reabsorbed, while the size of the endosperm relative to that of the whole caryopsis increases 20–30-fold (Leroux *et al.*, 2014). Residual mitotic activity persists in the peripheral cell layers, the aleurone and the sub-aleurone, until later (~20–25 DAP) (Kowles and Phillips, 1988).

Following the mitotic phase of endosperm development, in a process that starts at the center of the endosperm and progresses toward its periphery, cells asynchronously switch cell cycle mode from mitotic cell division to endoreduplication (also known as endocycle or endoreplication), which involves reiterated rounds of nuclear DNA synthesis in the absence of chromatin condensation, sister chromatid segregation, and cytokinesis (Kowles and Phillips, 1985; Larkins *et al.*, 2001; Sabelli, 2012a). Endoreduplication generates endosperm cells with increased DNA content (up to and in excess of 200C) that appears to contain uniform chromosomal complements (Bauer and Birchler, 2006). At any given time from around 10–12 DAP up to approximately 20 DAP, there is a ploidy gradient in the endosperm with inner cells (highly endoreduplicated) having the largest DNA contents and the outer cells (non- or lowly endoreduplicated) the smallest. This gradient is proportional to a corresponding gradient in cell size, with very large, inner endosperm cells and progressively smaller cells towards the periphery (Kowles and Phillips, 1988; Larkins *et al.*, 2001) (Fig. 10.1). The expansion in cell size accounts for most of endosperm growth (Vilhar *et al.*, 2002) and the increase in seed size during the endoreduplication phase. In addition to increases in cell and tissue volumes, temporal-spatial patterns of endosperm endoreduplication are virtually superimposed with those of starch and seed storage protein accumulation. The long-observed association between increased endosperm cell ploidy and synthesis, and deposition of storage compounds suggests a causal relationship between endoreduplication, cell and tissue growth, and storage metabolite accumulation (Kowles and Phillips, 1985; Larkins *et al.*, 2001; Kowles, 2009; Sabelli and Larkins, 2009b,c). However, definite evidence linking these processes has remained elusive, primarily because it has been experimentally challenging to disentangle endoreduplication from these other processes. In addition, PCD takes place in the starchy endosperm, noticeably from around 16 DAP, with a pattern that roughly follows that of endoreduplication and starch deposition (Domínguez and Cejudo, 2014; Sabelli, 2012a; Young and Gallie, 2000), further complicating our understanding of the precise role played by endoreduplication during endosperm development.

## 10.3 Molecular Regulation of Endosperm Cell Division and Endoreduplication: Unsettled Issues

The cell division cycle entails three important stages: (i) the G1/S-phase transition leading to DNA synthesis; (ii) the G2/M-phase transition leading to chromosome segregation; and (iii) the induction of cytokinesis (Dante *et al.*, 2014a). Both the G1/S- and G2/M-phase transitions are controlled by activities of different cyclin-dependent kinase (CDK) complexes made of catalytic kinase and regulatory cyclin subunits that control downstream processes through phosphorylation of a variety of protein substrates (Inzé and De Veylder, 2006). CDKs are themselves regulated by phosphorylation and dephosphorylation by upstream kinases and phosphatases, and by specific inhibitors known as CKIs. Cyclin abundance is controlled at the level of gene transcription and also by the ubiquitin-proteasome system (UPS). Periodic fluctuations in the activity of different CDK/CYC complexes drive cells through the unidirectional transitions that characterize the cell cycle (Fig. 10.1A).

CDKs can be classified into two major groups: CDKA, which are involved primarily in the regulation of the G1/S-phase transition but also the G2/M-phase transition; and CDKB, which appear to specifically control M-phase entry. Additional CDK types, known as CDK-activating kinases (CAKs), function upstream of and regulate CDKAs and CDKBs. In maize, members of the CDKA, CDKB, and cyclin families of genes and proteins have been identified and characterized during endosperm development in some detail (Sun *et al.*, 1999b; Leiva-Neto *et al.*, 2004; Sabelli *et al.*, 2013; Dante *et al.*, 2014b; Sabelli *et al.*, 2014), as well as a WEE1 kinase that inhibits CDK activity (Sun *et al.*, 1999a). Among other CDK-specific inhibitors are several members of the CKI family (Coelho *et al.*, 2005).

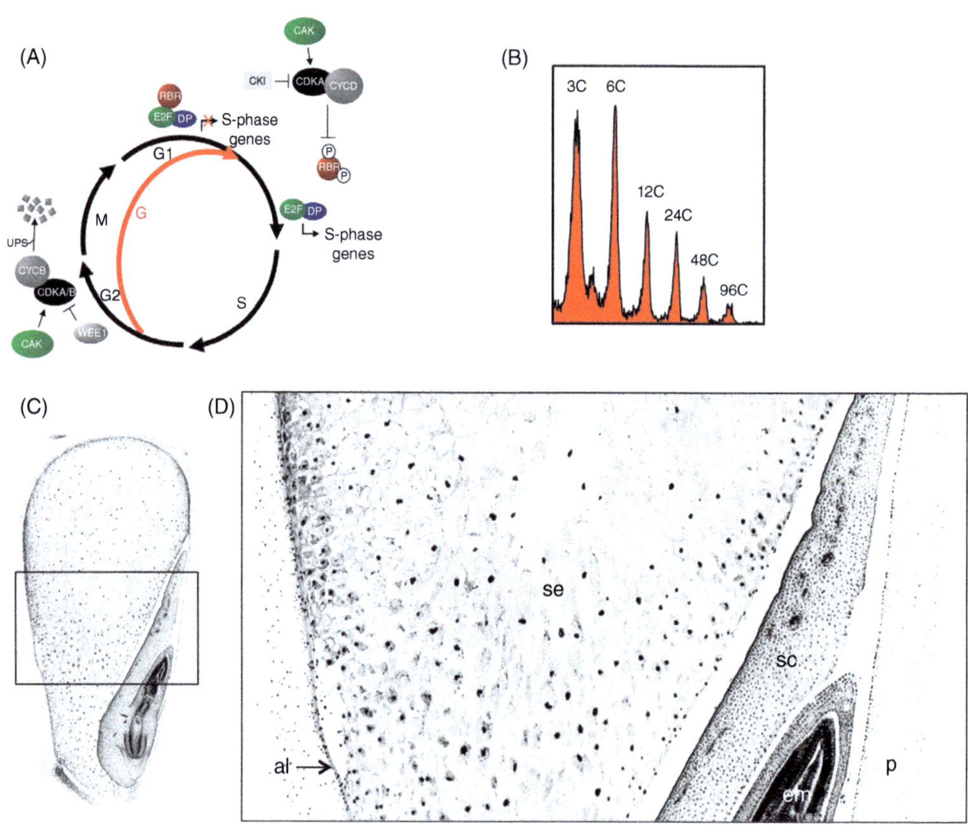



**Fig. 10.1.** Key features of cell cycle regulation and cell size patterns in maize endosperm. (A) Simplified diagram showing regulation of G1/S-phase and G2/M-phase transitions during the mitotic cell cycle by periodic fluctuations in the activities of specific CDK/CYC complexes. Upstream regulation by CAK, WEE1, and CKIs is shown. Rapid degradation of cyclin subunits by the UPS results in a sudden drop in CDK activity and exit from M-phase. The RBR pathway plays an important role in regulating E2F/DP-dependent gene expression of S-phase genes and is a target of CDK activity. The red arrow indicates the endoreduplication cell cycle, which is characterized by alternating G- and S-phases without intervening mitosis. (B) Typical flow-cytometric profile of 16-DAP endosperm showing discrete populations of endoreduplicated nuclei with ploidies greater than 6C. (C) Longitudinal median section of a 19-DAP kernel stained with 4′,6-diamidino-2-phenylindole (DAPI) to reveal nuclei. (D) Detail of boxed area in (C) showing a clear gradient in nuclear and cell sizes between the peripheral aleurone cell layer (al) and the inner starchy endosperm (se) cells. The pericarp (p), scutellum (sc), and embryo axis (em) are indicated. See main text for details. Panels (C) and (D) are modified from Sabelli *et al.* (2013).

Downstream of CDKA, the retinoblastoma-related (RBR) pathway plays a paramount role in controlling DNA synthesis, primarily by inhibiting E2F-related transcription factors required for expression of a plethora of S-phase genes (Inzé and De Veylder, 2006; Sabelli and Larkins, 2009a) (Fig. 10.1A). Maize possesses at least two RBR gene types, *RBR1* and *RBR3*, each with nearly identical paralogs, *RBR2* and *RBR4,* respectively (Grafi *et al.*, 1996; Sabelli *et al.*, 2005, 2009, 2013; Sabelli and Larkins, 2009a). In contrast with *Arabidopsis*, which possesses only one RBR gene with a clear inhibitory cell cycle role, the complexity of the RBR gene family in maize (and related cereals) suggests specialization of function among its members (Sabelli and Larkins, 2006). Indeed, it was shown that RBR1 and RBR3 have complementary expression patterns during endosperm development: RBR1 represses RBR3 expression (most likely through inhibition of E2F), and RBR3, surprisingly, could play a positive role in expression of MCM genes, DNA synthesis and cell cycle regulation (Sabelli *et al.*, 2005, 2009, 2013; Sabelli and Larkins, 2006).

With regard to how these genes and encoded proteins impact maize endosperm development, investigation has focused on studying gene expression patterns but also, more recently, on functional studies, particularly with regard to the endoreduplication phase. Consistent with an evolutionarily conserved paradigm, endoreduplication in the endosperm involves a downregulation of M-phase-specific CDK activity and associated programs, and sustained oscillations in S-phase-specific CDK activity. This would ensure reiterated rounds of complete DNA replication, followed by licensing and firing of origins of DNA replication in the absence of mitosis and cytokinesis (Grafi and Larkins, 1995; Dante *et al.*, 2014a). There is evidence that core cell cycle genes, such as *CDKA;1*, cyclins, and *RBR1*, affect cell proliferation and endoreduplication during maize endosperm development according to accepted models. However, uncharacteristic aspects of cell cycle regulation have also been revealed. For example, as mentioned, RBR1 represses expression of RBR3 (Sabelli *et al.*, 2005, 2013), and the relevance of this remains largely unexplained. RBR1, in addition, integrates different pathways and processes important for endosperm development and, through an unknown mechanism, appears to specifically couple nuclear DNA content to cell size. Also, although CDKA;1 and RBR1 function in the same pathway to control endoreduplication, they do not do so with regard to RBR1-dependent gene expression programs (Sabelli *et al.*, 2013). Furthermore, CDKA;1-associated kinase activity is negatively regulated by RBR1 through a feedback loop, providing an additional level of regulation of this pathway. Another largely unexplained observation is that endoreduplication is associated with reduced degradation of different types of cyclins by the UPS (Dante *et al.*, 2014b; Sabelli *et al.*, 2014).

Together, these observations suggest that the interplay between CDKs, cyclins, RBRs and the UPS during endosperm development is considerably more nuanced than previously thought. They also highlight areas where future research could make significant inroads towards understanding the role of cell cycle regulation in seed development.

## 10.4 Other Outstanding Questions

### 10.4.1 The relationship between the coenocytic phase and cellularization

In cereals such as maize, the rate of seed growth and its sink capacity are highly correlated with endosperm cell number and the number of the starch granules within. Both parameters, in turn, depend primarily on the mitotic phase of endosperm development (Brocklehurst, 1977; Reddy and Daynard, 1983; Jones *et al.*, 1996; Sabelli and Larkins, 2009b; Dante *et al.*, 2014a; Leroux *et al.*, 2014). This phase encompasses two distinct stages temporally separated by the period of endosperm cellularization/partition characterized by: (i) acytokinetic proliferation of the primary endosperm nucleus; and (ii) cell proliferation in which mitosis is coupled to cytokinesis. Nuclei generated in the coenocytic endosperm create the founder cells that go on to proliferate and eventually populate the whole endosperm. Thus, the extent of the coenocytic phase also correlates with endosperm and seed growth. Premature endosperm cellularization, one that occurs with fewer nuclei than normal, tends to result eventually in smaller kernels (Dante *et al.*, 2014a; Sekhon *et al.*, 2014), although the size of the coenocytic endosperm is inversely correlated with the number of nuclei at cellularization (Leroux *et al.*, 2014). Delayed cellularization, on the other hand, can be associated with larger endosperm size, but both cases of abnormal cellularization timing can lead to seed abortion. Although the corresponding information is lacking for maize, the timing of cellularization in rice endosperm appears to depend on correct expression patterns of core cell cycle genes, such as *E2F1*, *CYCB2;2*, and *CYCB1;1* (Guo *et al.*, 2010; Ishimaru *et al.*, 2013). Interestingly, there is considerable variation in the number of nuclei in the maize coenocyte, even within the same genotype (e.g. 128–512 nuclei in the B73 inbred, see Leroux *et al.*, 2014); thus, this feature does not appear to be under strict genetic control.

Studies mostly carried out in *Arabidopsis* and related species indicate that timing of endosperm cellularization is strongly dependent on epigenetic mechanisms and the relative proportion of maternal versus paternal genome complements or the proportion of parentally-expressed genes (Gehring and Satyaki, 2017; see also Chapter 9, this volume). Maternal genomic excess or a prevalence of maternally-expressed genes leads to precocious cellularization, whereas delayed or even absent cellularization could depend on paternal genomic excess and/or a prevalence of paternally-expressed genes. Imprinted genes involved in the regulation of endosperm cellularization have been identified in *Arabidopsis*, however, they are poorly conserved in maize and other species (Xin *et al.*, 2013; Bai and Settles, 2015); thus, it is doubtful whether *Arabidopsis* represents a reliable system for extrapolating conclusions concerning the regulation of endosperm cellularization in the grasses. Further insight into the role played by endosperm syncytial development in seed growth could conceivably be achieved through forward genetics experiments aimed at manipulating the rates of nuclear proliferation in the coenocyte and/or the onset of cellularization.

### 10.4.2 Is endoreduplication necessary for endosperm and seed development?

To the best of our knowledge and within the limits of sensitivity of a number of early assays, increased ploidies in the endosperm result entirely from complete duplication of the entire triploid genome, with each chromosome containing multiple chromatids (Kowles and Phillips, 1985, 1988; Kowles, 2009; Dante *et al.*, 2014a). There is no evidence for preferential DNA replication, underreplication, or alternative modes of cell cycle activity leading to different forms of polyploidy (Bauer and Birchler, 2006; Edgar *et al.*, 2014).

There is no indication the endosperm or seed can develop normally without endoreduplication, suggesting perhaps this cell cycle type is crucial to endosperm/seed development in some, still unclear, ways. Early investigations uncovered genotypes with greatly reduced or abolished endoreduplication, but they generally belong to the defective kernel (*dek*) mutant class and so do not undergo normal seed development (Kowles *et al.*, 1992). However, the exact function(s) of endoreduplication during maize endosperm development has not yet been established. Through reverse genetics experiments targeting CDKA;1 activity, the level of endoreduplication was significantly reduced, but the endosperm and the seed seemingly developed normally (Leiva-Neto *et al.*, 2004). More recently, downregulation of RBR1 in the *RBR1DS1* mutant resulted in increased cell proliferation and endoreduplication and yet in spite of substantial changes in gene expression, PCD activity, and nuclear and cell size patterns, both the endosperm and seed generally developed normally (Sabelli *et al.*, 2013). This is important, because seed dry mass and the rate of grain filling depend to a large extent on the number of starch granules and their relative size. Although starch granule number increases during the cell division phase of endosperm development and remains essentially constant thereafter, the increase in granule size is correlated with the endoreduplication/cell enlargement phase. Thus, it is surprising that the increases in cell number and endoreduplication in the *RBR1DS1* mutant were not reflected in enhanced grain mass or size. These observations suggest action of compensatory mechanisms that preserve tissue homeostasis, and they argue against a fundamental role for endoreduplication in endosperm development. But for reasons that cannot be ruled out—related to the pleiotropic effects of the CDKA;1–RBR1 pathway that potentially couple cell cycle activity with cell size control—this interpretation cannot be considered conclusive (Sabelli *et al.*, 2013). Consequently, several important questions concerning endosperm endoreduplication remain to be clarified:

### a) Does endoreduplication enable increased gene expression?

According to one long-standing theory, increased amounts of nuclear DNA resulting from endoreduplication potentially provide more DNA templates for upregulated transcription and gene expression. In addition, endoreduplicated chromosomes appear to be loosely polythenic (Bauer and Birchler, 2006), suggesting that relaxed chromatin could facilitate increased gene transcription. This could account for the dramatic increase in storage protein gene expression that occurs during mid-endosperm development (see Chapter 14, this volume). However, downregulation of CDKA;1 activity in the *CDKA1DN* mutant (which inhibited endoreduplication) (Leiva-Neto *et al.*, 2004) and that of RBR1 (which stimulated endoreduplication) (Sabelli *et al.*, 2013) did not provide evidence supporting this hypothesis. In the former, analysis of a selected pool of genes involved in seed storage compound accumulation did not show diminished transcriptional activity, while in the latter the quantity of proteins accumulated was actually inversely correlated with nuclear DNA content. These observations suggested that, on a whole-endosperm basis, manipulating the extent of endoreduplication has no proportional effect on gene expression, or protein synthesis and accumulation.

Why not? In the case of RBR1 downregulation, it appeared that on a DNA basis, the extra complements of endoreduplicated DNA were not as functionally active as wild type. Additionally, both nucleus and cell size were significantly reduced in spite of increased endoreduplication. In fact, the karyoplasmic ratios were virtually identical in wild type and *RBR1DS1* mutant endosperm. Although the results did not lead to any conclusive explanation, they prompted speculation that endoreduplicated chromatin was possibly more densely compacted in the *RBR1DS1* mutant and thus was less transcriptionally active than wild type. But why was cell size also decreased? Was this because cell size in endoreduplicated cells depends not on the amount of DNA but rather on nuclear size and the level of chromatin condensation? While these and other questions could be addressed through further analysis of the *CDKA1DN* and *RBR1DS1* mutants, there is the potential caveat that the DNA affected by their altered endoreduplication levels might not be functionally equivalent to the endoreduplicated DNA produced during normal endosperm development. Indeed, if there are pleiotropic effects of CDKA;1 and RBR1 on the regulation of chromatin structure and functionality, besides those specific to cell cycle regulation and endoreduplication in particular, then it could become prohibitive to conceptually disentangle these processes and extrapolate precise conclusions concerning the role of endoreduplication in endosperm development. One interesting corollary of the RBR1 downregulation investigation is that nuclear size, at least under certain conditions, might not directly reflect the amount of DNA. Thus, it would be sensible to use caution in deducing DNA content levels from measurements of nuclear size, a rather common practice, without the support of direct ploidy measurements.

### b) Does endoreduplication drive cell enlargement, or vice versa?

A correlation between nuclear DNA content and cell size has long been observed in numerous plant and animal systems, including endosperm cells in maize and related cereals (Larkins *et al.*, 2001; Kowles, 2009; Sabelli and Larkins, 2009b; Sabelli, 2012a; Edgar *et al.*, 2014). Endoreduplication is often believed to be essential for attainment of the final size and full functionality of certain differentiated cell types. Different mechanisms contribute to enlargement and growth of plant cells, which are typically vacuolated and possess a rigid cellulosic wall (Sugimoto-Shirasu and Roberts, 2003). Cell growth usually involves an increase in cytoplasmic mass, driven by macromolecule biosynthesis and accumulation. Cell expansion, instead, is an increase in cell volume, and is generally brought about by an increase in cell turgor pressure, expansion of the vacuole(s), relaxation of the cell wall and deposition of new cell wall material, and by related cytoskeleton rearrangements. Cell growth and expansion need not be mutually exclusive. So, are mechanical forces responsible for the endoreduplication and cell size patterns observed in the endosperm? In plants, mechanical forces originating in the rigid cell wall counteract the drive towards cell expansion upon increased turgor pressure and generally constrain cell size (Sablowski, 2016). But do they impinge on cell cycle activity and regulate endoreduplication, particularly in the endosperm?

Mature starchy endosperm cells are atypical in that they are virtually devoid of large vacuoles and possess very thin cell walls (Otegui, 2007; Burton and Fincher, 2014). Cells grow, to a limited degree, during the mitotic phase of development, which appears to be due mainly to a cytoplasmic mass increase. But this has very little impact on the size of the endosperm. However, these cells subsequently undergo endoreduplication and enlarge primarily through a process distinct from growth and/or expansion, as strictly defined above. Instead, they become filled with starch granules and storage protein bodies. Some experiments designed to shed light on the relationship between endoreduplication and cell and endosperm size, gene expression, and storage metabolite accumulation have been carried out. The downregulation of CDKA;1 and RBR1 affects endoreduplication levels on a tissue-wide basis in a manner coherent with the predicted cell cycle functions of these proteins (Leiva-Neto *et al.*, 2004; Sabelli *et al.*, 2013). However, there is no indication cell sizes were affected in a directly proportional way in these mutants. In the *RBR1DS1* mutant, starchy endosperm cells were significantly smaller, with smaller nuclei compared to wild type. The relationship between nuclear and cell size was not affected by RBR1 downregulation, which supports the karyoplasmic theory that states there is a fixed relationship between the size of the nucleus and that of the cell. However, ploidy increases in the mutant were practically uncoupled from nuclear and cell size. Thus, RBR1 appears to play a role in coupling DNA content to nuclear and cell sizes, possibly by regulating chromatin condensation. During endosperm development, a scenario in which active RBR1 helps retard the endoreduplication cell cycle while relaxing chromatin conformation, thus enabling high levels of gene expression, is conceivable and supported by a number of observations. First, RBR1 expression increases during mid-development of wild-type endosperm. Second, downregulation of RBR1 results in increased ploidy, but there is virtually no change in expression and accumulation of endosperm proteins relative to wild type, suggesting that additional endoreduplicated chromatin is transcriptionally inactive. Third, expression of the DNA methyltransferase *MET1* homolog, *DMT101*, is upregulated in the *RBR1DS1* mutant, which could contribute to gene silencing (Sabelli *et al.*, 2013). In fact, it is known that portions of the genome, particularly those containing genes preferentially or specifically expressed in the endosperm, are normally hypomethylated in both rice (Zemach *et al.*, 2010) and maize (Chen *et al.*, 2014; Wang *et al.*, 2015; Waters *et al.*, 2011). What remains to be determined are the genome-wide characterization of gene expression changes and chromatin modification patterns in the *RBR1DS1* mutant. Technologies, such as RNA-seq, the utilization of DNA methylation microarrays and MeDIP-seq can be used to attain a more complete picture of how endoreduplication in the *RBR1DS1* mutant affects gene expression and chromatin organization. It would also be helpful, albeit quite challenging, to couple these approaches with nuclear and/or cell sorting, based on DNA content, to increase sample homogeneity.

Analysis of the *RBR1DS1* mutant suggests endoreduplication could drive cell enlargement, but only if it is mediated by an increase in nuclear volume. If so, it would seem an increase in nuclear volume following the increase in ploidy is a "cell autonomous" process. But it is also possible non-autonomous cell processes are involved. Constraint by surrounding cells could be a factor, but if that were the case one would perhaps expect to find cells with the least DNA, the smallest nuclei, and the smallest volume, deep in the innermost area of the endosperm, where constraining mechanical forces from surrounding tissue would be the strongest, and the presence of a gradient for increasingly endoreduplicated and enlarged nuclei and cells towards the periphery of the endosperm. But exactly the opposite is observed.

The ploidy/nuclear size/cell size spatial patterns normally observed can perhaps be explained by assuming that, in a cell-autonomous manner, inner cells transition to the endoreduplication cycle earlier than outer cells and that, at any developmental time, they simply have more time to undergo additional rounds of endoreduplication and more enlargement. However, it is also possible cells of the relatively thick-walled aleurone and pericarp exert mechanical force on the adjacent inner cells, following a gradient that decreases centripetally. This, coupled to lower rigidity of the thin walls characteristic of starchy endosperm cells, would enable the innermost endosperm cells to reach larger sizes. Reduced cell numbers in the peripheral endosperm was shown to be associated with decreased cell volume, at any given ploidy, of inner starchy cells in the *miniature1* (*mn1*) mutant; this is presumably due to more severe mechanical constraints (Vilhar *et al.*, 2002). Indeed, one interesting experiment that could shed light on these alternative scenarios involves specifically stimulating cell proliferation (and/or cell elongation) in the aleurone and sub-aleurone layers and/or the seed coat, aiming to relieve mechanical constraints on inner endosperm cells. It is intriguing that as early as the cellularization/partitioning phase of endosperm development, and clearly before the onset of endoreduplication, a cell size gradient is observed, with larger inner cells than outer cells (Leroux *et al.*, 2014). The cause of this is not clear, and both physical constraints and developmental gradients could be at play. It is also puzzling that popcorn has small seeds and a disproportionately thick pericarp, while containing high-ploidy starchy endosperm cells. There is no information about spatial distribution patterns of cell size and endoreduplication in popcorn lines, but it is conceivable that physical constraints inhibit inner cell expansion, regardless of the amount of nuclear DNA in the cells (Dilkes *et al.*, 2002; Coelho *et al.*, 2007).

Does cell size determine ploidy level? In budding yeast, mechanisms that sense cell size are implicated in control of gene expression in polyploid cells (Wu *et al.*, 2010). In agreement with the karyoplasmic theory, the volume of endosperm cells appears to increase proportionally with nuclear DNA content. However, assuming endosperm cells are roughly isodiametric, there is a decrease in cell area/volume and nucleus area/volume ratios in endoreduplicated, large cells. Processes that depend on crosstalk between the nucleus and cytoplasm, such as nuclear uptake of metabolites necessary to carry out DNA and RNA syntheses, mRNA export, and nutrient influx into the cytoplasm from the surrounding environment to support mitosis, would be negatively affected by a decreased surface area/volume ratio. Conceivably, these limitations could reach a threshold where DNA synthesis could no longer be supported, thereby bringing endoreduplication to a halt. According to this view, cell size increase would ultimately undermine the biosynthetic processes that rely on ratios of nuclear and cell surface areas and their corresponding volumes. It is interesting in this respect to consider the case of the jelly-like locular tissue of the tomato fruit mesocarp, which envelops the seeds and undergoes extensive endoreduplication, and a dramatic increase in cell size. In these cells, the nuclear envelope is highly invaginated, thus creating a larger surface area that could enhance molecular trafficking with the surrounding cytoplasm, meeting the demands of the large nucleus and cell. Also, the nuclear membrane grooves appear to be densely populated by mitochondria, which could play a key role in linking endoreduplication to high metabolic output (Bourdon *et al.*, 2012). It would be interesting to investigate whether a similar situation characterizes endoreduplicated endosperm cells in maize and related cereals.

There are also additional confounding factors in considering the relationship between DNA content, the surface areas and volumes of the nucleus and the cell, and the biosynthetic potential of endoreduplicated endosperm cells. Although the karyoplasmic ratio is maintained in these cells as they increase in size, on a unit DNA basis the net amount of cytoplasm (and thus its biosynthetic capability) actually decreases as starch granules and storage protein bodies increasingly occupy the cell volume. The reprogramming of biosynthetic activity as endosperm cells transition to endoreduplication at the beginning of the grain filling period could indicate, on the one hand, that endoreduplication is a primary strategy to maximize carbon/energy sequestration away from costly processes such as mitosis, cytokinesis and cell wall deposition, or on the other hand, that it has built in mechanisms by which reiterated DNA synthesis cannot be supported beyond a certain threshold. Although most of the starch forming apparatus is located in the amyloplast (i.e. the developing starch granule), the key rate-limiting enzyme, adenosine diphosphate glucose pyrophosphorylase (AGPase), is largely cytoplasmic (Hannah, 2007; see also Chapter 12, this volume). Thus, all else being equal, starch accumulation could conceivably still be subject to the constraints mentioned above that are imposed by reduced cytoplasmic volumes.



Endoreduplication and associated endosperm cell enlargement precede the grain-filling phase of seed development, when metabolic resources are prioritized for synthesis and accumulation of starch and storage proteins. Having relatively few but large cells minimizes total surface area, compared with having many small cells. Thus, endoreduplication could be viewed in this context as an effective means to achieve cell enlargement and conserve energy/resources by limiting highly energy-dependent cellular processes, such as chromatin condensation, mitosis, and cytokinesis. Also, maize endosperm cell walls have relatively little cellulose and are remarkably thin (Otegui, 2007; Burton and Fincher, 2014), indicating that demand for cell wall biosynthesis is kept to a minimum. Interestingly, ATP levels are higher in inner endosperm cells and lower in peripheral cells, following a gradient similar to those of endoreduplication and cell size. Thus, endoreduplication could be part of an adaptive mechanism to maximize storage compound synthesis, especially given that the starchy endosperm is an hypoxic environment that theoretically could restrict metabolic output (Rolletschek *et al.*, 2005; Chapter 11, this volume).

A comparison with the emerging grass model, *Brachypodium distachyon*, could shed light on the interplay between cell cycle regulation, endoreduplication, cell expansion and the form in which carbohydrate is stored in the endosperm (Trafford *et al.*, 2013). *Brachypodium* endosperm cells, relative to those of its close relative, barley, are characterized by reduced expansion, low levels of starch accumulation, and extremely thickened walls; also, cell proliferation is reduced and endoreduplication is virtually lacking. The dramatically reduced expression of *CDKB1* and *CYCA3* in *Brachypodium* is consistent with its low level of cell cycle activity (Trafford *et al.*, 2013). In contrast with maize and related cereals, cell walls in *Brachypodium* endosperm contain large amounts of 1,3;1,4-β-glucan, the main carbohydrate storage component, and are the principal determinant of endosperm enlargement; expression of starch biosynthetic genes in *Brachypodium* is downregulated. Analysis of the differences between maize and *Brachypodium* suggests there might be mutually exclusive programs in grass endosperm linking carbohydrate metabolism, endoreduplication, and cell size. On the one hand, endoreduplication, the starch/1,3;1,4-β-glucan ratio, and cell size are all positively correlated in cereals like maize, whereas in *Brachypodium* high levels of

1,3;1,4-β-glucan compared to starch are associated with only minor increases in cell size and greatly reduced, if any, endoreduplication. It is not known whether the shift in carbohydrate storage from starch granules to cell wall polymer is responsible for differences in cell cycle and cell size patterns between maize and *Brachypodium* endosperm, or vice versa. Regardless, these observations suggest that pathways regulating carbohydrate metabolism, cell cycle, and cell size in the endosperm are intertwined. Indeed, in the *miniature1* (*mn1*) mutant, loss of the cell-wall-bound invertase INCW2, which is critically important during the mitotic phase of endosperm development, results in a substantial decrease in glucose level, cell division activity and endosperm/seed size (Vilhar *et al.*, 2002). It would be very interesting to study in detail the relationship between cell cycle regulation, cell size, and storage compound accumulation in maize endosperm. Stacking mutations that affect these processes could create the necessary genetic resources for such an undertaking.



Is there a role for core cell cycle genes and endoreduplication in the regulation of PCD in maize endosperm? It is well established that PCD in this tissue begins around 16 DAP (but the underlying molecular mechanisms are activated earlier in development), with a pattern that roughly recapitulates those of endoreduplication and starch accumulation (Sabelli, 2012a; Young and Gallie, 2000). Although this, of course, could simply underscore a developmental gradient between inner and outer endosperm cells, there is potential for a causative link. It has been established that the presence of ethylene induces DNA fragmentation and PCD in the starchy endosperm, and analysis of the *shrunken2* (*sh2*) mutant showed an abnormally high concentration of sugar (versus starch) that results in higher ethylene levels and hastens PCD (Young and Gallie, 2000). It is not known whether the relationship between

carbohydrate metabolism and PCD is physiological, although onset of endosperm PCD correlates with a noticeable decrease in the hexose/sucrose ratio. Sabelli *et al.* (2013) showed that downregulation of RBR1 stimulates PCD along with endoreduplication, seemingly in the absence of clear effects on starch deposition, but neither sugar nor hormone levels were determined in the mutant endosperm. Thus, although RBR1 appears to regulate endosperm cell death, whether carbohydrate metabolism, hormones, and/or endoreduplication play any role in the underlying mechanism remains an unresolved question.

The so-called DNA salvage hypothesis could explain a causative link between endoreduplication and PCD. According to this view, endoreduplication would provide the means to generate and store large amounts of DNA that could be converted, through PCD, into precursors to support cell division, embryo development, and growth of the germinating seedling (Kowles and Phillips, 1985; Larkins *et al.*, 2001). However, in spite of a noticeable correlation between the rate of PCD in the endosperm and growth of the embryo, direct evidence supporting this hypothesis is lacking (Kowles, 2009).

## 10.5   Conclusions

So, what have we learned thus far and where do we go from here? Not surprisingly, over the last two decades as research has delved deeper into seed development, investigators have progressively come to the realization that regulation of the cell cycle is not compartmentalized from that of cell size control and synthesis and accumulation of storage metabolites. Moreover, the endosperm processes described in this chapter occur in a highly diversified tissue that is composed of multiple cell types characterized by different physical, biochemical, and molecular features, rather than a monolithic storage compartment as perhaps initially thought. I have described a number of puzzling observations and unresolved questions. Reductionist approaches, while effective for entry-level investigations, are unlikely to provide unequivocal answers. Large amounts of data are being accumulated through so-called "omics" methodologies, but this information for the most part relates to whole-tissue or whole-seed analysis and could fail to provide the level of resolution required to reveal underlying mechanisms free from confounding factors. Thus, in future investigations it will be important to couple techniques that increase sample homogeneity with the "ad hoc" genetic resources and powerful computational analyses that are available. For example, sample uniformity can be significantly increased by sorting nuclei or cells according to specific criteria or by laser-assisted dissection of cell populations. Newly available techniques, such as CRISPR, can efficiently generate gene-specific mutations at an unprecedented rate, and thus overcome a long-standing technological bottleneck. Advanced imaging techniques, including the utilization of *in vivo* fluorescent markers, support development of precise mathematical models. These and other approaches will ultimately provide a more integrated understanding of the nuances of cell cycle and cell size regulation during seed development.

# References


Bai, F. and Settles, A.M. (2015) Imprinting in plants as a mechanism to generate seed phenotypic diversity. *Frontiers in Plant Science* 5, 780. DOI:10.3389/fpls.2014.00780

Bauer, M.J. and Birchler, J.A. (2006) Organization of endoreduplicated chromosomes in the endosperm of *Zea mays* L. *Chromosoma* 115, 383–394.

Bleckmann, A., Alter, S. and Dresselhaus, T. (2014) The beginning of a seed: regulatory mechanisms of double fertilization. *Frontiers in Plant Science* 5, 452. DOI:10.3389/fpls.2014.00452

Bourdon, M., Pirrello, J., Cheniclet, C., Coriton, O., Bourge, M., *et al.* (2012) Evidence for karyoplasmic homeostasis during endoreduplication and a ploidy-dependent increase in gene transcription during tomato fruit growth. *Development* 139, 3817–3826. DOI:10.1242/dev.084053

Brocklehurst, P.A. (1977) Factors controlling grain weight in wheat. *Nature* 266, 348–349.

Burton, R.A. and Fincher, G.B. (2014) Evolution and development of cell walls in cereal grains. *Frontiers in Plant Science* 5, 456. DOI:10.3389/fpls.2014.00456

Chen, J., Zeng, B., Zhang, M., Xie, S., Wang, G., *et al.* (2014) Dynamic transcriptome landscape of maize embryo and endosperm development. *Plant Physiology* 166, 252–264. DOI:10.1104/pp.114.240689

Coelho, C.M., Dante, R.A., Sabelli, P.A., Sun, Y., Dilkes, B.P., *et al.* (2005) Cyclin-dependent kinase inhibitors in maize endosperm and their potential role in endoreduplication. *Plant Physiology* 138, 2323–2336. DOI:10.1104/pp.105.063917

Coelho, C.M., Wu, S., Li, Y., Hunter, B., Dante, R.A., *et al.* (2007) Identification of quantitative trait loci that affect endoreduplication in maize endosperm. *Theoretical and Applied Genetics* 115, 1147–1162. DOI:10.1007/s00122-007-0640-z

Dante, R.A., Larkins, B.A. and Sabelli, P.A. (2014a) Cell cycle control and seed development. *Frontiers in Plant Science* 5, 493. DOI:10.3389/fpls.2014.00493

Dante, R.A., Sabelli, P.A., Nguyen, H.N., Leiva-Neto, J.T., Tao, Y., *et al.* (2014b) Cyclin-dependent kinase complexes in developing maize endosperm: evidence for differential expression and functional specialization. *Planta* 239, 493–509. DOI:10.1007/s00425-013-1990-1

Dilkes, B.P., Dante, R.A., Coelho, C. and Larkins, B.A. (2002) Genetic analyses of endoreduplication in *Zea mays* endosperm: evidence of sporophytic and zygotic maternal control. *Genetics* 160, 1163–1177.

Domínguez, F. and Cejudo, F.J. (2014) Programmed cell death (PCD): an essential process of cereal seed development and germination. *Frontiers in Plant Science* 5, 366. DOI:10.3389/fpls.2014.00366

Edgar, B.A., Zielke, N. and Gutierrez, C. (2014) Endocycles: a recurrent evolutionary innovation for post-mitotic cell growth. *Nature Reviews Molecular Cell Biology* 15, 197–210. DOI:10.1038/nrm3756

Gehring, M. and Satyaki, P.R. (2017) Endosperm and imprinting, inextricably linked. *Plant Physiology* 173, 143–154. DOI:10.1104/pp.16.01353

Grafi, G. and Larkins, B.A. (1995) Endoreduplication in maize endosperm – involvement of M-phase-promoting factor inhibition and induction of S-phase-related kinases. *Science* 269, 1262–1264.

Grafi, G., Burnett, R.J., Helentjaris, T., Larkins, B.A., DeCaprio, J.A., *et al.* (1996) A maize cDNA encoding a member of the retinoblastoma protein family: involvement in endoreduplication. *Proceedings of the National Academy of Sciences of the United States of America* 93, 8962–8967.

Guo, J., Wang, F., Song, J., Sun, W. and Zhang, X.S. (2010) The expression of *Orysa;CycB1;1* is essential for endosperm formation and causes embryo enlargement in rice. *Planta* 231, 293–303. DOI:10.1007/s00425-009-1051-y

Hannah, L.C. (2007) Starch formation in the cereal endosperm. In: Olsen, O.-A. (ed.) *Endosperm*. Springer, Berlin, Heidelberg, Germany, pp. 179–193. DOI:10.1007/7089_2007_116

Inzé, D. and De Veylder, L. (2006) Cell cycle regulation in plant development. *Annual Review of Genetics* 40, 77–105. DOI:10.1146/annurev.genet.40.110405.090431

Ishimaru, K., Hirotsu, N., Madoka, Y., Murakami, N., Hara, N., *et al.* (2013) Loss of function of the IAA-glucose hydrolase gene *TGW6* enhances rice grain weight and increases yield. *Nature Genetics* 45, 707–711. DOI:10.1038/ng.2612

Jones, R.J., Schreiber, B.M.N. and Roessier, J.A. (1996) Kernel sink capacity in maize: genotypic and maternal regulation. *Crop Science* 36, 301–306.

Kiesselbach, T.A. (1949) *The Structure and Reproduction of Corn*. University of Nebraska Press, Lincoln, Nebraska.

Kladnik, A., Chamusco, K., Dermastia, M. and Chourey, P. (2004) Evidence of programmed cell death in post-phloem transport cells of the maternal pedicel tissue in developing caryopsis of maize. *Plant Physiology* 136, 3572–3581. DOI:10.1104/pp.104.045195

Kowles, R.V. (2009) The importance of DNA endoreduplication in the developing endosperm of maize. *Maydica* 54, 387–399.

Kowles, R.V. and Phillips, R.L. (1985) DNA amplification patterns in maize endosperm nuclei during kernel development. *Proceedings of the National Academy of Sciences of the United States of America* 82, 7010–7014.

Kowles, R.V. and Phillips, R.L. (1988) Endosperm development in maize. *International Review of Cytology* 112, 97–136.

Kowles, R.V., McMullen, D, Yerk, G., Phillips, R.L., Kraemer, S., *et al.* (1992) Endosperm mitotic activity and endoreduplication in maize affected by defective kernel mutations. *Genome* 35, 68–77.

Larkins, B.A., Dilkes, B.P., Dante, R.A., Coelho, C.M., Woo, Y.M., *et al.* (2001) Investigating the hows and whys of DNA endoreduplication. *Journal of Experimental Botany* 52, 183–192.

Leiva-Neto, J.T., Grafi, G., Sabelli, P.A., Dante, R.A., Woo, Y.M., *et al.* (2004) A dominant negative mutant of cyclin-dependent kinase A reduces endoreduplication but not cell size or gene expression in maize endosperm. *Plant Cell* 16, 1854–1869. DOI:10.1105/tpc.022178

Leroux, B.M., Goodyke, A.J., Schumacher, K.I., Abbott, C.P., Clore, A.M., *et al.* (2014) Maize early endosperm growth and development: from fertilization through cell type differentiation. *American Journal of Botany* 101, 1259–1274. DOI:10.3732/ajb.1400083

Lopes, M.A. and Larkins, B.A. (1993) Endosperm origin, development, and function. *Plant Cell* 5, 1383–1399.

Lur, H.S. and Setter, T.L. (1993) Endosperm development of maize defective-kernel (dek) mutants. Auxin and cytokinin levels. *Annals of Botany* 72, 1–6. DOI:10.1006/anbo.1993.1074

Monjardino, P., Machado, J., Gil, F.S., Fernandes, R. and Salema, R. (2007) Structural and ultrastructural characterization of maize coenocyte and endosperm cellularization. *Canadian Journal of Botany* 85, 216–223.

Olsen, O.-A. (2001) Endosperm development: cellularization and cell fate specification. *Annual Review of Plant Physiology and Plant Molecular Biology* 52, 233–267. DOI:10.1146/annurev.arplant.52.1.233

Olsen, O.-A. (2004) Nuclear endosperm development in cereals and *Arabidopsis thaliana*. *Plant Cell* 16, 214–227. DOI:10.1105/tpc.017111

Otegui, M. (2007) Endosperm cell walls: formation, composition, and functions. In: Olsen, O.-A. (ed.) *Endosperm*. Springer, Berlin, Heidelberg, Germany, pp. 159–177. DOI:10.1007/7089_2007_113

Radchuk, V., Weier, D., Radchuk, R., Weschke, W. and Weber, H. (2011) Development of maternal seed tissue in barley is mediated by regulated cell expansion and cell disintegration and coordinated with endosperm growth. *Journal of Experimental Botany* 62, 1217–1227. DOI:10.1093/jxb/erq348

Randolph, L.F. (1936) Developmental morphology of the caryopsis in maize. *Journal of Agricultural Research* 53, 881–916.

Reddy, V.M. and Daynard, T.B. (1983) Endosperm characteristics associated with rate of grain filling and kernel size in corn. *Maydica* 28, 339–355.

Rolletschek, H., Koch, K., Wobus, U. and Borisjuk, L. (2005) Positional cues for the starch/lipid balance in maize kernels and resource partitioning to the embryo. *The Plant Journal* 42, 69–83. DOI:10.1111/j.1365-313X.2005.02352.x

Sabelli, P.A. (2012a) Replicate and die for your own good: endoreduplication and cell death in the cereal endosperm. *Journal of Cereal Science* 56, 9–20.

Sabelli, P.A. (2012b) Seed development: a comparative overview on biology of morphology, physiology, and biochemistry between monocot and dicot plants. In: Agrawal, G.K. and Rakwal, R. (eds.) *Seed Development: OMICS Technologies toward Improvement of Seed Quality and Crop Yield.* Springer, Dordrecht, The Netherlands, pp. 3–25. DOI:10.1007/978-94-007-4749-4_1

Sabelli, P.A. (2014) Cell cycle regulation and plant development: a crop production perspective. In: Pessarakli, M. (ed.) *Handbook of Plant and Crop Physiology.* CRC Press, Boca Raton, Florida, pp. 3–32.

Sabelli, P.A and Larkins, B.A. (2006) Grasses like mammals? Redundancy and compensatory regulation within the retinoblastoma protein family. *Cell Cycle* 5, 352–355. DOI:10.4161/cc.5.4.2428

Sabelli, P.A. and Larkins, B.A. (2009a) Regulation and function of retinoblastoma-related plant genes. *Plant Science* 177, 540–548.

Sabelli, P.A. and Larkins, B.A. (2009b) The contribution of cell cycle regulation to endosperm development. *Sexual Plant Reproduction* 22, 207–219. DOI:10.1007/s00497-009-0105-4

Sabelli, P.A. and Larkins, B.A. (2009c) The development of endosperm in grasses. *Plant Physiology* 149, 14–26. DOI:10.1104/pp.108.129437

Sabelli, P.A., Dante, R.A., Leiva-Neto, J.T., Jung, R., Gordon-Kamm, W.J., *et al.* (2005) RBR3, a member of the retinoblastoma-related family from maize, is regulated by the RBR1/E2F pathway. *Proceedings of the National Academy of Sciences of the United States of America* 102, 13005–13012. DOI:10.1073/pnas.0506160102

Sabelli, P.A., Hoerster, G., Lizarraga, L.E., Brown, S.W., Gordon-Kamm, W.J., *et al.* (2009) Positive regulation of minichromosome maintenance gene expression, DNA replication, and cell transformation by a plant retinoblastoma gene. *Proceedings of the National Academy of Sciences of the United States of America* 106, 4042–4047. DOI:10.1073/pnas.0813329106

Sabelli, P.A. Liu, Y., Dante, R.A., Lizarraga, L.E., Nguyen, H.N., *et al.* (2013) Control of cell proliferation, endoreduplication, cell size, and cell death by the retinoblastoma-related pathway in maize endosperm. *Proceedings of the National Academy of Sciences of the United States of America* 110, E1827–E1836. DOI:10.1073/pnas.1304903110

Sabelli, P.A., Dante, R.A., Nguyen, H.N., Gordon-Kamm, W.J. and Larkins, B.A. (2014) Expression, regulation and activity of a B2-type cyclin in mitotic and endoreduplicating maize endosperm. *Frontiers in Plant Science* 5, 561. DOI:10.3389/fpls.2014.00561

Sablowski, R. (2016) Coordination of plant cell growth and division: collective control or mutual agreement? *Current Opinion in Plant Biology* 34, 54–60. DOI:10.1016/j.pbi.2016.09.004

Sekhon, R.S., Hirsch, C.N., Childs, K.L., Breitzman, M.W., Kell, P., *et al.* (2014) Phenotypic and transcriptional analysis of divergently selected maize populations reveals the role of developmental timing in seed size determination. *Plant Physiology* 165, 477–478. DOI:10.1104/pp.114.235424

Sugimoto-Shirasu, K. and Roberts, K. (2003) "Big it up": endoreduplication and cell-size control in plants. *Current Opinion in Plant Biology* 6, 544–553.

Sun, Y.J., Dilkes, B.P., Zhang, C., Dante, R.A., Carneiro, N.P., *et al.* (1999a) Characterization of maize (*Zea mays* L.) Wee1 and its activity in developing endosperm. *Proceedings of the National Academy of Sciences of the United States of America* 96, 4180–4185.

Sun, Y.J., Flannigan, B.A. and Setter, T.L. (1999b) Regulation of endoreduplication in maize (*Zea mays* L.) endosperm. Isolation of a novel B1-type cyclin and its quantitative analysis. *Plant Molecular Biology* 41, 245–258.

Trafford, K., Haleux, P., Henderson, M., Parker, M., Shirley, N.J., *et al.* (2013) Grain development in *Brachypodium* and other grasses: possible interactions between cell expansion, starch deposition, and cell-wall synthesis. *Journal of Experimental Botany* 64, 5033–5047. DOI:10.1093/jxb/ert292

Vilhar, B., Kladnik, A., Blejec, A., Chourey, P.S. and Dermastia, M. (2002) Cytometrical evidence that the loss of seed weight in the *miniature1* seed mutant of maize is associated with reduced mitotic activity in the developing endosperm. *Plant Physiology* 129, 23–30. DOI:10.1104/pp.001826

Wang, P., Xia, H., Zhang, Y., Zhao, S., Zhao, C., *et al.* (2015) Genome-wide high-resolution mapping of DNA methylation identifies epigenetic variation across embryo and endosperm in maize (*Zea may*). *BMC Genomics* 16, 21. DOI:10.1186/s12864-014-1204-7

Waters, A.J., Makarevitch, I., Eichten, S.R., Swanson-Wagner, R.A., Yeh, C.-T., *et al.* (2011) Parent-of-origin effects on gene expression and DNA methylation in the maize endosperm. *Plant Cell* 23, 4221–4233. DOI:10.1105/tpc.111.092668

Wu, C.-Y., Rolfe, P.A., Gifford, D.K. and Fink, G.R. (2010) Control of transcription by cell size. *PLOS Biology* 8, e1000523. DOI:10.1371/journal.pbio.1000523

Xin, M., Yang, R., Li, G., Chen, H., Laurie, J., *et al.* (2013) Dynamic expression of imprinted genes associates with maternally controlled nutrient allocation during maize endosperm development. *Plant Cell* 25, 3212–3227. DOI:10.1105/tpc.113.115592

Young, T.E. and Gallie, D.R. (2000) Programmed cell death during endosperm development. *Plant Molecular Biology* 44, 283–301.

Zemach, A., Kim, M.Y., Silva, P., Rodrigues, J.A., Dotson, B., *et al.* (2010) Local DNA hypomethylation activates genes in rice endosperm. *Proceedings of the National Academy of Sciences of the United States of America* 107, 18729–18734. DOI:10.1073/pnas.1009695107

# 11 Central Metabolism and Its Spatial Heterogeneity in Maize Endosperm

Hardy Rolletschek[1], Ljudmilla Borisjuk[1], Tracie A. Hennen-Bierwagen[2] and Alan M. Myers[2,*]

[1]*Leibniz Institute of Plant Genetics and Crop Plant Research (IPK), Department of Molecular Genetics, Gatersleben, Germany;* [2]*Roy J. Carver Department of Biochemistry, Biophysics, and Molecular Biology, Iowa State University, Ames, Iowa, USA*

## 11.1   Introduction

This chapter addresses bioenergetic considerations of the metabolic processes in maize kernels by which sugars are converted to starch, protein, and other metabolites. Central metabolism in this context is divided into modules: (i) hexose and sucrose supply; (ii) hexose phosphorylation; (iii) ATP production; (iv) starch biosynthesis; (v) amino acid biosynthesis; (vi) storage protein synthesis; and (vii) lipid biosynthesis. The components of each node are a group of enzymes and the genes encoding them, so queries of RNA transcripts and quantitative proteomics reflect these metabolic pathways. Flux analysis is relatively well developed for maize endosperm and provides information about rates of metabolic interconversions in particular nodes. Connected metabolic pathways can be proposed based on these considerations and models tested by perturbing them.

The ultimate product of these metabolic processes is grain, which is a large component of our food supply. Production of maize and other cereals is tremendously important, considering the worldwide scale of agriculture and the fact that the energy supply for endosperm metabolism is renewable. Maize is a highly advanced experimental system, and kernel biology is amenable to genetic manipulation to achieve beneficial outcomes. Knowledge about maize endosperm development is applicable to other cereals as well.

Constraints for modeling metabolism in maize endosperm result from several factors, including cellular heterogeneity, which flux analysis typically does not consider. Metabolic outputs vary in different regions of the endosperm, so some aspects of models may need to be developed and/or adapted for cells in different regions. It is important to note that maize endosperm develops in a hypoxic environment, meaning that small changes in $O_2$ concentration can affect fluxes, for example through the citric acid cycle. Further complication arises from subcellular compartmentalization of metabolic pathways. Glycolysis is essentially duplicated in the amyloplast stroma and the cytosol. Some oxidative pentose phosphate pathway (oxPPP) reactions are duplicated as well, and citric acid cycle reactions are, apparently, present in both the amyloplast

*Corresponding author e-mail: albergo@mac.com

   

stroma and mitochondrial matrix. In the few instances tested, plastid and cytosolic pathways do not complement each other (Muñoz-Bertomeu *et al.*, 2009; Spielbauer *et al.*, 2013), so metabolite pools in subcellular compartments are not necessarily in equilibrium. This brings into consideration transport mechanisms that move metabolites across membranes.

We do not address every aspect of kernel metabolism and refer the reader to Chapters 12, 14, and 15 for discussions of starch biosynthesis, protein metabolism, and sugar supply. This chapter addresses energy metabolism and how it relates to storage-compound deposition. Our focus is the grain fill stage from 10 to 30 days after pollination (DAP) when starch and protein accumulation rates are maximal. During this period starchy endosperm (SE) cells, which have completed cellularization and endoreduplication, are poised for high-level gene expression (see Chapter 3). Their metabolism is central to our discussion, but consideration to metabolic distinctions of other endosperm cell types is given. Our major conclusions are: (i) anabolic and catabolic fluxes in the SE are balanced by allosteric enzyme regulation to achieve optimum seed physiology; (ii) $O_2$ availability limits ATP yield and affects central metabolic pathways; and (iii) the endosperm is a metabolically and spatially heterogeneous tissue. Among potential signals for flux control are ATP/ADP/AMP ratios, pyrophosphate (PPi) levels, and the redox state of a cell.

## 11.2  Current Understanding of the System

### 11.2.1  Metabolic fuel—the starting point of metabolism in SE cells

There remains much to learn about sucrose utilization during kernel biomass accumulation, despite the fundamental importance of this process and extensive previous investigations. The reducing hexoses, glucose (Glc), and fructose (Frc) are primary fuels for SE central metabolism, and the

possibility exists that the non-reducing disaccharide sucrose (Suc) can also serve this function. This subject is addressed in Chapters 5 and 15.

Photosynthate is supplied to kernels as Suc that is transported through the plant vasculature (Fig. 11.1A; see Chapter 15). This was shown by exposing leaves to $^{14}CO_2$ and quantifying the label in various metabolites in developing SE and surrounding maternal tissues (Shannon, 1968, 1972). The plant's vasculature terminates in the pedicel, which in turn is in contact with basal endosperm transfer layer (BETL) cells (Felker and Shannon, 1980). It is important to note that this tissue interface is between two genetically distinct individuals. Sugars appear to be transported symplastically within maternal tissue (Felker and Shannon, 1980) prior to crossing into the endosperm by passive efflux (Porter *et al.*, 1985; Griffith *et al.*, 1987). Pulse-labeling showed some or all of the Suc delivered from phloem is hydrolyzed to Glc and Frc prior to entry into SE, and all three sugars are approximately equally abundant in basal regions of kernels (Shannon, 1972; Griffith *et al.*, 1987; Schmalstig and Hitz, 1987). Sugars move between maternal and filial tissue through a narrow interface that contains the BETL, where cell wall-attached invertase (CWIN) is exposed on the cell surface (Shannon and Dougherty, 1972; Chourey *et al.*, 2006; Chapter 5, this volume). This is the only location for metabolite exchange between maternal and endosperm tissues, in contrast to other species where transfer can occur between the pericarp and endosperm. Transport across the BETL involves hydrolysis of Suc by CWIN to generate Glc and Frc, intracellular import of the hexoses into BETL cells, and then export into the apoplastic space (Sosso *et al.*, 2015). Sugars move through the endosperm tissue in the apoplast (Shannon, 1972), where they access transporters for cytosolic entry. Some transporters have been defined, but others likely remain to be identified (see Chapter 15, this volume). Pulse-labeling showed some or all Glc and Frc that enters the endosperm is resynthesized into Suc, presumably after cellular import.

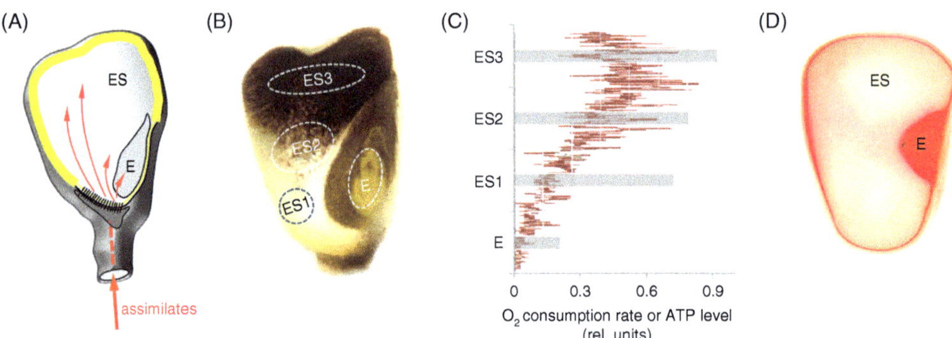



**Fig. 11.1.** Assimilate supply and metabolic heterogeneity in developing maize kernels. (A) Drawing indicates the route for assimilate supply toward endosperm (ES) and embryo (E); yellow line indicates a diffusion barrier restricting gas exchange (Radchuk and Borisjuk, 2014). (B) Starch distribution. Gradient in starch accumulation revealed by I₂/KI stain (Rolletschek *et al.*, 2005); (C) Relative ATP distribution and respiration rates. Red line indicates gradient in ATP level measured longitudinally across the endosperm (Rolletschek *et al.*, 2005); bars indicate respiratory activity in embryo and three endosperm regions as indicated in panel B; respiration was measured using planar sensor foils (unpublished results; method in Tschiersch *et al.*, 2012); (D) Lipid distribution. Sudan B stain reveals storage oils concentrated in aleurone and in embryo, and detectable stain throughout SE tissue (Rolletschek *et al.*, 2005).

Genetic analyses indicate that delivery of hexose to SE is critical for normal kernel development. Kernel CWIN can almost be completely eliminated by the *mn1-* mutation, which causes about 65% reduction in endosperm biomass (Cheng *et al.*, 1996). Mutation of the SWEET4c transporter that moves Glc and Frc through the BETL into the SE apoplast also conditions a major biomass reduction (Sosso *et al.*, 2015). Thus, normal SE development depends to a large extent on hexoses supplied by CWIN and the SWEET4c transporter.

The hexose requirement may not be absolute, however, because appreciable residual kernel biomass accumulates in mutants deficient in CWIN and SWEET4c. Potential explanations for this are as follows: First, a second gene encoding CWIN is expressed in the basal endosperm and supplies approximately 1% residual activity in *mn1-* mutants (Chourey *et al.*, 2006), which may be sufficient for remnant biomass accumulation. Second, Suc as well as hexoses could enter the apoplast, be transported into SE cells, and enter central metabolism. Consistent with this explanation, labeling studies, including use of a non-hydrolysable Suc analog, demonstrated Suc entry to SE tissue without hydrolysis (Cobb and Hannah, 1986;

Schmalstig and Hitz, 1987), and into SE cytosol (Felker and Goodwin, 1988). Future genetic analyses of double mutants that are entirely deficient in kernel CWIN will resolve this issue. Learning precisely which sugars are available to SE cells is important for understanding central metabolism, because there are different bioenergetic constraints for Suc and hexoses, and primary fuel distribution may contribute to metabolic heterogeneity.

Intracellular resynthesis of Suc from Glc and Frc could be a means of driving passive hexose transport from the apoplast into the cell (Chapter 5). As shown by mutants lacking sucrose synthase (SUS) activity, this is another metabolic process that is dispensable to some extent (McCarty *et al.*, 1986; Chourey *et al.*, 1998; Li *et al.*, 2013). SUS regulates an initial step in intracellular Suc catabolism (Cobb and Hannah, 1988); however, mutants of this enzyme exhibit about 50% residual biomass. This implies that resynthesis of Glc and Frc into Suc is not obligate for SE biomass accumulation and hexose can directly serve this purpose.

Collectively, data regarding carbohydrate supply are consistent with the hypothesis that wild-type SE has access to Glc,

Frc, and Suc in the apoplastic space and any of these sugars can serve as the primary source for substantial amounts of starch and protein. Additional experiments are needed to test this hypothesis, because direct evidence for specific sugar concentrations in the SE apoplast is lacking and processes that occur in mutant lines may not operate in non-mutant tissue.

Transport of substantial amino groups from maternal tissue into SE is required for amino acid and storage protein biosynthesis. A great deal of information is available regarding amino group transport in dicots (Tegeder, 2014) and maize maternal tissues (Liseron-Monfils *et al.*, 2013). The exact carriers involved in delivery to maize endosperm and the metabolic processes that extract amino groups have not been identified.

### 11.2.2 Bioenergetics overview

The nature of metabolic flux between anabolic and catabolic pathways in endosperm tissues is an open question. The minimal ATP requirement for storage-compound accumulation, calculated from dry weight, is 2–3 mmol ATP per kernel (Fig. 11.2). The amount of Glc units needed to generate this ATP depends on the efficiency of energy capture, which can vary from ~2 mol ATP/mol Glc when operating glycolytically, to ~36 mol ATP/mol Glc in a fully oxidative mode. This value affects the availability of Glc units for incorporation into starch and protein, which account for 80–90% of mature grain dry weight. These ATP estimates imply that the ratio of Glc units entering anabolic versus catabolic pathways varies between ~30:1 and 1:1, depending on the extent of oxidative and non-oxidative metabolism. How ATP is generated in SE is not fully known; central to this question is the hypoxic environment of the tissue, as discussed in the following section. Other unknowns are the ATP used in metabolic recycling reactions and cellular maintenance, neither of which are considered in these estimates of the anabolic:catabolic ratio.

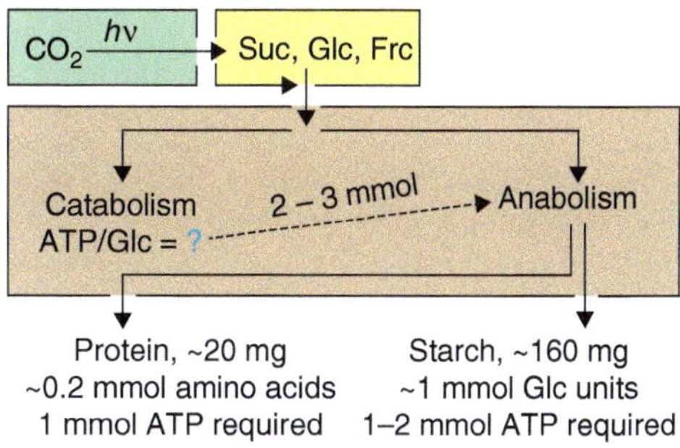


| Efficiency = | anabolic hexose / catabolic hexose | Metabolic mode |
|---|---|---|
| | 30:1 | fully oxidative |
| | 1:1 | fully glycolytic |

**Fig. 11.2.** Minimum ATP requirements for storage-compound deposition. The estimate for storage protein includes 4 mol ATP/mol peptide bonds formed and 1 mol ATP/mol amino acids synthesized. The estimate for starch is 1 mol ATP/mol glucose units incorporated starting from Suc, or 2 mol ATP/mol glucose units incorporated starting from hexose. Efficiency estimates consider the number of hexoses used to produce starch or protein ("anabolic hexose") compared to the number used to produce ATP ("catabolic hexose") when cells operate in particular metabolic modes. Green, yellow, and brown boxes represent maternal photosynthetic tissue, the SE apoplast, and the SE interior (symplast), respectively.

### 11.2.3 Metabolic heterogeneity

SE is often viewed as a homogeneous tissue; however, this is an oversimplification because different regions have distinct developmental and metabolic programs/networks. Endosperm cellularization occurs 6–9 DAP and precedes differentiation of its several tissues, including the aleurone, BETL, embryo-surrounding region (ESR), and SE cells that store starch and storage protein (see Chapter 3). SE cells increase in number and volume as growth ensues. They fill with storage compounds from the top of the kernel downward, as evident from the gradient of starch accumulation (Fig. 11.1B). The region below SE tissue performs a major role in solute, water, and nutrition acquisition from the BETL (see Chapter 5). Likewise, the ESR and conducting tissue within the central endosperm have specialized functions (see Chapter 3). Metabolic heterogeneity of these different tissues is recognizable at distinct levels:

**1.** The gradient distribution of starch accumulation (Fig. 11.1B).
**2.** The spatial pattern of vitreousness and non-vitreous (floury) mature endosperm (see Chapters 3 and 14).
**3.** Lipid partitioning between floury and vitreous regions (Gayral *et al.*, 2015).
**4.** ATP gradients within the SE that are developmentally dependent (Fig. 11.1C) (Rolletschek *et al.*, 2005). These gradients generally relate to cell size and starch storage. The steady-state ATP concentration varies twofold across the endosperm.
**5.** Gradients in respiratory activity across endosperm and embryo (Fig. 11.1C). These correlate with steady-state ATP levels in respective tissues and indicate distinct, tissue-specific capacities for mitochondrial respiratory fluxes.
**6.** The small size of peripheral SE cells and their nuclei compared with larger cells and nuclei in the interior. This variation occurs along a seemingly continuous gradient and is the consequence of endoreduplication (see Chapter 10). Peripheral SE cells in kernels relatively late in development (28 DAP) contain clearly defined mitochondria and ER, in addition to starch granules and protein bodies (Fig. 11.3), and thus differ from interior SE cells that experience programmed cell death (PCD) at earlier stages.
**7.** The spatial pattern of PCD in the central endosperm recapitulates that of endoreduplication and starch accumulation (Young *et al.*, 1997; Young and Gallie, 2000; Sabelli, 2012; Domínguez and Cejudo, 2014) (Chapter 10, this volume).

Some of the above reflect spatially distinct metabolic states (1–5), whereas others (6–7) are indirect indicators of metabolic heterogeneity. Endoreduplication poses specific metabolic demands, e.g. biosynthesis of pentoses and nucleotides, and this is fueled by metabolic output. However, the metabolic pathways that terminate with storage protein and starch synthesis commence after endoreduplication is complete and can be considered separately.

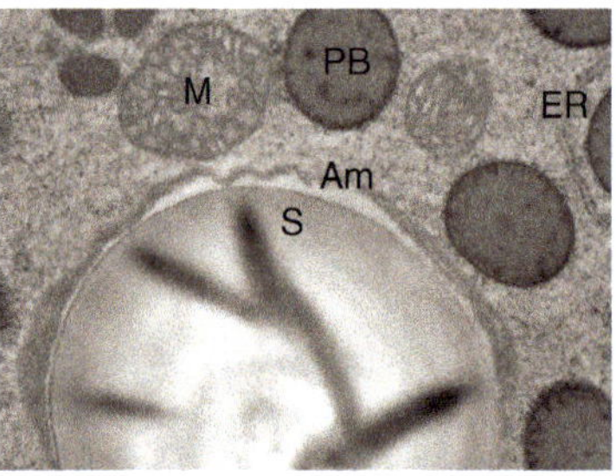


**Fig. 11.3.** TEM of peripheral SE cell from a kernel harvested 28 DAP (Myers *et al.*, 2011). M, mitochondrion; PB, protein body; ER, endoplasmic reticulum; S, starch granule; Am, amyloplast membrane. The starch granule is 3 µm diameter.

Some aspects of metabolic heterogeneity in maize endosperm are beginning to be unraveled and are expected to provide novel cues about kernel development. Study of metabolic heterogeneity requires experimental tools that allow spatially resolved analysis, e.g. laser dissection, mass spectrometry imaging, magnetic resonance imaging, and metabolic modeling tools for *in silico* studies. Application of these tools will impact maize research in unpredictable ways. As an example, it would interesting to learn whether there is spatial heterogeneity within the endosperm of the transporters that move hexose or sucrose from the apoplast into SE and other cells.

### 11.2.4 Phosphorylation for entry into metabolism; PPi metabolism

The disaccharide Suc, and the hexoses Glc and Frc, enter metabolism via different pathways with different energetic requirements. Glc and Frc must first be converted to Glc-6-P or Frc-6-P by hexokinase, of which several isoforms could be present. This requires energy input, i.e. for each hexose utilized, 1 ATP is hydrolyzed to ADP. Hexose-phosphate pools can use Glc-1-P for generation of UDPGlc (for cell wall deposition and other metabolic conversions) or ADPGlc (for addition of glucosyl units to starch) (see Chapter 12), or Glc-6-P and Frc-6-P for entry into glycolysis and subsequent ATP generating reactions. The fact that the

$O_2$ supply is rate limiting (see following section) implies that ATP generation reactions are proceeding at suboptimal levels, so the ATP yield per mole of hexose phosphate entering glycolysis is not known. In contrast to hexoses, Suc can be converted to two moles of hexose-phosphate using inorganic pyrophosphate (PPi) without net ATP hydrolysis. This pathway includes sucrose synthase, UDPglucose pyrophosphorylase (UGPase), hexokinase, and nucleotide diphosphate kinase:

$$
\begin{aligned}
\text{Suc} + \text{UDP} &\longrightarrow \text{Frc} + \text{UDPGlc} \\
\text{UDPGlc} + \text{PPi} &\longrightarrow \text{Glc-1-P} + \text{UTP} \\
\text{Frc} + \text{ATP} &\longrightarrow \text{Frc-6-P} + \text{ADP} \\
\underline{\text{UTP} + \text{ADP}} &\underline{\longrightarrow \text{UDP} + \text{ATP}} \\
\text{Suc} + \text{PPi} &\longrightarrow \text{Frc-6-P} + \text{Glc-1-P}
\end{aligned}
$$

Sucrose utilization is thus energetically more efficient than the hexose pathway as long as there is a supply of PPi, which is abundant in plant cytosol (Edwards *et al.*, 1984; Mohanty and ap Rees, 1992; Plaxton, 1996). For endosperm, the best estimate of PPi comes from non-aqueous fractionation of barley tissue, where the cytosolic PPi concentration was 118 μM, compared to the plastidial concentration of 18 μM (Tiessen *et al.*, 2012). PPi is generated by aminoacyl-tRNA synthetases at 1 mol per mol amino-acid residue incorporated to protein, and also during every cycle of nucleic acid synthesis. Also, the ADPGlc pyrophosphorylase (AGPase) reaction that directs hexose into starch releases one mol PPi per mol Glc incorporated (see Chapter 12). Cereal endosperm AGPase is predominantly cytosolic, so the PPi generated in either starch or protein synthesis collects as a single pool in the cytosol.

Variation in PPi homeostasis was proposed as a control mechanism for metabolic fluxes, in particular in hypoxic tissue. When $O_2$ availability in roots is rapidly curtailed by flooding, the abundance of two PPi-metabolizing enzymes, specifically PPi-dependent phosphofructokinase (PFP) and pyruvate phosphate dikinase (PPDK), increase rapidly (Huang *et al.*, 2008). They are proposed to increase net ATP yield in the central metabolism by affecting PPi pools and shifting to a PPi-utilizing mode of sucrose degradation. PPDK is abundant in developing maize endosperm (Mechin *et al.*, 2007; Walley *et al.*, 2013), and this may be a constitutive adaptation to hypoxia, whether through effects on PPi homeostasis or via other potential mechanisms that remain to be identified.

Hypoxic conditions of flooded roots and developing endosperm could be comparable, although seed tissue is permanently $O_2$-limited whereas roots experience induced hypoxia. SE metabolism should be viewed as an organ-specific system that is permanently acclimated to low $O_2$ concentration, rather than a tissue that responds rapidly to induced hypoxia. Nonetheless, specific $O_2$-related signals likely regulate genetic and metabolic responses in SE and elsewhere in the kernel using hypoxia signaling mechanisms shared by other tissues.

Precisely how PPDK or PFP contribute to steady-state PPi concentration is unknown. PPi will have a substantial production rate in highly metabolically active tissue such as SE. One consumption mechanism is the PPi-dependent proton pump that moves protons into the vacuolar space (Martinoia *et al.*, 2007). Multiple forms of this transporter are expressed in maize endosperm (Walley *et al.*, 2013), where it is expected to maintain cytosol pH. The high rate of net synthesis and potential high rate of degradation implies a substantial flux rate for PPi conversions. PPDK and PFP are reversible reactions, so they can function both to generate and remove PPi and thus affect the fluxes in various ways (Stitt, 1990).

In summary, developing SE cells are plentifully supplied with Glc, Frc, and potentially Suc, which they access using plasma membrane transporters. The hexoses and the disaccharide have different pathways for entry into either anabolism for starch accumulation or catabolism for ATP supply and precursor conversions. The energetic considerations for these two pathways differ, and they may be spatially separated. PPi homeostasis could play a role in determining flux through the sucrose pathway. Utilization of these two different modes of entering metabolism could contribute to metabolic heterogeneity and thus provide overall advantage to the system.

### 11.2.5 Oxygen status

Modeling endosperm central metabolism must take into account the fact that the main stages of kernel development are oxygen limited (Rolletschek *et al.*, 2005). Microsensor measurements show that mean $O_2$ concentration in SE immediately inside the pericarp is only 1.4% (~3.8 µM) of atmospheric saturation, with local minima below 0.1% saturation. More than twofold increases in external oxygen levels were necessary to saturate the respiratory demands of the growing kernel. The fact that $O_2$ is not saturating implies that the ATP yield per glucosyl unit entering glycolysis cannot be known with certainty, and accordingly that the ATP supply may limit metabolic processes. This is likely considering that increasing the internal $O_2$ concentration resulted in elevated ATP levels in SE. Hypoxia appears to be normal for developing seeds of our major crops (Borisjuk and Rolletschek, 2009), and this is particularly pronounced in cereals and especially pervasive in maize kernels.

Generally, the hypoxic state comes about whenever the capacity for oxygen diffusion is restricted and the oxygen concentration falls below the level required for mitochondrial respiration. Intense respiratory metabolism inside SE, i.e. the high metabolic flux around the citric acid cycle/mitochondrial electron transport chain (mETC), is probably one of the causes of oxygen deficiency. There is also evidence of a diffusion barrier at the SE boundary that is potentially created by the presence of the aleurone layer and/or the seed cuticle (Fig. 11.1A). Hindrance of gas exchange is generally known to be mediated by lipid-rich membranes in some seed tissues (Rolletschek *et al.*, 2007), and cutin and lipidous layers occur in the outermost layers of the maize kernel (Fig. 11.1D). Even tiny lipid-rich membranes enveloping the grain can severely hamper gas exchange. High resolution X-ray imaging of oilseeds demonstrated that storage parenchyma have small and poorly interconnected, gas-filled pore spaces, further impeding gas diffusion (Cloetens *et al.*, 2006; Verboven *et al.*, 2013). There are no equivalent studies on cereals, but similar conditions are likely to exist in maize endosperm. Oxygen levels in the maize embryo are significantly higher than in endosperm, another example of metabolic heterogeneity. Together, limited oxygen diffusion and a high $O_2$ consumption rate create a low steady-state $O_2$ level that varies within specific regions of the endosperm.

How SE catabolism provides sufficient ATP for starch and protein synthesis is an open question in endosperm biology. In many species, oxygen deprivation first limits oxidative phosphorylation in the mETC and subsequently ATP production and flux through the citric acid cycle. Major metabolic adjustments follow, owing in part to altered ATP/ADP ratios, activation of signal transduction pathways initiated by low-oxygen sensors, and subsequent transcriptional responses (van Dongen and Licausi, 2015). One predicted change is greater flux through glycolytic pathways. Transcriptome and proteome analyses of developing SE show that, in general, glycolytic enzymes are present at higher levels than citric acid cycle enzymes, which is consistent with a shift towards glycolysis in a low-oxygen environment (Prioul *et al.*, 2008; Walley *et al.*, 2013). In a glycolytic mode, recycling NADH to NAD+ in SE does not conform to canonical pathways, because neither lactate nor ethanol accumulate during grain fill. Based on the high abundance of sorbitol in SE, sorbitol dehydrogenase was proposed to catalyze NADH oxidation by reducing fructose to sorbitol (de Sousa *et al.*, 2008; Walley *et al.*, 2013), but this hypothesis remains to be completely tested.

The hypoxic nature of developing endosperm likely signals multiple adaptive mechanisms. Low $O_2$ responses are too numerous to discuss in detail here; however, they can include the following: (i) NO signaling and coordinated regulation of cytochrome oxidase (reducing mETC activity and hence oxidative phosphorylation) (Benamar *et al.*, 2008); (ii) reduced flux towards reactive oxygen species (Simontacchi *et al.*,

1995); (iii) potential connection to PCD mediated by changes in the ROS pathways and/or ethylene (Igamberdiev *et al.*, 2014); (iv) potential connection to cell shape determination (as shown by a gradient of smaller cells with reduced starch content towards the periphery of the tissue where $O_2$ concentration is highest); (v) potential changes in sugar supply at the interface between maternal and filial tissue, as shown for wheat (van Dongen *et al.*, 2004); and (vi) potential changes in the state of redox regulation through thioredoxin (Wong *et al.*, 2003). Obviously, a great deal is unknown regarding signal transduction pathways operating during SE development and grain fill, including those initiated by signals other than oxygen status and interactions between multiple pathways that respond to distinct signals.

Hypoxia could be proposed a priori to reduce seed metabolite storage owing to less efficient ATP production and the consequences of reduced assimilate uptake due to reduced ATP-driven transport. On the other hand, a selective advantage might arise from the hypoxic condition of the seed. One possibility is that hypoxia reduces the levels of reactive oxygen species that would otherwise arise. Hypoxia could also be beneficial by creating local metabolic adjustments. The latter has been demonstrated for the developing (hypoxic) endosperm of barley, where an appreciable level of locally established alanine aminotransferase activity limits accumulation of fermentation products and eventually improves the efficiency of nitrogen and energy utilization in the grain (Rolletschek *et al.*, 2011). Such metabolic compartmentation is likely to occur in maize endosperm as well, and this might provide a mechanism to ensure metabolic flexibility and contribute to its high carbon conversion efficiency (Alonso *et al.*, 2011). Hypoxic constraints are not uniformly experienced and any imbalances are corrected *in vivo* by local metabolic adjustments. An important future task is to unravel the mechanisms involved. This will greatly enhance our understanding of the regulatory principles governing the central metabolism in the maize kernel and provide novel avenues

for its improvement/manipulation. Specifically in this regard, storage-compound accumulation and possibly kernel size might be increased by more efficient ATP production.

### 11.2.6 Metabolic recycling reactions and metabolic flux analysis

Additional complexity in the bioenergetics of maize SE results from metabolic recycling reactions (also discussed in Chapter 12), where breakdown of some hexoses through partial glycolysis is followed by resynthesis of intermediates back into hexose phosphate, which is then incorporated into starch. This requires ATP expenditure, and the more recycling that occurs the lower the ratio of Glc units that eventually end up in starch and protein compared with ATP production.

Maize kernels can develop *ex planta* on cob sections in synthetic medium, which allows $^{14}C$- or $^{13}C$-labeling in a carbon-site-specific manner (Gengenbach, 1977; Gengenbach and Jones, 1994). If the hexose incorporates directly into the starch, then labeling with $^{13}C_1$-Glc results in the isotope at the $C_1$ position. In contrast, one such study in wheat endosperm found that 15% of the Glc units in starch were labeled at the $C_6$ position (Keeling *et al.*, 1988). This requires that some portion of the Glc be derived from the glycolytic intermediates, dihydroxyacetone phosphate (DHAP), and glyceraldehyde-3-phosphate (Gly-3-P). Interconversion of these two compounds changes the position of the labeled carbon when they are resynthesized back into Glc-1-P, resulting in the $^{13}C_6$ label. The route of resynthesis could conceivably involve parts of gluconeogenesis, the pentose phosphate pathway, and/or other metabolic pathways (Ettenhuber *et al.*, 2005; Spielbauer *et al.*, 2006; Alonso *et al.*, 2011). There is a debate about the percentage of Glc residues in starch that have been recycled, rather than incorporated directly, but in all instances examined there was clear evidence for some degree of catabolism to trioses and resynthesis of Glc-1-P. Strong support for recycling comes from a mutant defective in the plastid oxPPP that exhibited substantially reduced starch accumulation (Spielbauer *et al.*, 2013).

The *ex planta* ear culture method is amenable to metabolic flux analysis using combined $^{14}$C- and $^{13}$C- labeling, with metabolite tracking by GC-MS and NMR (Alonso *et al.*, 2011). Using computer modeling, these data predict that 30–50% of the total ATP produced in SE is used for purposes other than starch and protein synthesis. Further, these studies demonstrated up to 47% of the total ATP is used for substrate cycling. These are rough estimates based on computational assumptions; however, repeated flux analyses using this approach make it possible to test and refine the models. The metabolic flux pathways varied substantially between endosperm and embryo, as expected considering the many differences between their tissues, including biomass composition and steady-state $O_2$ level. Metabolic flux analyses of the SE and embryo suggest substantial contributions of plastids, which underscores unknowns resulting from dual localization of metabolic enzymes.

### 11.2.7 System perturbation—What happens to metabolism when conditions change?

Models of SE metabolism make predictions that can be tested by monitoring the effects of perturbations. Rolletschek *et al.* (2005) increased $O_2$ partial pressure during SE development and found changes in adenylate energy status, ATP gradients, metabolite levels, and assimilate partitioning. These results are consistent with $O_2$ as rate limiting and emphasize the fine balance between catabolic reactions that produce ATP and anabolic pathways that utilize ATP.

Other approaches to block specific metabolic steps use enzyme inhibitors or loss of function mutations affecting enzymes in specific pathways. For example, when several mutant lines were isotopically labeled in the *ex planta* culture system (Spielbauer *et al.*, 2006, 2013), the degree of recycling through trioses and back to glucose was increased when cytosolic AGPase

was eliminated. In other instances, mutation of specific metabolic steps did not affect $^{13}$C labeling patterns (Spielbauer *et al.*, 2006) or other measurable aspects of metabolism. These included single mutations of (i) sucrose synthase; (ii) the BT1 plastid membrane ADPGlc transporter (see Chapter 12); (iii) CWIN in the BETL; and (iv) several starch assembly enzymes. Thus, it appears that the SE metabolic system is adaptable to changes in many different reactions. The consequences of only a few mutations have been tested in this way, but many could be analyzed in culture and this could be a fruitful avenue for future investigation. Analyzing double mutant combinations could also be informative.

The stability of the SE metabolic network apparently extends to the cellular and molecular level. Certain mutations in cell cycle regulation were shown to affect the number and size of SE cells, the degree of endoreduplication, and the pattern of programmed cell death (Sabelli *et al.*, 2013; Chapter 10, this volume). Despite the nature of these global perturbations, neither kernel biomass, in terms of kernel weight, nor storage protein content are significantly altered.

Clearly many questions remain about how metabolism in SE adjusts to signals that maintain homeostasis or substantially alter the metabolic state. First-order responses for modulation of metabolism likely include allosteric enzyme regulation. Energy charge, i.e. the ratio of ATP to ADP + AMP, is well known as a regulator of many controlling enzymes in glycolysis and the citric acid cycle, and it is likely to be tightly controlled in SE owing to low $O_2$ availability. PPi availability may also be regulated and contribute to ATP production efficiency. Redox regulation is also important for multiple enzymes active in SE, and this could vary substantially in heterogeneous regions of the tissue. Beyond these immediate metabolic responses are those that require transcriptional reprogramming, and for the most part these remain to be characterized. Yet another likely means of adjusting kernel metabolism in defined microenvironments or in response to external stimuli is post-translational modification. Many of the enzymes involved in the metabolic pathways discussed here, including the BT1 transporter, one of the sucrose synthase isoforms, AGPase, and several others are phosphorylated (Duncan *et al.*, 2006; Walley *et al.*, 2013). This and other post-translational modifications in maize kernels are areas requiring further investigation.

Overall, the results of these perturbations suggest that SE has a fine balance between oxidative metabolism, involving the citric acid cycle and mETC, and glycolysis where far less ATP per Glc is generated. Reverse genetics, likely using CRISPR genome editing together with SE-specific RNAi, will be useful in continuing these investigations and researching a great variety of enzyme deficiencies for global changes in metabolite fluxes.

## 11.3 Future Directions

We are at the very early stages of understanding metabolic fluxes in maize kernels, and knowledge gained from future research addressing central metabolic processes will further the long-term goal of manipulating metabolism for specific outcomes. These goals include increased yield, improved nutrition, adaptation to environmental stress, systems engineering for alternative bioproducts etc. Advances in our understanding of approaches to achieve these ends require some of the following:

### 11.3.1 High resolution characterization of cellular spatial heterogeneity within kernels

Experimental tools and instrumentation for advanced metabolomics studies need to be developed to provide spatially-resolved analyses. One emerging approach is to dissect specific regions of the kernel and obtain detailed biomass composition, followed by flux balance analysis (FBA) that predicts metabolic pathway activities that could generate the respective biomass compositions and thus provide spatially resolved flux maps. Initial FBA models for maize embryo and endosperm are available (Seaver *et al.*, 2015). While this approach neglects developmental aspects, it can uncover local metabolic networks and heterogeneity within the kernel. Based on this approach, researchers can establish links between cell cycle regulation, gene expression programs, and specific metabolic networks fueling the respective demands. It could also help address how the sink strength of particular kernel regions is determined, e.g. what mechanisms regulate partitioning of the sugar supply between embryo and endosperm?

### 11.3.2 Expanded application of metabolic flux analysis

Much of our current description of metabolic states relies on transcriptome and proteome data, along with steady-state levels of metabolites; however, these parameters are often poorly correlated with metabolic fluxes (Schwender *et al.*, 2014; Walley *et al.*, 2013). A better understanding of central metabolism will require greater knowledge of fluxes, although this information is technically challenging and difficult to obtain. To date, such studies on maize kernels are limited (Alonso *et al.*, 2010, 2011). Future advances will require expanded application of the *ex planta* ear development system, use of mutants and transgenic genome modification in this system, and development of experimental approaches suitable for whole plant analyses.

### 11.3.3 Transcriptional networks

Transcriptional programming of metabolic enzymes and related regulatory factors has received less attention in maize kernels than other plant tissues. Identification of transcription factors and their targets and how they respond to developmental and environmental cues merit further investigation. Included in these studies is how hypoxia influences global regulation of gene expression in SE and other kernel tissues.

In maize, G4 motif-containing nuclear genes presumably play key roles in energy metabolism, hypoxic acclimation, and nutrient signaling pathways (Andorf *et al.*, 2014). Further investigations of functional links between G4 motifs in the maize genome and genes coupled to energy status and/or hypoxia are warranted.

### 11.3.4 Assimilate supply, maternal–filial interaction, and filial–filial relationships

This chapter focuses on metabolic processes associated with the primary metabolic fuel in SE cells, i.e., Suc, Glc, and Frc. A global understanding of kernel metabolism requires knowledge of assimilate supply to the maternal tissues that house the ovule, interactions between the maternal and filial tissues, and interactions between adjacent ovules that can affect seed size, kernel abortion, and mature kernel number. For example, are there differences in tissue heterogeneity between individual kernels depending on ear position or pollination efficiency? Other questions arise regarding assimilate supply during the diurnal cycle. Diurnal rhythms in source leaf metabolism are well characterized; however, there is limited knowledge about them in sink organs, including maize kernels. One study showed that ATP/ADP/AMP ratios in maize SE endosperm vary with the diurnal cycle, yet the levels of ADPGlc and UDPGlc do not (Scott, 2000). This implies adjustments in metabolism depending on assimilate supply or other consequences of the diurnal cycle. In barley grains, diurnal shifts in the transcriptome, assimilate import, and central metabolic pathways have been demonstrated (Mangelsen *et al.*, 2010; Rolletschek *et al.*, 2015). Additional research is required to understand how maize kernel metabolism is affected by this natural cycle.

Another poorly understood aspect of assimilate transport relates to the finding that phloem transport of carbohydrate toward maize ovaries is coupled with water potential/water availability (Mäkelä *et al.*, 2005). Similar observations were made for developing barley grains (Rolletschek *et al.*, 2015). This suggests water and assimilate uptake into the sink structure, i.e. the maize kernel, are tightly coupled by an unknown mechanism(s) (see Chapter 15).

Potential effects of SE metabolism on ovary development also require further attention. This is illustrated by results of a study in which a transgenic AGPase with altered catalytic properties was expressed in SE and found to increase yield under certain environmental conditions (Hannah *et al.*, 2012). Rather than affecting metabolism within the SE, however, the transgene was found to affect seed number by increasing the frequency at which ovary primordia develop into mature seeds. The connection between a key enzyme in starch synthesis and seed development is obscure, and further study of this phenomenon is likely to be informative.

### 11.3.5 Genetic manipulation of the degree of hypoxia

Unraveling the effect of hypoxia on kernel metabolism and development can be addressed with mutations that alter the structure/composition of the outermost layers of the kernel, e.g. in cutin or suberin. Such plants might be altered in kernel gas exchange characteristics, allowing effects of increased endogenous oxygen levels to be assessed. The kernels might show changes in starch and protein accumulation, lipid content, partitioning of assimilates between the SE and the embryo. Mutants with altered gas exchange at the surface of the kernel could also help unravel hypoxia-induced signaling processes. For example, ethylene induces PCD in maize endosperm (Young *et al.*, 1997; Sabelli and Larkins, 2009) and is produced in response to hypoxia (Mustroph *et al.*, 2010). Whether or not there is a link between hypoxia, ethylene, and PCD could be examined by altering the steady-state $O_2$ level of SE in mutant lines.

# References

Alonso, A.P., Dale, V.L. and Shachar-Hill, Y. (2010) Understanding fatty acid synthesis in developing maize embryos using metabolic flux analysis. *Metabolic Engineering* 12, 488–497.

Alonso, A.P., Val, D.L. and Shachar-Hill, Y. (2011) Central metabolic fluxes in the endosperm of developing maize seeds and their implications for metabolic engineering. *Metabolic Engineering* 13, 96–107.

Andorf, C.M., Kopylov, M., Dobbs, D., Koch, K.E., Stroupe, M.E., Lawrence, C.J. and Bass, H.W. (2014) G-quadruplex (G4) motifs in the maize (*Zea mays* L.) genome are enriched at specific locations in thousands of genes coupled to energy status, hypoxia, low sugar, and nutrient deprivation. *Journal of Genetics and Genomics* 41, 627–647.

Benamar, A., Rolletschek, H., Borisjuk, L., Avelange-Macherel, M.-H., Curien, G., *et al.* (2008) Nitrite-nitric oxide control of mitochondrial respiration at the frontier of anoxia. *Biochimica et Biophysica Acta* 1777, 1268–1275.

Borisjuk, L. and Rolletschek, H. (2009) The oxygen status of the developing seed. *New Phytologist* 182, 17–30.

Cheng, W.H., Taliercio, E.W. and Chourey, P.S. (1996) The miniature1 seed locus of maize encodes a cell wall invertase required for normal development of endosperm and maternal cells in the pedicel. *Plant Cell* 8, 971–983.

Chourey, P.S., Taliercio, E.W., Carlson, S.J. and Ruan, Y.-L. (1998) Genetic evidence that the two isozymes of sucrose synthase present in developing maize endosperm are critical, one for cell wall integrity and the other for starch biosynthesis. *Molecular and General Genetics* 259, 88–96.

Chourey, P.S., Jain, M., Li, Q.B. and Carlson, S.J. (2006) Genetic control of cell wall invertases in developing endosperm of maize. *Planta* 223, 159–167.

Cloetens, P., Schlenker, M. and Lerbs-Mache, S. (2006) Quantitative phase tomography of *Arabidopsis* seeds reveals intercellular void network. *Proceedings of the National Academy of Sciences of the United States of America* 103, 14626–14630.

Cobb, B.G. and Hannah, L.C. (1986) Sugar utilization by developing wild-type and *shrunken-2* maize kernels. *Plant Physiology* 80, 609–611.

Cobb, B.G. and Hannah, L.C. (1988) *Shrunken-1* encoded sucrose synthase is not required for sucrose synthesis in the maize endosperm. *Plant Physiology* 88, 1219–1221.

de Sousa, S.M., Paniago Mdel, G., Arruda, P. and Yunes, J.A. (2008) Sugar levels modulate sorbitol dehydrogenase expression in maize. *Plant Molecular Biology* 68, 203–213.

Domínguez, F. and Cejudo, F.J. (2014) Programmed cell death (PCD): an essential process of cereal seed development and germination. *Frontiers in Plant Science* 5, 366.

Duncan, K.A., Hardin, S.C. and Huber, S.C. (2006) The three maize sucrose synthase isoforms differ in distribution, localization, and phosphorylation. *Plant Cell Physiology* 47, 959–971.

Edwards, J., ap Rees, T., Wilson, P.M. and Morrell, S. (1984) Measurement of the inorganic pyrophosphate in tissues of *Pisum sativum* L. *Planta* 162, 188–191.

Ettenhuber, C., Spielbauer, G., Margl, L., Hannah, L.C., Gierl, A., *et al.* (2005) Changes in flux pattern of the central carbohydrate metabolism during kernel development in maize. *Phytochemistry* 66, 2632–2642.

Felker, F.C. and Goodwin, J.C. (1988) Sugar uptake by maize endosperm suspension cultures. *Plant Physiology* 88, 1235–1239.

Felker, F.C. and Shannon, J.C. (1980) Movement of C-labeled assimilates into kernels of *Zea mays* L. III. An anatomical examination and microautoradiographic study of assimilate transfer. *Plant Physiology* 65, 864–870.

Gayral, M., Bakan, B., Dalgalarrondo, M., Elmorjani, K., Delluc, C., *et al.* (2015) Lipid partitioning in maize (*Zea mays* L.) endosperm highlights relationships among starch lipids, amylose, and vitreousness. *Journal of Agricultural and Food Chemistry* 63, 3551–3558.

Gengenbach, B.G. (1977) Development of maize caryopsis resulting from *in vitro* pollination. *Planta* 134, 91–93.

Gengenbach, B.G. and Jones, R. (1994) *In vitro* culture of maize kernels. In: Freeling, M. and Walbot, V. (eds.) *The Maize Handbook*, Springer-Verlag, Berlin.

Griffith, S.M., Jones, R.J. and Brenner, M.L. (1987) *In vitro* sugar transport in *Zea mays* L. kernels. I. Characteristics of sugar absorption and metabolism by developing maize endosperm. *Plant Physiology* 84, 467–471.

Hannah, L.C., Futch, B., Bing, J., Shaw, J.R., Boehlein, S., *et al.* (2012) A *shrunken-2* transgene increases maize yield by acting in maternal tissues to increase the frequency of seed development. *Plant Cell* 24, 2352–2363.

Huang, S., Colmer, T.D. and Millar, A.H. (2008) Does anoxia tolerance involve altering the energy currency towards PPi? *Trends in Plant Science* 13, 221–227.

Igamberdiev, A.U., Stasolla, C. and Hill, R.D. (2014) Low oxygen stress, nonsymbiotic hemoglobins, NO, and programmed cell death. In: van Dongen, J.T. and Licausi, F. (eds.) *Low Oxygen Stress in Plants*. Springer-Verlag, Vienna.

Keeling, P.L., Wood, J.R., Tyson, R.H. and Bridges, I.G. (1988) Starch biosynthesis in developing wheat grain: evidence against the direct involvement of triose phosphates in the metabolic pathway. *Plant Physiology* 87, 311–319.

Li, J., Baroja-Fernández, E., Bahaji, A., Muñoz, F.J., Ovecka, M., *et al.* (2013) Enhancing sucrose synthase activity results in increased levels of starch and ADP-glucose in maize (*Zea mays* L.) seed endosperms. *Plant Cell Physiology* 54, 282–294.

Liseron-Monfils, C., Bi, Y.M., Downs, G.S., Wu, W., Signorelli, T., *et al.* (2013) Nitrogen transporter and assimilation genes exhibit developmental stage-selective expression in maize (*Zea mays* L.) associated with distinct cis-acting promoter motifs. *Plant Signaling & Behavior* 8, e26056.

Mäkelä, P., McLaughlin, J.E. and Boyer, J.S. (2005) Imaging and quantifying carbohydrate transport to the developing ovaries of maize. *Annals of Botany* 96, 939–949.

Mangelsen, E., Wanke, D., Kilian, J., Sundberg, E., Harter, K. and Jansson, C. (2010) Significance of light, sugar, and amino acid supply for diurnal gene regulation in developing barley caryopses. *Plant Physiology* 153, 14–33.

Martinoia, E., Maeshima, M. and Neuhaus, H.E. (2007) Vacuolar transporters and their essential role in plant metabolism. *Journal of Experimental Botany* 58, 83–102.

McCarty, D.R., Shaw, J.R. and Hannah, L.C. (1986) The cloning, genetic mapping, and expression of the constitutive sucrose synthase locus of maize. *Proceedings of the National Academy of Sciences of the United States of America* 83, 9099–9103.

Mechin, V., Thevenot, C., Le Guilloux, M., Prioul, J.L. and Damerval, C. (2007) Developmental analysis of maize endosperm proteome suggests a pivotal role for pyruvate orthophosphate dikinase. *Plant Physiology* 143, 1203–1219.

Mohanty, B. and ap Rees, T. (1992) Demonstration and measurement of inorganic pyrophosphate in potato tubers. *Potato Research* 35, 195–198.

Muñoz-Bertomeu, J., Cascales-Minaña, B., Mulet, J.M., Baroja-Fernández, E., Pozueta-Romero, J., *et al.* (2009) Plastidial glyceraldehyde-3-phosphate dehydrogenase deficiency leads to altered root development and affects the sugar and amino acid balance in Arabidopsis. *Plant Physiology* 151, 541–558.

Mustroph, A., Lee, S.C., Oosumi, T., Zanetti, M.E., Yang, H., *et al.* (2010) Cross-kingdom comparison of transcriptomic adjustments to low-oxygen stress highlights conserved and plant-specific responses. *Plant Physiology* 152, 1484–1500.

Myers, A.M., James, M.G., Lin, Q., Yi, G., Stinard, P.S., Hennen-Bierwagen, T.A. and Becraft, P.W. (2011) Maize *opaque5* encodes monogalactosyldiacylglycerol synthase and specifically affects galactolipids necessary for amyloplast and chloroplast function. *Plant Cell* 23, 2331–2347.

Plaxton, W.C. (1996) The organization and regulation of plant glycolysis. *Annual Review of Plant Physiology and Plant Molecular Biology* 47, 185–214.

Porter, G.A., Knievel, D.P. and Shannon, J.C. (1985) Sugar efflux from maize (*Zea mays* L.) pedicel tissue. *Plant Physiology* 77, 524–531.

Prioul, J.L., Mechin, V., Lessard, P., Thevenot, C., Grimmer, M., *et al.* (2008) A joint transcriptomic, proteomic and metabolic analysis of maize endosperm development and starch filling. *Plant Biotechnology Journal* 6, 855–869.

Radchuk, V. and Borisjuk, L. (2014) Physical, metabolic and developmental functions of the seed coat. *Frontiers in Plant Science* 5, 510.

Rolletschek, H., Koch, K., Wobus, U. and Borisjuk, L. (2005) Positional cues for the starch/lipid balance in maize kernels and resource partitioning to the embryo. *Plant Journal* 42, 69–83.

Rolletschek, H., Borisjuk, L., Sánchez-García, A., Gotor, C., Romero, L.C., Martínez-Rivas, J.M. and Mancha, M. (2007) Temperature-dependent endogenous oxygen concentration regulates microsomal oleate desaturase in developing sunflower seeds. *Journal of Experimental Botany* 58, 3171–3181.

Rolletschek, H., Melkus, G., Grafahrend-Belau, E., Fuchs, J., Heinzel, N., *et al.* (2011) Combined noninvasive imaging and modeling approaches reveal metabolic compartmentation in the barley endosperm. *Plant Cell* 23, 3041–3054.

Rolletschek, H., Grafahrend-Belau, E., Munz, E., Radchuk, V., Kartausch, R., *et al.* (2015) Metabolic architecture of the cereal grain and its relevance to maximize carbon use efficiency. *Plant Physiology* 169, 1698–1713.

Sabelli, P.A. (2012) Replicate and die for your own good: endoreduplication and cell death in the cereal endosperm. *Journal of Cereal Science* 56, 9–20.

Sabelli, P.A. and Larkins, B.A. (2009) The development of endosperm in grasses. *Plant Physiology* 149, 14–26.

Sabelli, P.A., Liu, Y., Dante, R.A., Lizarraga, L.E., Nguyen, H.N., *et al.* (2013) Control of cell proliferation, endoreduplication, cell size, and cell death by the retinoblastoma-related pathway in maize endosperm. *Proceedings of the National Academy of Sciences of the United States of America* 110, E1827–E1836.

Schmalstig, J.G. and Hitz, W.D. (1987) Transport and metabolism of a sucrose analog (1'-fluorosucrose) into *Zea mays* L. endosperm without invertase hydrolysis. *Plant Physiology* 85, 902–905.

Schwender, J., Konig, C., Klapperstuck, M., Heinzel, N., Munz, E., *et al.* (2014) Transcript abundance on its own cannot be used to infer fluxes in central metabolism. *Frontiers in Plant Science* 5, 668.

Scott, M.P. (2000) Diurnal and developmental changes in levels of nucleotide compounds in developing maize endosperms. *Plant, Cell and Environment* 23, 1281–1286.

Seaver, S.M., Bradbury, L.M., Frelin, O., Zarecki, R., Ruppin, E., Hanson, A.D. and Henry, C.S. (2015) Improved evidence-based genome-scale metabolic models for maize leaf, embryo, and endosperm. *Frontiers in Plant Science* 6, 142.

Shannon, J.C. (1968) Carbon-14 distribution in carbohydrates of immature *Zea mays* kernels following $^{14}CO_2$ treatment of intact plants. *Plant Physiology* 43, 1215–1220.

Shannon, J.C. (1972) Movement of $^{14}$C-labeled assimilates into kernels of *Zea mays* L. I. Pattern and rate of sugar movement. *Plant Physiology* 49, 198–202.

Shannon, J.C. and Dougherty, C.T. (1972) Movement of $^{14}$C-labeled assimilates into kernels of *Zea mays* L. II. Invertase activity of the pedicel and planceno-chalazal tissues. *Plant Physiology* 49, 203–206.

Simontacchi, M., Caro, A. and Puntarulo, S. (1995) Oxygen-dependent increase of antioxidants in soybean embryonic axes. *The International Journal of Biochemistry & Cell Biology* 27, 1211–1229.

Sosso, D., Luo, D., Li, Q.-B., Sasse, J., Yang, J., *et al.* (2015) Seed filling in domesticated maize and rice depends on SWEET-mediated hexose transport. *Nature Genetics* 47, 1489–1493.

Spielbauer, G., Margl, L., Hannah, L.C., Romisch, W., Ettenhuber, C., *et al.* (2006) Robustness of central carbohydrate metabolism in developing maize kernels. *Phytochemistry* 67, 1460–1475.

Spielbauer, G., Li, L., Romish-Margl, L., Do, P.T., Fouquet, R., *et al.* (2013) Chloroplast-localized 6-phosphogluconate dehydrogenase is critical for maize endosperm starch accumulation. *Journal of Experimental Botany* 64, 2231–2242.

Stitt, M. (1990) Fructose-2,6-bisphosphate as a regulatory molecule in plants. *Annual Review of Plant Physiology and Plant Molecular Biology* 41, 153–185.

Tegeder, M. (2014) Transporters involved in source to sink partitioning of amino acids and ureides: opportunities for crop improvement. *Journal of Experimental Botany* 65, 1865–1878.

Tiessen, A., Nerlich, A., Faix, B., Hummer, C., Fox, S., *et al.* (2012) Subcellular analysis of starch metabolism in developing barley seeds using a non-aqueous fractionation method. *Journal of Experimental Botany* 63, 2071–2087.

Tschiersch, H., Liebsch, G., Borisjuk, L., Stangelmayer, A. and Rolletschek, H. (2012) An imaging method for oxygen distribution, respiration and photosynthesis at a microscopic level of resolution. *New Phytologist* 196, 926–936.

van Dongen, J.T. and Licausi, F. (2015) Oxygen sensing and signaling. *Annual Review of Plant Biology* 66, 345–367.

van Dongen, J.T., Roeb, G.W., Dautzenberg, M., Froehlich, A., Vigeolas, H., Minchn, P.E.H. and Geigenberger, P. (2004) Phloem import and storage metabolism are highly coordinated by the low oxygen concentrations within developing wheat seeds. *Plant Physiology* 135, 1809–1821.

Verboven, P., Herremans, E., Borisjuk, L., Helfen, L., Ho, Q.T., *et al.* (2013) Void space inside the developing seed of *Brassica napus* and the modelling of its function. *New Phytologist* 199, 936–947.

Walley, J.W., Shen, Z., Sartor, R., Wu, K.J., Osborn, J., Smith, L.G. and Briggs, S.P. (2013) Reconstruction of protein networks from an atlas of maize seed proteotypes. *Proceedings of the National Academy of Sciences of the United States of America* 110, E4808–E4817.

Wong, J.H., Balmer, Y., Cai, N., Tanaka, C.K., Vensel, W.H., Hurkman, W.J. and Buchanan, B.B. (2003) Unraveling thioredoxin-linked metabolic processes of cereal starchy endosperm using proteomics. *FEBS Letters* 547, 151–156.

Young, T.E. and Gallie, D.R. (2000) Programmed cell death during endosperm development. *Plant Molecular Biology* 44, 283–301.

Young, T.E., Gallie, D.R. and DeMason, D.A. (1997) Ethylene-mediated programmed cell death during maize endosperm development of wild-type and *shrunken2* genotypes. *Plant Physiology* 115, 737–751.

# 12 Starch Biosynthesis in Maize Endosperm

L. Curtis Hannah* and Susan Boehlein

*Program in Plant Molecular and Cellular Biology and Horticultural Sciences, University of Florida, Gainesville, Florida, USA*

## 12.1 Introduction

Starch constitutes approximately 70% of the maize kernel and provides energy for the germinating seed, allowing embryo growth and development until the seedling is photosynthetically active. Starch provides approximately 50% of calories for humans and other animals and is used to manufacture many products, including biofuels. With incipient climate change and adverse environmental conditions, plant scientists are challenged to find ways to enhance starch synthesis in order to meet the needs of a growing human population.

Because of the importance of starch and the availability of seminal mutants affecting its biosynthesis, our knowledge of this process is robust. Here we describe our current understanding of this process in maize endosperm, the primary storage site for starch in the kernel. Our understanding of this process is incomplete, and questions are identified that serve as a guide for those interested in investigating starch synthesis.

## 12.2 Maize Endosperm Starch

Starch is deposited in endosperm cells as insoluble granules in amyloplast stroma. Starch consists of two relatively simple glucose homopolymers, linear amylose and branched amylopectin. In amylose, glucose residues are almost exclusively linked through α-1,4 glycosidic bonds; amylopectin has the same backbone, but differs by the presence of α-1,6 branch points that comprise ~3–5% of its glycosidic bonds (Fig. 12.1). The non-randomly placed, or clustered, side branches in amylopectin form double helices between pairs of linear chains, and this extends into higher-order packing that gives rise to the semi-crystalline structure of starch granules. Amylose molecules typically contain several hundred to several thousand glucose residues, whereas amylopectin contains hundreds of thousands of glucose molecules. The linear chain length population of amylopectin ranges from approximately 7 to 50 residues. In structural terms, amylopectin is the same as glycogen found in bacterial, fungal, and mammalian cells; however, amylopectin has approximately one half the

*Corresponding author e-mail: lchannah@ufl.edu

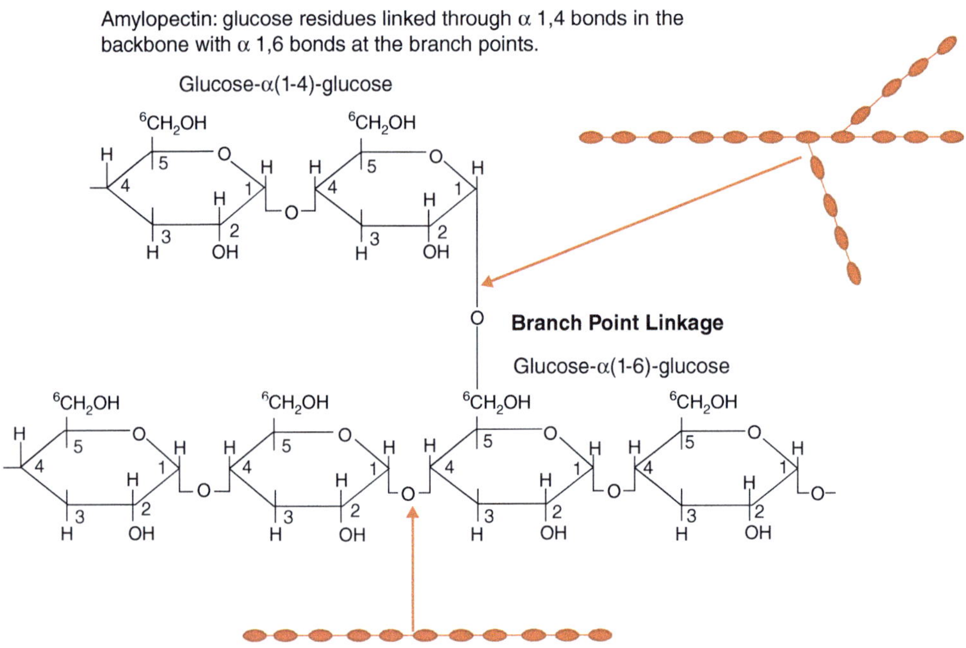



Fig. 12.1. Structures of amylose and amylopectin.

branch frequency as glycogen, and they are clustered rather than dispersed. This accounts for the defining difference: glycogen is soluble and amylopectin, assembled within starch granules, is not. The insolubility of starch drives the conversion of sugars to starch. It is interesting to speculate what plants would be like if amylopectin were soluble.

The simplicity of starch structure contrasts with the complexity of its synthesis. An illustration of some of the enzymatic steps involved in this process is shown in Fig. 12.2 and details are explained in a series of reviews (Hannah, 2007; Preiss 2009; Jeon *et al.*, 2010; Keeling and Myers, 2010) and below. Sucrose entering the endosperm is cleaved initially by sucrose synthase to form fructose and UDP-glucose. A series of reactions catalyzed by UDP-glucose pyrophosphorylase, phosphoglucomutase, phosphoglucoisomerase, and hexose kinases give rise to phosphorylated glucose and fructose residues. The majority of the resulting glucose-1-phosphate then cycles through the oxidative pentose phosphate pathway before it is used by ADP-glucose pyrophosphorylase (AGPase) in the cytosol to synthesize the starch precursor, ADP-glucose (ADPG). ADPG then enters the amyloplast where starch synthases, starch branching enzymes, and starch debranching enzymes are used to synthesize starch. These steps are addressed in more detail in Chapter 11 in this volume.

## 12.3 Unexpected Complexity in the Starch Biosynthetic Pathway

Several enzymatic steps in this pathway are more convoluted than researchers expected. The first surprise was that sucrose is synthesized twice. When it enters the seed, sucrose is rapidly cleaved by invertase into glucose and fructose; it is then resynthesized into sucrose. This discovery was based on classic pulse chase experiments (Shannon, 1968). Maize plants with developing ears were fed $^{14}CO_2$ and the amounts of labeled glucose, fructose, sucrose, and

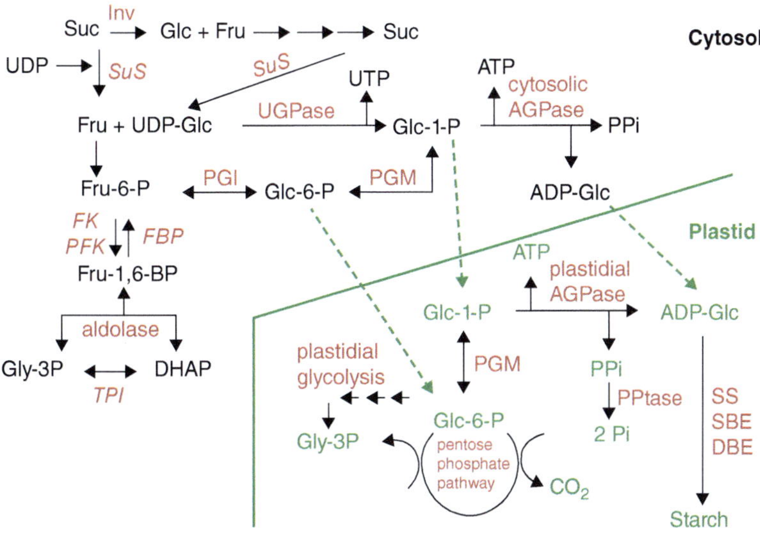


**Fig. 12.2.** Selected aspects of the starch biosynthetic pathway in the maize endosperm. Suc, sucrose; Fru, fructose; Glc, glucose; Glc-1-P, glucose 1-phosphate; Glc-6-P, glucose 6-phosphate; Fru-6-P, fructose 6-phosphate; Fru-1,6-BP, fructose 1,6 bisphosphate; Gly-3-P, glycerol 3-phosphate; DHAP, dihydroxyacetone phosphate; PPi, pyrophosphate; Pi, inorganic phosphate; ADP-Glc, ADP glucose; Inv, invertase; SuS, sucrose synthase; UGPase, UDP-glucose pyrophosphorylase; AGPase, ADP-glucose pyrophosphorylase; FK, fructokinase; PFK, phosphofructokinase; TPI, triose phosphate isomerase; PK, pyruvate kinase; PPtase pyrophosphatase; PGM, phosphoglucomutase; PGI, phosphoglucoisomerase; FBP, fructose 1,6 bisphosphatase; SS, starch synthase; SBE, starch branching enzyme and DBE, debranching enzyme.

starch were monitored over a chase period. While $^{14}$C was rapidly chased from glucose and fructose into starch, labeled sucrose increased during the early time points. This and other observations were interpreted to mean sucrose was degraded and then resynthesized as it moved from the leaf to the endosperm. In agreement with this hypothesis, it is now known that mutation of either invertase or sucrose synthase reduces kernel starch content. Cheng *et al.* (1996) showed cell wall invertase, primarily at the base of the kernel, is greatly reduced in maize mutants of the *miniature-1* (*Mn1*) locus. Loss of *Mn1* function increases sucrose in the seed and reduces starch content. However, *mn1* mutants have pleiotropic effects; hence a straightforward interpretation of their phenotypes is difficult.

There is clear evidence for a second sucrose cleavage reaction. Chourey and Nelson (1976, 1979) showed that the major endosperm sucrose synthase is encoded by the *shrunken-1* (*Sh1*) locus. Based on kinetic

considerations, this enzyme is thought to function in the direction of sucrose degradation. Consistent with the idea that sucrose synthase is actually degrading sucrose, a cross between two particular *sh1* mutants unexpectedly produced a plump (non-mutant) kernel. Enzyme assays of the heterozygote, exhibiting interallelic complementation, showed elevated activity in the direction of sucrose cleavage, but not in synthesis. In addition and in agreement with this observation, Cobb and Hannah (1988) found that wild-type and *shrunken-2* (*Sh2*) mutant kernels exhibit superior growth and development when grown on sucrose rather than glucose. Kernels lacking *Sh1* function have the opposite pattern: they exhibit superior development on reducing sugars. This latter observation supports the hypothesis that the *Sh1*-encoded sucrose synthase cleaves sucrose and supplementation of reducing sugars bypasses the defect in this mutant.

Given the structures of amylose and amylopectin, it was assumed amylose is the substrate for amylopectin synthesis. However, discovery that mutation at the *waxy* locus greatly reduces amylose and has little effect on amylopectin, coupled with work of Nelson and Rines (1962) who showed the *waxy* gene encodes a granule-bound starch synthase, led to the hypothesis that amylose and amylopectin are synthesized via parallel pathways.

Another unexpected observation regarding starch synthesis in maize endosperm is that that glycoside bond hydrolysis must occur to generate wild-type starch levels. This insight came from analysis of the *sugary1* mutants used extensively in the first sweet corn varieties. Loss of *sugary1* (*Su1*) function leads to elevated levels of sucrose, decreased amylopectin and an increase in the more highly branched and soluble polymer, phytoglycogen. Analysis of the *Su1* gene (James *et al.*, 1995) revealed that it encodes an isoamylase that hydrolyzes α-1,6 bonds in amylopectin. Hence, wild-type starch synthesis requires activities of starch synthase, starch branching enzyme, and starch debranching activity. The role of starch debranching activity in attaining normal levels of starch synthesis is currently unknown.

Yet another unexpected finding was that most of the glucose residues in sucrose are not incorporated directly into starch. Rather, they are first metabolized to triose phosphates and then resynthesized into glucose before starch synthesis. This was first demonstrated in wheat (Keeling *et al.*, 1988) and subsequently shown in maize from experiments in which developing kernels were fed two forms of glucose, including one in which each of the six carbons was labeled with $^{13}$C (Ettenhuber *et al.*, 2005; Spielbauer *et al.*, 2006). Starch from mature kernels was isolated and the location of $^{13}$C in glucose molecules was determined. If glucose were directly incorporated into starch, one would expect the labeled glucose to contain $^{13}$C at all six positions. However, the majority contained $^{13}$C at carbons 1 and 2 and 3 or 4 and 5 and 6. Less than 20% had labels matching the starting molecule. Furthermore, decreasing the rate of ADPG incorporation into starch by loss of *Sh2* and *brittle2* (*Bt2*), which encode

AGPase, reduced the amount of glucose incorporated directly into starch. This labeling pattern is consistent with synthesis of triose phosphates from $^{13}$C glucose, followed by glucose resynthesis as precursor steps in starch synthesis. The authors proposed glycolytic cleavage of glucose and passage through the pentose phosphate pathway as important steps in starch synthesis. Interestingly, loss of *brittle1* (*Bt1*) function did not alter the pattern of glucose labeling. *Bt1* encodes the ADPG transporter that is important for movement of ADPG into the amyloplast (above), so this indicates the likelihood of extensive metabolic plasticity in developing endosperm.

Genetic evidence for the importance of the pentose phosphate pathway in starch synthesis came with the discovery that the maize *Pgd3* gene encodes a chloroplast-localized 6-phosphogluconate dehydrogenase (Spielbauer *et al.*, 2013). Loss of this enzyme leads to a reduction in endosperm starch comparable to that found with severe mutant alleles of *Sh2*, *Bt1*, and *Bt2*.

Finally, it was surprising that the majority of endosperm AGPase is located in the cytosol, while in the leaf, tuber, and other tissues the enzyme is in the amyloplast stroma. AGPase synthesizes the glucose donor for starch, ADPG, and is thought to be a key regulatory enzyme for starch synthesis. A hint that the protein might be cytosolic came from the absence of a typical chloroplast transit sequence (Giroux and Hannah, 1994). Definitive evidence for its location came from Shannon *et al.* (1996, 1998), who showed that ADPG concentrations increase 16-fold in *bt1* mutants. *Bt1* encodes a protein with sequence similarity to a transporter protein (Sullivan *et al.*, 1991). Furthermore, Shannon *et al.* (1996) showed the *bt1* mutant-induced increase in ADPG does not occur when cytosolic AGPase is removed by incorporation of a *sh2* mutation. Hence, it was concluded that BT1 transports ADPG from the cytosol into the amyloplast. Independent evidence of this was furnished by Shannon *et al.* (1998) and Denyer *et al.* (1996) who showed the vast majority of AGPase is in the cytosol. Interestingly, Huang *et al.* (2014) reported that endosperm starch levels were reduced approximately 7% when the AGPase small subunit expressed primarily in the leaf was genetically removed.

There is no obvious reason why AGPase is located in different subcellular compartments in different plant tissues. The enzyme is usually allosteric, and perhaps the cytosolic location in the endosperm creates a more effective environment for regulation. However, analysis of metabolites in the cytosol and plastid of developing barley seed (Tiessen *et al.*, 2012) showed that while concentrations of the four reactants and two known activators of the enzyme, 3-phosphoglyceric acid and fructose-6-phosphate, are higher in the cytosol, the ratio of metabolite concentrations in the two locations is approximately equal.

It is interesting to note that cereal endosperm AGPases exhibit different heat stability, kinetic, and allosteric properties (reviewed in Boehlein *et al.*, 2013). The rice enzyme is heat stable, while those from maize, barley, and wheat are heat labile. The wheat and barley AGPases are almost recalcitrant to 3-PGA activation and Pi inhibition, but the maize and rice enzymes are responsive to the allosteric effectors. In addition, maize AGPase shows little Pi inhibition in the absence of 3-PGA, while the wheat and rice enzymes exhibit Pi inhibition that can be reversed by 3-PGA.

Accordingly to Tiessen *et al.* (2012), the Pi concentration in barley endosperm cytosol is high, 27 mM, a value some 1000-fold greater than typical $K_i$ values for plastid AGPases. If this concentration is common for cereal endosperms, then mechanisms to circumvent Pi inhibition of AGPase could be exploited to increase starch synthesis. Barley AGPase is not inhibited by Pi, while the maize endosperm enzyme has biphasic inhibition, with 50% inhibition possible regardless of other conditions.

## 12.4 Important Known Unknowns in the Starch Biosynthetic Pathway

While much is known about the starch biosynthetic pathway, there are many important questions yet to answer, including the following:

### 12.4.1 The origin of the first glycosidic bond

While the role of starch synthases, branching enzymes, and debranching enzymes in starch synthesis is clear, what primes the first glycosidic bond? In contrast to what is known in some other organisms, there is generally little information about what happens in plants, maize in particular. Perhaps the best-characterized reaction involves muscle glycogenin-I, a self-glucosylating enzyme that creates a maltosaccharide chain using UDP-glucose (Lomako *et al.*, 2004).

Boyer and Preiss (1979) reported an unprimed starch synthesis reaction in maize endosperm using a preparation of starch synthase termed "isoform I." Plants normally have four soluble starch synthases, SS-I, -II, -III and -IV, as well as a granule-bound starch synthase. Boyer and Preiss (1979) reported their SS-I enzyme preparation contained an anhydroglucose moiety, allowing for the possibly it could serve as a primer. This enzyme was not purified to homogeneity, so it is unknown whether the glucose moiety was covalently linked to starch synthase. Previously, Singh *et al.* (1995) reported an autocatalytic self-glucosylating enzyme in sweet corn that uses UDP-glucose. This enzyme bears no resemblance to muscle glycogenin, and it binds the sugar to a different amino acid. The role of this protein in starch biosynthesis, if any, was not reported.

Germane to the initiation of starch synthesis in maize endosperm are studies by Szydlowski *et al.* (2009) that characterized SS-III and SS-IV mutants of Arabidopsis. Loss of SS-IV reduced the number of starch granules per chloroplast from five or six to one. The single granule was large, as total starch content was reduced only 40%. Loss of SS-III had no effect on granule number, but no granules were produced in a SS-III/IV double mutant. These data suggest a role for SS-III and -IV in controlling starch granule number. The authors asked whether SS-III and/or SS-IV could initiate starch synthesis by expressing His-tagged forms in *Escherichia coli* and purifying the enzymes. Following electrophoresis, an in-gel assay without primer showed that SS-III, but not SS-IV, could synthesize a polyglucan. Both enzymes could elongate glucose polymers when glycogen or amylopectin was present. These data are in accord with the idea that one or both of the enzymes can initiate starch synthesis. It will be interesting to learn whether either of them contains a covalently attached glucose polymer. Certainly, SS-III does not serve as an initiator of starch synthesis, since starch granule number is not affected in a SS-III single mutant. It is possible that SS-IV can prime starch synthesis, but this was not detectable with iodine staining; more sensitive assays might address this question.

While it is assumed a primer is necessary for starch synthesis, results of Ral *et al.* (2004) suggest one might not be needed. These investigators studied starch synthesis in a unicellular green alga, *Ostreococcus tauri*, the "tiniest eukaryote" with one of the smallest genomes. A typical chloroplast in this alga has a single starch granule. The investigators found that the granule actually divides into two small granules. If this is common in other organisms, a primer for starch synthesis would not be necessary.

### 12.4.2 The starch biosynthetic enzyme complex

Typically, the role of specific enzymes is evaluated after they are purified to the greatest extent possible and parameters such as $K_m$, heat stability, effects of inhibitors and activators, and $K_{cat}$ (with pure enzymes) are measured. This approach can be misleading for starch synthesis, because many of the biosynthetic enzymes occur in a complex. How inclusion of an enzyme in a complex influences its kinetic and other properties must be determined to draw valid inferences concerning its properties *in vivo*.

Following observations with wheat (reviewed in Tetlow *et al.*, 2008), Hennen-Bierwagen *et al.* (2009) showed that SS-IIa and SS-III and starch branching enzymes (SBE) IIa and IIb are present in a phosphorylation-dependent 670-kD complex in the amyloplast. Mutational loss of any one of the four enzymes blocked its formation. Furthermore, a ~ 300-kD complex was found that contained all the enzymes, except SS-III. Another complex was described that contained SS-III, both forms of pyruvate orthophosphate dikinase, the *Sh1*-encoded sucrose synthase, and the large and small subunits of AGPase. Interestingly, starch phosphorylase was also found in the complex from wheat but not maize endosperm.

While virtually all of the SS-III protein was in the 670-kD complex and most of the SS-IIa protein was in the ~300-kD complex, the relative amounts of other proteins in these complexes is not known. Gel fractionation of endosperm enzyme preparations revealed virtually all the active AGPase is ~240-kD, the size expected for the heterotetrameric enzyme (Hannah and Nelson, 1975), so at least with this enzyme only a small portion of AGPase is in the ~300-kD complex.

Since several of the important starch biosynthetic enzymes reside in a complex in the amyloplast, this could provide an unanticipated form of regulation. The presence of one protein might influence the enzymatic activity of another. For example, Boyer and Preiss (1981) reported loss of SS-III activity via mutation at the *dull-1* locus concomitantly reduced SBE-IIa activity. This supports the hypothesis that SBE-IIa is physiologically most active when present in the complex. It is also possible that formation of the complex blocks degradation of the individual proteins.

The discovery of these enzyme complexes raises many questions. Which fraction of each of the various starch biosynthetic enzymes resides in a complex? Some proteins reported to be in amyloplast-localized complexes normally reside in the cytosol. Are they simply contaminants trapped during amyloplast isolation, and if not, how do they enter the amyloplast? How do we relate data describing the activity of purified enzymes or data from expressing genes encoding these enzymes in heterogeneous systems? Do unique biochemical and physiological functions arise from multi-enzyme complexes? Do either of the 670-kD or ~300-kD complexes give rise to the clustered α-1,6 branch points in amylopectin?

### 12.4.3 The limiting reaction in starch biosynthesis

Perhaps the most important "known unknown" about starch biosynthesis is the rate-limiting enzymatic step. Knowledge of the regulation of this reaction could potentially allow increased starch synthesis, and in turn, larger grain yields. Virtually all previous relevant research focused on unique aspects of endosperm starch synthesis, namely the synthesis of ADPG, ADPG transport from the cytosol into the amyloplast, and starch polymerization. Evidence favoring each of these steps in cereal seeds has been reported.

In maize, ADPG is synthesized by AGP-ase, an allosteric enzyme consisting of two identical small and two identical large subunits. The small subunit is encoded by *brittle-2* (*Bt2*) and the large subunit is encoded by *shrunken-2* (*Sh2*). Giroux *et al.* (1996) used transposon mutagenesis to modify a motif affecting the holoenzyme's allosteric regulation. The result was an enzyme less sensitive to phosphate inhibition, and maize seeds expressing it exhibited an 11–17% increase in seed weight in some genetic backgrounds. Wang *et al.* (2007) expressed a similar allosterically-modified version of *E. coli* AGP-ase, *glgC16*, in maize and increased seed weight 13–25%. Using a different approach, Li *et al.* (2011) overexpressed *Sh2* and *Bt2* and increased maize seed weight by 15%.

Hannah *et al.* (2012) reported one of the largest yield increases in maize seed conditioned by an altered AGPase. They expressed a modified AGPase with enhanced heat stability and reduced phosphate sensitivity and increased seed yield up to 68%. The gene also conditioned comparable yield increases in wheat and rice (Smidansky *et al.*, 2002, 2003). Surprisingly, while the gene was expressed in the endosperm, its effect on seed yield was manifest in the plant by enhancing the probability an ovary developed into a seed. The seed increases occurred if daily high temperatures 4 days after pollination exceeded 33°C. Elevated nighttime temperatures might also be important. Interestingly, a variant of the small subunit of AGPase exhibits identical properties. Altered AGPases that enhance yield were reported for potato (Stark *et al.*, 1992), wheat (Smidansky *et al.*, 2002), rice (Smidansky *et al.*, 2003; Sakulsingharoj *et al.*, 2004), and Arabidopsis (Obana *et al.*, 2006), suggesting this approach has broad utility.

Transport of ADPG into the amyloplast could limit starch synthesis. Tiessen *et al.* (2012) used non-aqueous conditions to isolate metabolites from the cytosol and amyloplasts of developing barley endosperm. The concentration in the plastid of the four AGPase reaction components, i.e. ATP, glucose-1-P, ADPG, and PPi, ranged from 10–16% of their concentration in the cytosol. Concentrations in the cytosol were in the range of their respective $K_m$ values; however in this regard, the products of the AGPase reaction, ADPG and PPi, were much higher than the concentrations of the substrates, ATP and G1P. They concluded that transport of ADPG into the amyloplast, rather than ADPG synthesis, is the limiting step for starch biosynthesis.

Recent studies in rice endosperm (Cakir *et al.*, 2016) point to the enzymatic transfer of glucose from ADPG to starch as the rate-limiting biosynthetic step. These investigators showed expression of an altered *E. coli* AGPase, *glgc-TM*, in rice endosperm produced a 40% increase in ADPG; however, the mature seed weight increased only 11%. They enhanced transport of ADPG into the amyloplast by overexpression of the maize BT1 protein. With both altered AGPase and overexpression of BT1, the increase in seed weight was similar to that with just the *E. coli* AGPase transgene, even when plants were grown under high $CO_2$. Consequently, they concluded that the enzymatic process of starch polymerization is rate limiting.

To summarize, studies with maize endosperm suggest ADPG synthesis is rate limiting for starch synthesis; whereas in barley ADPG transport into the amyloplast is limiting; in rice neither synthesis of

ADPG nor its transport into the amyloplast is limiting, which implies that starch polymerization is the rate-limiting reaction. Perhaps the vastly different kinetic, allosteric and stability properties of cytosolic AGPases in the different cereal endosperms explain these observations.

Bahaji *et al.* (2014) suggest AGPase is, in fact, not important for starch synthesis. They argue that the ADPG used in starch synthesis is synthesized primarily by sucrose synthase, and the AGPase in the amyloplast salvages glucose for starch resynthesis. Their model does not account for a cytosolic AGPase, the fact that allosterically enhanced AGPases increase starch synthesis, and the fact that attempts to detect starch turnover in endosperm amyloplasts failed to find any turnover.

The above experimental approaches are based on the premise, or prejudice, that biochemistry unique to starch synthesis must be rate limiting in the endosperm. This might or might not be true. We recently employed an approach that requires no knowledge of the biochemistry associated with a particular enzyme. It simply requires a recessive, loss-of-function mutation that conditions an easily scorable seed phenotype. Many such mutants are available from large transposon-mutagenized maize populations (for example *TUSC* ( Mena *et al.*, 1996); *Ac/Ds* (Cowperthwaite *et al.*, 2002); and *UniformMu* (McCarty *et al.*, 2005; Settles *et al.*, 2007)). The rationale for this approach is as follows: virtually all loss-of-function mutants are recessive. In other words, one dose of the wild-type allele is sufficient to produce a wild-type phenotype. For example, all the mutants discussed above, *sh1, sh2, bt1, su1*, etc., are recessive, because one functional allele conditions a plump rather than a shrunken kernel. But at the gene expression level (i.e. the functional cognate transcript, protein, and/or enzyme), a dosage effect is almost always observed; hence, there is an excess of gene product. Examples where enzymatic activity is not proportional to the amount of protein are exceptional, as in the case of allosteric enzymes. But without this or a similar phenomenon, if an enzymatic step in a pathway is normally rate limiting, then reducing its activity should produce a change in the amount of product.

Since the maize seed is approximately 70% starch, maize kernel weight is a good indicator of starch content. Hence, seed weight of kernels expressing different numbers of functional alleles can be monitored. Plants heterozygous for a loss-of-function mutation are crossed reciprocally with homozygous wild-type individuals. When the heterozygote is used as the ear parent, sibling kernels are produced in a 1:1 ratio on the ear with one or three doses of a functional allele expressed in the endosperm. The reciprocal cross produces kernels with two or three doses of the functional allele expressed in the endosperm. Individual kernels are weighed and genotyped, and if an enzymatic step is rate limiting, seeds with less than three functional alleles are expected to be lighter than kernels with three functional alleles. This approach can also be used to test specific hypotheses. For example, the conclusion drawn from the barley experiments pointing to ADPG transport as the limiting step for starch synthesis was tested in maize by determining whether *bt1* alleles exhibit a dosage effect. Kernels having one or two doses of the functional *Bt1* allele were no lighter than kernels having three doses of the *Bt1* allele, consistent with the view that ADPG transport into the amylopast is not rate limiting in the maize endosperm.

In conclusion, while many aspects of the starch biosynthetic pathway in maize and other cereal endosperms are understood, there are many important unanswered questions. It is our hope that this overview stimulates others to pursue the questions we have identified.

## Acknowledgment

Work in the authors' lab was supported by National Institute of Food and Agriculture (2010-04228). We thank Tom Okita, Alan Myers, Tracie Hennen-Bierwagen, Bill Tracy, and Brian Larkins, as well as our colleagues at the University of Florida for valuable discussions. Finally, we give immense thanks to the inventors of Google Scholar.

# References


Bahaji, A., Li, J., Sánchez-López, A.M., Baroja-Fernández, E., Muñoz, F.J., *et al.* (2014) Starch biosynthesis: its regulation and biotechnological approaches to improve crop yields. *Biotechnology Advances* 32, 87–106.

Boehlein, S.K., Shaw, J.R., McCarty, D.R., Hwang, S.-K., Stewart, J.D. and Hannah, L.C. (2013) The potato tuber, maize endosperm and a chimeric maize-potato ADP-glucose pyrophosphorylase exhibit fundamental differences in Pi inhibition. *Archives of Biochemistry and Biophysics* 537, 210–216.

Boyer, C.D. and Preiss, J. (1979) Properties of citrate-stimulated starch synthesis catalyzed by starch synthase-I of developing maize kernels. *Plant Physiology* 64, 1039–1042.

Boyer, C.D. and Preiss, J. (1981) Evidence for independent genetic control of the multiple forms of maize endosperm branching enzymes and starch synthases. *Plant Physiology* 67, 1141–1145.

Cakir, B., Shiraishi, S., Tuncel, A., Matsusaka, H., Satoh, R., *et al.* (2016) Analysis of the rice ADP-glucose transporter (OsBT1) indicates the presence of regulatory processes in the amyloplast stroma that control ADP-glucose flux into starch. *Plant Physiology* 170, 1271–1283.

Cheng, W.H., Taliercio, E.W. and Chourey, P.S. (1996) The *Miniature1* seed locus of maize encodes a cell wall invertase required for normal development of endosperm and maternal cells in the pedicel. *Plant Cell* 8, 971–983.

Chourey, P.S. and Nelson, O.E. (1976) The enzymatic deficiency conditioned by *shrunken 1* mutations in maize. *Biochemical Genetics* 14, 1041–1055.

Chourey, P.S. and Nelson, O.E. (1979) Interallelic complementation at the *sh*-locus in maize at the enzyme level. *Genetics* 91, 317–325.

Cobb, B.G. and Hannah, L.C. (1988) *Shrunken-1* encoded sucrose synthase is not required for sucrose synthesis in the maize endosperm. *Plant Physiology* 88, 1219–1221.

Cowperthwaite, M., Park, W., Xu, Z.N., Yan, X.H., Maurais, S.C. and Dooner, H.K. (2002) Use of the transposon Ac as a gene-searching engine in the maize genome. *Plant Cell* 14, 713–726.

Denyer, K., Dunlap, F., Thorbjornsen, T., Keeling, P. and Smith, A.M. (1996) The major form of ADP-glucose pyrophosphorylase in maize endosperm is extra-plastidial. *Plant Physiology* 112, 779–785.

Ettenhuber, C., Spielbauer, G., Margl, L., Hannah, L.C., Gierl, A., *et al.* (2005) Changes in flux pattern of the central carbohydrate metabolism during kernel development in maize. *Phytochemistry* 66, 2632–2642.

Giroux, M.J. and Hannah, L.C. (1994) ADP-glucose pyrophosphorylase in *shrunken-2* and *brittle-2* mutants of maize. *Molecular and General Genetics* 243, 400–408.

Giroux, M.J., Shaw, J., Barry, G., Cobb, B.G., Greene, T., Okita, T. and Hannah, L.C. (1996) A single gene mutation that increases maize seed weight. *Proceedings of the National Academy of Sciences of the United States of America* 93, 5824–5829.

Hannah, L.C. (2007) Starch formation in the cereal endosperm. In: Olsen, O.-A. (ed.) *Endosperm: Developmental and Molecular Biology.* Springer, Berlin, Heidelberg, Germany, pp. 179–193.

Hannah, L.C. and Nelson, O.E. (1975) Characterization of adenosine-diphosphate glucose pyrophosphorylases from developing maize seeds. *Plant Physiology* 55, 297–302.

Hannah, L.C., Futch, B., Bing, J., Shaw, J.R., Boehlein, S., *et al.* (2012) A *shrunken-2* transgene increases maize yield by acting in maternal tissues to increase the frequency of seed development. *Plant Cell* 24, 2352–2363.

Hennen-Bierwagen, T.A., Lin, Q., Grimaud, F., Planchot, V., Keeling, P.L., James, M.G. and Myers, A.M. (2009) Proteins from multiple metabolic pathways associate with starch biosynthetic enzymes in high molecular weight complexes: a model for regulation of carbon allocation in maize amyloplasts. *Plant Physiology* 149, 1541–1559.

Huang, B.Q., Hennen-Bierwagen, T.A. and Myers, A.M. (2014) Functions of multiple genes encoding ADP-glucose pyrophosphorylase subunits in maize endosperm, embryo, and leaf. *Plant Physiology* 164, 596–611.

James, M.G., Robertson, D.S. and Myers, A.M. (1995) Characterization of the maize gene *sugary1*, a determinant of starch composition in kernels. *Plant Cell* 7, 417–429.

Jeon, J.-S., Ryoo, N., Hahn, T.-R., Walia, H. and Nakamura, Y. (2010) Starch biosynthesis in cereal endosperm. *Plant Physiology and Biochemistry* 48, 383–392.

Keeling, P.L. and Myers, A.M. (2010) Biochemistry and genetics of starch synthesis. *Annual Review of Food Science and Technology* 1, 271–303.

Keeling, P.L., Wood, J.R., Tyson, R.H. and Bridges, I.G. (1988) Starch biosynthesis in developing wheat-grain – evidence against the direct involvement of triose phosphates in the metabolic pathway. *Plant Physiology* 87, 311–319.

Li, N., Zhang, S., Zhao, Y., Li, B. and Zhang, J. (2011) Over-expression of AGPase genes enhances seed weight and starch content in transgenic maize. *Planta* 233, 241–250.

Lomako, J., Lomako, W.M. and Whelan, W.J. (2004) Glycogenin: the primer for mammalian and yeast glycogen synthesis. *Biochimica et Biophysica Acta* 1673, 45–55.

McCarty, D.R., Settles, A.M., Suzuki, M., Tan, B.C., Latshaw, S., *et al.* (2005) Steady-state transposon mutagenesis in inbred maize. *Plant Journal* 44, 52–61.

Mena, M., Ambrose, B.A., Meeley, R.B., Briggs, S.P., Yanofsky, M.F. and Schmidt, R.J. (1996) Diversification of C-function activity in maize flower development. *Science* 274, 1537–1540.

Nelson, O.E. and Rines, H.W. (1962) The enzymatic deficiency in waxy mutant of maize. *Biochemical and Biophysical Research Communications* 9, 297–300.

Obana, Y., Omoto, D., Kato, C., Matsumoto, K., Nagai, Y., *et al.* (2006) Enhanced turnover of transitory starch by expression of up-regulated ADP-glucose pyrophosphorylases in *Arabidopsis thaliana*. *Plant Science* 170, 1–11.

Preiss, J. (2009) Biochemistry and molecular biology of starch biosynthesis. In: BeMiller, J.N. and Whistler, R.L. (eds.) *Starch: Chemistry and Technology* (3rd edn.). Academic Press, Cambridge, Massachusetts, pp. 83–148.

Ral, J.P., Derelle, E., Ferraz, C., Wattebled, F., Farinas, B., *et al.* (2004) Starch division and partitioning: a mechanism for granule propagation and maintenance in the picophytoplanktonic green alga *Ostreococcus tauri* (1 w). *Plant Physiology* 136, 3333–3340.

Sakulsingharoj, C., Choi, S.B., Hwang, S.K., Edwards, G.E., Bork, J., *et al.* (2004) Engineering starch biosynthesis for increasing rice seed weight: the role of the cytoplasmic ADP-glucose pyrophosphorylase. *Plant Science* 167, 1323–1333.

Settles, A.M., Holding, D.R., Tan, B.C., Latshaw, S.P., Liu, J., *et al.* (2007) Sequence-indexed mutations in maize using the UniformMu transposon-tagging population. *BMC Genomics* 8, 116. DOI:10.1186/1471-2164-8-116

Shannon, J.C. (1968) Carbon-14 distribution in carbohydrates of immature *zea mays*. Kernels following $^{14}CO_2$ treatment of intact plants. *Plant Physiology* 43, 1215–1220.

Shannon, J.C., Pien, F.M. and Liu, K.C. (1996) Nucleotides and nucleotide sugars in developing maize endosperms – synthesis of ADP-glucose in *brittle-1*. *Plant Physiology* 110, 835–843.

Shannon, J.C., Pien, F.M., Cao, H.P. and Liu, K.C. (1998) Brittle-1, an adenylate translocator, facilitates transfer of extraplastidial synthesized ADP-glucose into amyloplasts of maize endosperms. *Plant Physiology* 117, 1235–1252.

Singh, D.G., Lomako, J., Lomako, W.M., Whelan, W.J., Meyer, H.E., Serwe, M. and Metzger, J.W. (1995) Beta-glucosylarginine – a new glucose protein bond in a self-glucosylating protein from sweet corn. *FEBS Letters* 376, 61–64.

Smidansky, E.D., Clancy, M., Meyer, F.D., Lanning, S.P., Blake, N.K., Talbert, L.E. and Giroux, M.J. (2002) Enhanced ADP-glucose pyrophosphorylase activity in wheat endosperm increases seed yield. *Proceedings of the National Academy of Sciences of the United States of America* 99, 1724–1729.

Smidansky, E.D., Martin, J.M., Hannah, L.C., Fischer, A.M. and Giroux, M.J. (2003) Seed yield and plant biomass increases in rice are conferred by deregulation of endosperm ADP-glucose pyrophosphorylase. *Planta* 216, 656–664.

Spielbauer, G., Margl, L., Hannah, L.C., Roemisch, W., Ettenhuber, C., *et al.* (2006) Robustness of central carbohydrate metabolism in developing maize kernels. *Phytochemistry* 67, 1460–1475.

Spielbauer, G., Li, L., Roemisch-Margl, L., Phuc Thi, D., Fouquet, R., *et al.* (2013) Chloroplast-localized 6-phosphogluconate dehydrogenase is critical for maize endosperm starch accumulation. *Journal of Experimental Botany* 64, 2231–2242.

Stark, D.M., Timmerman, K.P., Barry, G.F., Preiss, J. and Kishore, G.M. (1992) Regulation of the amount of starch in plant-tissues by ADP glucose pyrophosphorylase. *Science* 258, 287–292.

Sullivan, T.D., Strelow, L.I., Illingworth, C.A., Phillips, R.L. and Nelson, O.E. (1991) Analysis of maize *brittle-1* alleles and a defective *suppressor-mutator*-induced mutable allele. *Plant Cell* 3, 1337–1348.

Szydlowski, N., Ragel, P., Raynaud, S., Mercedes Lucas, M., Roldan, I., *et al.* (2009) Starch granule initiation in *Arabidopsis* requires the presence of either class IV or class III starch synthases. *Plant Cell* 21, 2443–2457.

Tetlow, I.J., Beisel, K.G., Cameron, S., Makhmoudova, A., Liu, F., *et al.* (2008) Analysis of protein complexes in wheat amyloplasts reveals functional interactions among starch biosynthetic enzymes. *Plant Physiology* 146, 1878–1891.

Tiessen, A., Nerlich, A., Faix, B., Huemmer, C., Fox, S., *et al.* (2012) Subcellular analysis of starch metabolism in developing barley seeds using a non-aqueous fractionation method. *Journal of Experimental Botany* 63, 2071–2087.

Wang, Z., Chen, X., Wang, J., Liu, T., Liu, Y., Zhao, L. and Wang, G. (2007) Increasing maize seed weight by enhancing the cytoplasmic ADP-glucose pyrophosphorylase activity in transgenic maize plants. *Plant Cell Tissue and Organ Culture* 88, 83–92.

# 13  Maize Kernel Oil Content


**Bo Shen* and Keith Roesler**

*Agricultural Biotechnology, DuPont-Pioneer, Johnston, Iowa, USA*


## 13.1  Introduction

Although maize is a staple food crop that contributes a large percentage of calories to human diets in a few countries, its major use worldwide is for animal feed. One approach to improve the metabolizable energy of maize for feed applications involves increasing kernel oil content, because oil has the highest energy density. The average kernel oil content in commodity maize is ~4.5% on a dry weight basis; each kilogram of oil contains 9400 calories, which is 2.25 times greater than that of starch on a weight basis. Several feeding trials using high-oil maize for poultry, hogs, and dairy cattle have shown increased growth rates and feed efficiency (Perry, 1988). In addition to direct use of maize grain in feed, distillers dried grains with solubles (DDGS), a co-product of ethanol production that contains 10% oil, is an economically valuable feed ingredient for livestock.

Aside from its importance in feed applications, maize oil is also a valuable co-product derived from the approximately 10–15% of maize grain processed for human consumption and industrial uses. In the wet milling process, oil is extracted from embryos after they are separated from the starch-rich endosperm. Maize is not considered an oilseed crop because of its relatively low oil content, but it should qualify because of the large volume of maize production (USDA, 2017): the total amount of maize oil produced in the U.S. in 2015 was ~4.7 billion pounds, second only to soybean oil at 21.7 billion pounds. The fatty acid profile of maize oil, with about 16% saturated fatty acids and 84% unsaturated fatty acids, including 11% palmitic (16:0), 2% stearic (18:0), 24% oleic (18:1), 61.9% linoleic (18:2), and 0.7% linolenic (18:3), is desirable for human consumption. The low level of linolenic acid enhances its stability for cooking and storage, in contrast to soybean oil which contains 7% linolenic acid and is less stable. In addition, maize oil is tasteless and odorless and contains high amounts of antioxidants, such as phytosterols and vitamin E. The high antioxidant content further increases oil stability and nutritional value (Weber, 1987).

Triacylglycerol (TAG) is the predominant storage lipid in maize kernels, but other lipids are also present, including phospholipids, glycolipids, phytosterols, and free fatty acids. Refined maize oil contains 98.8% TAG. In a typical maize kernel, the embryo accounts for ~9% of the weight and contains


*Corresponding author e-mail: bo.shen@pioneer.com

about 85% of the total lipid, with 12% in the aleurone and 3% in the starchy endosperm. The oil concentration in the embryo ranges from 25 to 32% by weight, depending on growth environments. The endosperm lipid is found mainly in the aleurone layer and ranges from 0.8 to 2%. Maize starchy endosperm does not accumulate TAG. Half of the lipids associated with starch are free fatty acids and lysophospholipid (Weber, 1987).

## 13.2  Biochemical Pathways of Oil Biosynthesis

Triacylglycerol (TAG) is synthesized from glycerol-3-phosphate and fatty acyl-CoA in the endoplasmic reticulum (ER) as shown in Fig. 13.1. Fatty acids are first synthesized from acetyl-CoA in the plastid through reactions catalyzed by acetyl-CoA carboxylase and the fatty acid synthase complex. Fatty acids are exported from the plastid and converted to acyl-CoA by acyl-CoA synthetase on the outer plastid membrane. Both the acyl-CoAs exported from the plastid and those elongated in the ER contribute to the pool for TAG biosynthesis (Ohlrogge and Browse, 1995). In the ER, TAGs are synthesized by stepwise acylation of glycerol-3-phosphate via the Kennedy pathway. First, fatty acyl moieties are added to the sn-1 and sn-2 positions of glycerol-3-phosphate by glycerol-3-phosphate acyltransferase (GPAT) and lyso-phosphatidic acid acyltransferase (LPAAT), respectively, to form phosphatidic acid. Phosphatidic acid is then hydrolyzed by phosphatidate phosphatase (PAP) to yield diacylglycerol (DAG). Diacylglycerol acyltransferase (DGAT) catalyzes the final step in TAG production by transferring the last acyl group from acyl-CoA to DAG to form TAG (Voelker and Kinney, 2001). Recent studies have revealed greater complexity in TAG synthesis than suggested by the Kennedy pathway, with editing and exchange of acyl groups occurring throughout TAG synthesis, and with DAG derived from both *de novo* synthesis and conversion of membrane lipid phosphatidylcholine (Bates and Browse, 2012).

Three different types of DGAT genes have been identified in plants. Types 1 and 2 are predicted to be integral membrane proteins with different numbers of transmembrane domains; they are localized in different regions of the ER and share almost no sequence identity (Lardizabal *et al.*, 2001; Lung and Weselake, 2006; Shockey *et al.*, 2006). Genetic analysis has demonstrated that a type1

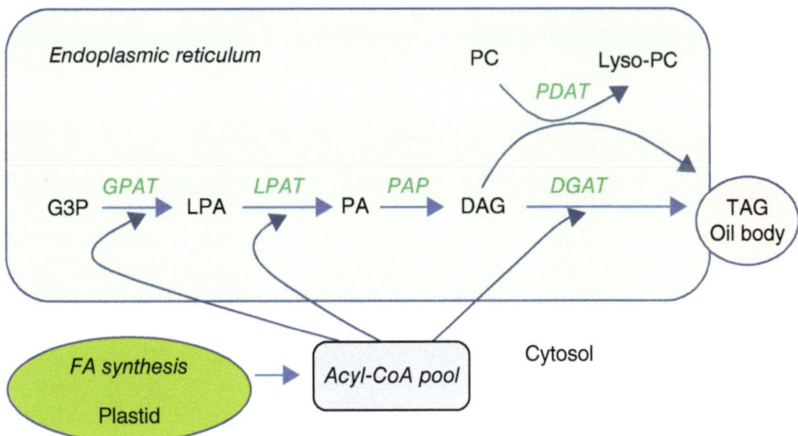


Fig. 13.1. TAG synthesis pathway. Substrate abbreviations: G3P, glycerol-3-phosphate; LPA, lyso-phosphatidic acid; PA, phosphatidic acid; DAG, diacylglycerol; TAG, triacylglycerol; PC, phosphatidylcholine; Lyso-PC, lyso-phosphatidylcholine. Enzymatic reactions are in green color: *GPAT*, acyl-CoA:G3P acyltransferase; *LPAT*, acyl-CoA:LPA acyltransferase; *PAP*, PA phosphatase, *DGAT*, acyl-CoA:DAG acyltransferase; *PDAT*, phospholipid:DAG acyltransferase. Complicated acyl editing and PC-DAG interconversion are described in the review article (Bates and Browse, 2012).

DGAT is the major enzyme catalyzing TAG production in *Arabidopsis*, whereas type 2 DGAT appears to be responsible for TAG synthesis in plant species with unusual fatty acid composition, such as castor bean ricinoleic acid (Zou *et al.*, 1999; Kroon *et al.*, 2006; Burgal *et al.*, 2008). In addition to type 1 and 2 DGATs, a soluble form of DGAT has been identified in developing peanut cotyledons (Saha *et al.*, 2006). In contrast to DGAT which uses acyl-CoA as a substrate, phospholipid:diacylglycerol acyltransferase (PDAT) directly transfers fatty acids from phosphatidylcholine (PC) to DAG to produce TAG (Zhang *et al.*, 2009). The relative flux through DGAT and PDAT is unclear in most oilseed plants. While DGAT1 is the major enzyme for TAG biosynthesis in developing seeds and in leaves during senescence, PDAT plays an important role in TAG biosynthesis in growing leaves (Zhang *et al.*, 2009). Overexpression of PDAT1 significantly increases TAG accumulation in *Arabidopsis* leaves, but does not affect TAG accumulation in seeds. PDAT may also play an overlapping role in TAG biosynthesis in pollen and vegetative tissues (Zhang *et al.*, 2009; Fan *et al.*, 2013). Finally, TAGs are stored in seeds in specialized structures termed oil bodies. Each oil body contains a triacylglycerol matrix surrounded by a monolayer of phospholipids embedded with structural oleosin proteins (Ting *et al.*, 1996; Hsieh and Huang, 2004).

## 13.3 Genetic Control of Oil Biosynthesis

High-oil maize lines have been successfully developed through breeding and selection. Three well-characterized high-oil populations are Illinois High-oil (IHO), Alexho Synthetic, and Beijing High-oil (BHO). IHO was developed by recurrent selection in the Illinois Long-term Selection Experiments. In 1896, 24 ears were chosen for high oil content out of 163 ears from the open-pollinated variety Burr's white. After more than 100 years of selection, by the 100th cycle kernel oil content increased from 4% to 20% (Dudley and Lambert, 2004; Moose *et al.*, 2004). Alexho Synthetic was developed by mixing 43 open-pollinated maize varieties and using

a single kernel oil selection approach. By increasing both embryo size and embryo oil concentration, after 27 selection cycles kernel oil content reached 21%. The seed size, however, was significantly reduced (Lambert *et al.*, 2004). BHO was developed from Zhongzong No. 2 synthetic by single kernel NMR selection for oil content. After 17 cycles of selection, BHO reached a kernel oil content of 15.5% with only a slight reduction in kernel weight (Song and Chen, 2004).

Multiple studies using mapping populations derived from the above three high-oil sources were conducted to understand the genetic basis of kernel oil content. Maize kernel oil is a typical quantitative trait controlled by a large number of quantitative trait loci (QTLs). In populations derived from IHO x Illinois Low Oil (ILO), more than 50 QTLs accounted for ~50% of the genetic variance. The QTL effect estimates are small and largely additive (Laurie *et al.*, 2004; Clark *et al.*, 2006). Oil content is a function of kernel weight, embryo size and oil concentration. A recombinant inbred population was developed from a cross between B73 and BHO line By804 and used to map QTLs affecting kernel oil content and composition. Twelve QTLs affecting kernel oil content, nine QTLs affecting embryo oil concentration, and ten QTLs affecting embryo/endosperm ratio were detected across overlapping chromosome regions (Yang *et al.*, 2012). To identify major QTLs for map-based cloning, Zheng *et al.* (2008) generated a mapping population from a cross between the normal oil line, PH09B, and the high-oil line, ASKC28IB1. Eight QTLs affecting kernel oil content and four QTLs affecting embryo oil concentration were identified (Zheng *et al.*, 2008). A QTL on chromosome 7 affects kernel oil by increasing embryo size at the expense of the endosperm. But a high-oil kernel with a large embryo and small endosperm is not desirable, as this can result in reduced grain yield. A favorable trait for high-oil would increase embryo oil concentration without affecting either the embryo or seed size. A QTL consistently detected on chromosome 6 between markers PHI077 and UMC1014 had a modest effect on embryo oil concentration. This QTL, *qHO6*, was cloned and shown to encode a type 1 DGAT. The gene affects the fatty acid profile, as well as embryo oil concentration. The high-oil *DGAT 1-2* allele increased oleic acid (18:1) content by 61.3% and reduced linoleic acid (18:2) by 24.1% (Zheng *et al.*, 2008). *qHO6* corresponds to the ln1 locus previously identified as a major QTL controlling oleic acid content in the IHO X ILO population (Alrefai *et al.*, 1995). The high-oil *DGAT1-2* allele contains a phenylalanine at position 469 (F469) that is absent in the normal *DGAT1-2* allele. The *DGAT1-2* allele from the high-oil line was shown to produce higher enzymatic activity when compared to the *DGAT1-2* allele from the normal oil line. The F469 residue is found in all teosinte lines and is the ancestral allele whereas the allele without F469 is a recent mutant selected by domestication or breeding (Zheng *et al.*, 2008). To understand when the high-oil allele was lost, *DGAT1-2* was sequenced from a set of maize landrace populations. The high-oil *DGAT1-2* allele was present in the majority of Northern Flint and Southern Dent populations, but had disappeared in five out of eight Corn Belt Dent open-pollinated populations as well as in most early inbred lines. Loss of the high-oil *DGAT1-2* allele could have been caused by the genetic drift that occurred in the early 1900s when only a few Corn Belt Dent populations were selected for inbred development (Chai *et al.*, 2012). Alternatively, the low oil *DGAT1-2* allele may have been coupled with other favorable agronomic traits, such as high starch content (Cook *et al.*, 2012), and therefore was genetically fixed in early breeding. In contrast to the successful cloning of *qHO6*, we were unable to clone the embryo oil concentration QTLs, QTL8 at 29–110 cM on chromosome 8 and QTL9 at 64–132 cM on chromosome 9. The effects of these QTLs on seed oil content and embryo oil concentration were detected in BC3S3 but disappeared at BC5S2, suggesting that the impacts of QTL8 and QTL9 are dependent on other genes in the ASK28 background or that there are multiple small effect QTLs clustered in the region. Once separated, the individual QTL effect is too small to be detected.

Recently, genome-wide association analysis has identified additional loci that are associated with maize kernel oil biosynthesis.

In an association mapping analysis of 553 inbred lines, a single QTL for oleic acid content was identified on chromosome 4. A fatty acid desaturase, *fad2*, mapped within 2 kilobases of the marker associated with elevated oleic acid levels and is the probable causative gene. Surprisingly, *qHO6* was not detected as a strong QTL affecting oleic acid content in the same population (Beló *et al.*, 2008). In the nest association mapping population, 22 oil QTLs were detected and *DGAT1-2* was a major oil QTL (Cook *et al.*, 2012). In addition to *DGAT1-2* and *FAD2*, additional candidate genes, including *ACP*, *LACS*, *WRI1a*, and *COPII*, have been identified from another genome-wide association study (Li *et al.*, 2013). The function of these candidate genes in oil biosynthesis needs to be validated by transgenic overexpression or downregulation.

## 13.4  Regulation of Oil Biosynthesis

In higher plants, the biosynthesis of seed storage compounds is usually coupled with seed development. *Leafy cotyledon1* (*LEC1*) is a master regulator for seed maturation. In *Arabidopsis*, *LEC1* encodes a *HAP3* subunit of the CCAAT-binding factor (Lotan *et al.*, 1998). The *lec1* mutation reduces seed desiccation tolerance and oil accumulation, and ectopic expression of *LEC1* leads to the formation of embryo-like structures containing oil and storage proteins in leaves. The effect of *LEC1* on seed oil accumulation can be attributed in part to its positive regulation of genes involved in carbon and lipid metabolism. Overexpression of *LEC1* upregulates genes involved in: (i) fatty acid biosynthesis, such as acetyl-CoA carboxylase and acyl carrier protein; (ii) fatty acid modification, such as stearoyl-ACP desaturase and *FAD2*; and (iii) oil biosynthesis, such as DGAT and oleosin. In addition, multiple genes involved in sucrose synthesis and transport, glycolysis, cell growth, and cell development are also upregulated by overexpressing *LEC1* (Mu *et al.*, 2008). *LEC1* could regulate lipid biosynthesis through downstream transcription factors, rather than directly binding to promoters of lipid pathway genes, as no evidence supports direct interactions. The maize ortholog of the *Arabidopsis LEC1* gene was identified and shown to have a similar impact on oil accumulation (Shen *et al.*, 2010). Downregulation of *ZmLEC1*, however, did not affect embryo development, embryo oil content, or seed germination, possibly due to functional redundancy of maize *LEC1*. Overexpression of *ZmLEC1* in the embryo increased seed oil content by 35%, but this had negative effects on seed germination and leaf growth (Shen *et al.*, 2010). Promoter optimization of *LEC1* expression was not successful in uncoupling the seed oil increase from the poor germination and growth phenotypes.

In addition to *LEC1*, other master regulators, such as *LEC2*, *FUS3*, and *GL2*, play important roles in *Arabidopsis* seed development and oil accumulation, but their roles in regulating maize oil accumulation are not clear (Santos-Mendoza *et al.*, 2008; Shen *et al.*, 2006). Furthermore, oil QTL analysis in maize did not detect an association of oil variation with maize *LEC1*, *LEC2*, *GL2*, or *FUSCA3* (Li *et al.*, 2013). It is unknown whether orthologs of these *Arabidopsis* regulators have a similar function.

*Wrinkled1* (*WRI1*) is a key transcription factor downstream of *LEC1* that controls seed oil accumulation in *Arabidopsis*. The *WRI1* transcription factor contains two AP2 domains. In *Arabidopsis*, the *wri1* mutant shows a wrinkled seed phenotype with reduced oil accumulation. Overexpression of *WRI1* increased the seed oil content in *Arabidopsis*, but negatively impacted seed germination in the presence of sucrose (Cernac and Benning, 2004). Expression of *WRI1* is upregulated by *LEC1*, *LEC2*, and *FUS3*, and the gene is a direct target of *LEC2* (Baud *et al.*, 2007; Santos-Mendoza *et al.*, 2008). *WRI1* directly regulates expression of genes involved in late glycolysis and fatty acid biosynthesis, whereas *LEC1* does not. RNA profiling of the *wri1* mutant and overexpression lines identified *WRI1* target genes, including pyruvate kinase and pyruvate dehydrogenase which are involved in glycolysis, and acetyl CoA carboxylase, ACP, ketoacyl-ACP synthase III, and acyl-ACP thioesterase, all of which are involved in fatty acid biosynthesis. In contrast to *LEC1*, *WRI1* does not affect expression of

oil biosynthesis genes, such as DGAT and oleosin (Baud *et al.*, 2009; To *et al.*, 2012). The AW box is a consensus *WRI1* DNA-binding motif identified from genes encoding a subunit of pyruvate kinase, acetyl-CoA carboxylase, acyl carrier protein, and ketoacyl-acyl carrier protein synthase, and it is enriched in the promoters of fatty acid pathway genes. *WRI1* has been shown to bind directly to the AW box *in vitro,* and mutating the AW box abolishes *WRI1*-mediated activation in protoplast assays (Baud *et al.*, 2009; Maeo *et al.*, 2009). Two maize homologues with unique expression patterns, *ZM-WRI1a* and *ZM-WRI1b*, have been identified and both complement the *Arabidopsis wri1* mutant phenotype, confirming their role in fatty acid biosynthesis (Pouvreau *et al.*, 2011). Overexpression of *Zm-WRI1* using an embryo-specific promoter increased kernel oil content by 36.6% without affecting germination, flowering time, or plant height. Expression of *ZmWrRI1* is upregulated by *ZmLEC1*. Protein content of the embryo is not affected by *WRI1* overexpression, but starch content is reduced by 60% in transgenic embryos, indicating carbon flow from starch to oil biosynthesis. Yield trials of transgenic maize overexpressing *WRI1* showed no significant yield penalty when compared to nulls (Shen *et al.*, 2010). Maize *WRI1* shows a conserved function in regulating late glycolysis and parts of the fatty acid biosynthesis pathway and is also involved in regulating native variation of oil content in association mapping populations (Pouvreau *et al.*, 2011; Li *et al.*, 2013). In addition to *Arabidopsis* and maize, *WRI* has been implicated in the control of oil biosynthesis in *Brassica napus* and oil palm and has emerged as a ubiquitous regulator of oil biosynthesis in plants (Ma *et al.*, 2013; Li *et al.*, 2015).

## 13.5 Overexpression of DGAT Genes to Increase Maize Oil and Oleic Acid Content

Numerous studies have achieved increased seed oil content by overexpression of DGAT genes (Jako *et al.*, 2001; and many others).

Considering only maize examples, embryo preferred expression of the *ZmDGAT1-2* high-oil allele (described previously) resulted in increases in oil and oleic acid concentrations of up to 41% and 107%, respectively (Zheng *et al.*, 2008). In comparison, expression of the normal DGAT allele using the identical promoter showed less extreme increases in oil and oleic acid content. These results demonstrate that a kinetically superior DGAT allele is more effective at increasing oil content in transgenic maize; the recombinant high-oil allele also had greater DGAT specific activity in yeast microsomal membrane assays than did the normal oil allele. In another maize example, embryo preferred expression of fungal type 2 DGAT genes resulted in increases in kernel oil and oleic acid contents of up to 26% and 18%, respectively, while reductions in DAG concentrations were also observed in developing transgenic embryos (Oakes *et al.*, 2011). Neither of these maize studies observed significant yield decreases with the high-oil lines, although the latter reported decreased kernel starch content. Collectively, these examples demonstrate that ectopic expression of either type 1 or type 2 DGAT genes from diverse organisms can increase maize kernel oil content. The increased oleic acid content often observed with DGAT overexpression plants has been attributed either to increased DGAT activity pulling DAG toward TAG synthesis and away from the phosphatidylcholine-based *FAD2* and *FAD3* desaturation reactions, or alternatively, to the substrate specificities of the DGAT being overexpressed (Zheng *et al.*, 2008; Zhang *et al.*, 2013; Roesler *et al.*, 2016).

## 13.6 Questions that Remain to be Answered

### 13.6.1 How can maize kernel oil content be further increased?

#### *Engineer DGAT*

Would engineering DGAT for improved kinetic properties or increased protein stability allow even greater maize oil content increases, beyond what was achieved with the high-oil *ZmDGAT1-2* allele? A site-directed mutagenesis study identified one amino acid substitution, S197A, out of several tested in the type 1 DGAT from *Tropaeolum majus* that increased DGAT specific activity in yeast microsomal membrane assays (Xu *et al.*, 2008). Both the mutated and the wild-type *Tropaeolum* DGATs increased oil content of transgenic *Arabidopsis* seeds, although the performance of each was difficult to compare as they were tested in different experiments. As an alternative to rational engineering, error-prone PCR was used to create *Brassica napus* type 1 DGAT variants with multiple mutations, followed by high throughput screening for high oil content in *Saccharomyces cerevisiae* (Siloto *et al.*, 2009). This study identified DGAT variants with an increased ability to synthesize oil in yeast, but efficacy in transgenic plants was not reported. To increase soybean oil content, DNA shuffling technology was used to create libraries of *Corylus americana* (American hazelnut) type 1 DGAT variants, followed by screening for high oil content in *Saccharomyces cerevisiae* (Roesler *et al.*, 2016). Sequencing the top 20 variants revealed amino acid substitutions at 63 different DGAT positions, with any one variant containing from 2 to 17 substitutions. Kinetic studies revealed that the most improved variant, CaDGAT1-C11, had an approximately fourfold decrease in the $S_{0.5}$ value for oleoyl-CoA and a threefold improvement in $V_{max}$. This variant was more effective than wild-type DGAT at increasing oil content of soybean somatic embryos. Improved soybean DGAT variants were also designed based on amino acid substitutions observed in promising *Corylus* DGAT variants, with the most effective one increasing oil content three percentage points (16% relative increase) in field-grown soybeans. This study demonstrated that: (i) wild-type DGATs are far from optimal for high-oil production (i.e. engineered variants are much more effective); (ii) DGAT affinity for the acyl-CoA substrate correlates strongly with oil content and is an important determinant of oil content in both soybeans and yeast; and (iii) beneficial amino acid substitutions identified in one plant type 1 DGAT variant can be successfully used to improve the function of that variant from a different species. Based on these observations, it seems likely that DGAT engineering could further increase maize oil content by: (i) using a similar DNA shuffling approach to directly improve the high-oil *ZmDGAT1-2* allele; (ii) by expressing in maize the most effective engineered dicot DGAT variants available, preferably following optimization for a maize codon bias; or (iii) by designing improved variants of the ZmDGAT1-2 high-oil allele to include beneficial amino acid substitutions already observed in promising *Corylus* or soybean DGAT variants. In support of this last approach, increased activity of *ZmDGAT1-2* variants was observed in yeast microsomal membrane assays following amino acid substitutions that were chosen based on the results of the DGAT DNA shuffling study (Roesler *et al.*, 2012).

### Engineer WRI1

Could maize *WRI1* function be improved by protein engineering or expression optimization? The *Arabidopsis WRI1* protein contains three intrinsically disordered regions that play specific roles in its function or post-translational modification. A putative PEST motif was identified in the intrinsically disordered region at the C-terminus. Deletion of this motif, or mutations in putative phosphorylation sites in the motif, increased the stability of the protein. The stabilized *WRI1* yielded greater oil increases in transgenic plant tissues than did the wild-type *WRI1* (Ma *et al.*, 2015). In maize *WRI1*, a PEST motif is likewise found in the intrinsically disordered region at the C-terminus, and similar modifications could be engineered to assess effects on maize kernel oil content. To maximize the oil increase achieved by wild-type *WRI1* overexpression in maize, several promoters have been tested. Of the seven embryo promoters studied, a strong embryo-specific oleosin promoter resulted in the greatest increase in oil, while expression of *WRI1* using the globulin-1 promoter had only a marginal effect (Shen *et al.*, 2010). These results show that WRI1 temporal and

tissue-specific expression, as well as level of expression, are critical for the high-oil phenotype.

### Stack WRI1 and DGAT and block oil degradation

*WRI1* increases expression of glycolysis and fatty acid biosynthesis pathway genes but does not affect expression of oil assembly genes (Baud *et al.*, 2009). Could a stack of *WRI1* and DGAT upregulate both the fatty acid and oil biosynthesis pathways? Using a transient tobacco leaf assay, co-expression of *WRI1* and DGAT resulted in oil levels exceeding those expected from an additive effect, reaching as high as 2.5% oil on a dry weight basis, approximately fivefold greater than values achieved with either gene alone (Vanhercke *et al.*, 2013). In *Arabidopsis*, co-expression of *WRI1* and *DGAT* combined with suppression of the triacylglycerol lipase *SUGAR-DEPENDENT 1* resulted in both greater seed oil content and greater seed mass, compared with overexpression of each gene individually (van Erp *et al.*, 2014). Similarly, in maize a genetic stack of *ZmWRI1* and *ZmDGAT1-2* led to increased oil content compared with overexpression of the individual genes (Fig. 13.2a). The molecular stack increased kernel oil content by as much as 50% (Fig. 13.2b). Although seed weight loss was not detected in transgenic lines expressing either *ZmWRI1* or *Zm-DGAT1-2*, an average seed weight loss of 3.4% was observed in transgenic lines expressing the molecular stack. Whether a small seed weight loss will lead to a yield penalty is not known. Clearly, combining *WRI1* and *DGAT* had an additive effect on maize oil content.

### Stack TAG assembly genes

Genes encoding all three acyltransferases of the Kennedy pathway, plus a GPDH gene, were co-expressed in tobacco seed, resulting in an oil increase of about 14.5% (Liu *et al.*, 2015). This increase was approximately equal to the sum of oil increases from individual genes. A similar approach has not yet been tried in maize, but could

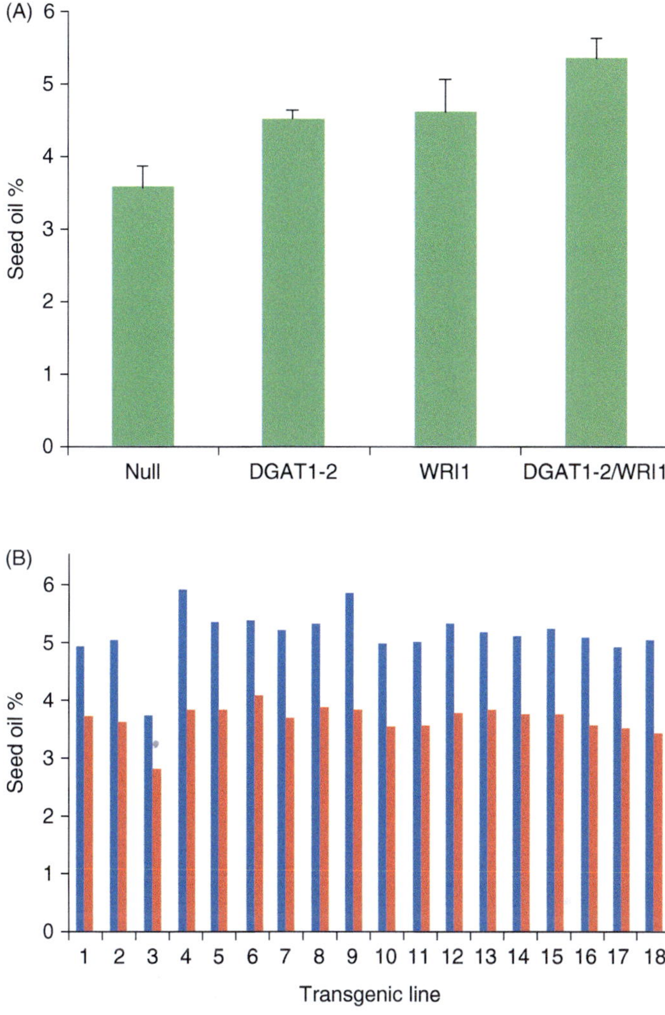

**Fig. 13.2.** Seed oil content of transgenic maize lines expressing *ZmDGAT1-2, ZmWRI1*, and the *ZmDGAT1-2/ZmWRI1*. Both *ZmDGAT1-2* and *ZmWRI1* were expressed under the control of the embryo-specific *Oleosin* promoter. Seed oil content is based on air-dried seed weight. (A) Transgenic *ZmDGAT1-2* line was crossed to *ZmWRI1* line to generate the *ZmDGAT1-2/ZmWRI1* stack genetically, (B) seed oil content in transgenic lines from a molecular stack of *ZmDGAT1-2* and *ZmWRI1*.

potentially give higher maize oil contents than previously observed.

### Increase embryo size

Selection of high-oil by breeding programs led to a 67% increase in embryo size in the ASK high-oil line. A few QTLs affecting embryo size have been identified, but none of them have been cloned (Yang *et al.*, 2012; Zheng *et al.*, 2008). Rice *Giant Embryo* (*GE*) mutants have large embryos at the expense of endosperm. *GE* encodes a cytochrome p450 protein (CYP78A13) essential for controlling embryo size (Nagasawa *et al.*, 2013). Its overexpression in transgenic maize seeds with an embryo-preferred *oleosin* promoter reduced the embryo/endosperm ratio by 23%.

However, downregulation of the maize *GE* homolog did not result in increased embryo size, possibly due to only partial downregulation or functional redundancy of the maize *GE* genes. *ZmGE2* shares 70% amino acid identity with rice *GE* and contains a transposable element insertion associated with a QTL affecting embryo/endosperm ratio (Zhang *et al.*, 2012). Further work to knock out *ZmGE2* using CRISPR/CAS9 technology would confirm whether *ZmGE2* is critical for controlling the embryo/endosperm ratio in maize.

Another creative way to increase embryo size is by using double embryo seed. The maize ear meristem produces an upper and a lower floret, with the lower floret usually aborting in early development, resulting in one mature kernel in each floret. Expression of a bacterial *isopentenyl transferase* (*IPT*) gene using a senescence inducible promoter (SAG12) blocked the abortion of the lower floret and resulted in one kernel with two fused seeds. The kernel contained two viable embryos and endosperms with an increased ratio of embryo to endosperm. Double embryo maize kernels contain more oil and more high quality protein (Young *et al.*, 2004). We also identified a spontaneous double embryo mutant (Fig. 13.3c,d,e) with phenotypes that varied in different environments. The gene responsible for this mutant has not yet been cloned, so it is not clear if the double embryo is caused by a change in cytokinin concentration in the floret, as implied by the *IPT* results of the Young *et al.* study. Double embryo maize not only improves nutritional value, but also reduces seed costs, because the grower needs to plant half the normal amount of seed to get the same plant population density. Large embryos can cause difficulties with current wet milling technology, but this would not be relevant for maize used in feed applications.



Maize endosperm typically consists of a central mass of starchy endosperm cells, a single surrounding layer of aleurone, and a basal layer of transfer cells (Olsen, 2001).

In the endosperm, only the aleurone cells produce TAG, with none detected in the starchy endosperm. Engineering multiple layers of aleurone cells could increase endosperm oil content. Interestingly, the maize landrace Mexican Coroico has as many as 4–5 layers of aleurone cells (Fig. 13.3 A,B), resulting in a doubling of endosperm oil content. In addition to oil, the aleurone cell layer is rich in beneficial antioxidants such as phytosterols. The genetic basis for the multiple layers of aleurone cells in Coroico is not understood, and attempts to breed for multiple aleurone layers were unsuccessful. Significant progress has been made in understanding the molecular pathway controlling aleurone cell differentiation and development, with key genes such as CR4, SAL1, and DEK1 having been identified (see Chapter 6). Overexpression of CR4 resulted in two layers of aleurone cells, but unfortunately this was insufficient to have a large impact on endosperm oil content. It remains to be seen whether manipulation of other genes in this pathway can generate multiple layers of aleurone cells.

Surprisingly, long-term recurrent selection for high-oil in maize has not resulted in multiple aleurone cell layers or oil accumulation in the starchy endosperm (Lambert *et al.*, 2004). High-oil lines have only a single layer of aleurone cells, albeit with increased oil concentration, and oil bodies have not been detected in starchy endosperm cells from these high-oil lines. In contrast, recurrent selection for high-oil in oat (*Avena sativa*) led to oil accumulation in the starchy endosperm (Peterson and Wood, 1997). Microscopy analysis showed that oil is most abundant in the sub-aleurone cells and cells in the vicinity of the scutellum and embryo, with oil accumulating primarily in the early stages of endosperm development (Banas *et al.*, 2007; Heneen *et al.*, 2009). Further research is required to understand how oat produces oil in the endosperm proper, but maize cannot.

Similar to maize starchy endosperm, plant vegetative tissues, such as leaves and stems, typically do not accumulate oil. A recent review article summarized the current understanding of oil synthesis, turnover,

**Fig. 13.3.** Aleurone cell layer and double embryo mutant phenotypes. (A) Arrow marks the single aleurone cell layer in wild-type; (B) multiple aleurone layers in Mexican Coroico; (C) cross section of immature double embryo (arrows) seed; (D) germination of double embryo seed; (E) two seedling plants resulting from double embryo seed.

storage, and function in leaves, and discussed recent advances in metabolic engineering efforts to increase oil accumulation in leaves (Xu and Shanklin, 2016). Co-expression of *WRI1*, DGAT1, and OLEOSIN genes in tobacco using constitutive promoters resulted in dramatically increased oil accumulation in the leaf tissue by up to 15% of dry weight without severely impacting plant development. Microscopy analysis confirmed accumulation of oil droplets in leaf mesophyll cells (Vanhercke *et al.*, 2014). Furthermore, co-expression of the same three genes in sorghum, with simultaneous co-suppression of either ADP-glucose pyrophosphorylase or a subunit of the peroxisomal ABC transporter, increased oil content in leaves and stems 95-fold and 43-fold, respectively (Zale *et al.*, 2016). Metabolic engineering of high biomass crops producing oil in vegetative tissues has the potential to create new feedstocks for biofuel production. Overexpression of *ZmWRI1* using the maize endosperm-specific 19-KD zein promoter, however, did not result in a significant increase in maize seed oil content (Shen *et al.*, 2010). Considering the above observations with vegetative tissues, the failure of *ZmWRI1* overexpression alone to increase oil in the starchy endosperm may have been due to insufficient expression of genes involved in oil biosynthesis and oil body formation. Co-expression of *WRI1* with DGAT1-2 and oleosin genes could be tested to influence oil content in maize starchy endosperm. In addition to DGAT, PDAT could play an important role in oil biosynthesis in growing leaves and young floral tissues. Further tests of PDAT with *WRI1* and oleosin are needed to understand the limiting factors for oil biosynthesis in the maize starchy endosperm.

### 13.6.2   Is it possible to develop a high-oil maize hybrid without decreased grain yield?

Breeding for high-oil maize has been successful because the oil trait has high heritability and is easily measured. However, despite several feeding trials showing improved livestock feed efficiencies, no high-oil maize hybrids have been widely adopted. Producers lose interest in high-oil maize if the premium for high-oil grain is insufficient to compensate for yield loss and the cost for identity preservation. Some high-oil hybrids have shown a 10–15% yield reduction with poor agronomic traits (Lambert *et al.*, 2004). While the energy needed for oil biosynthesis will always be greater than that for starch, does high oil content inevitably result in yield reduction? Alternatively, can the plant increase resources in response to sink demand and accommodate the increased energy requirement for oil storage? Lambert *et al.* studied whether a high-oil maize pollinator could reduce grain yield losses by a "topcross" strategy. Four normal oil hybrids were pollinated by either a high-oil pollinator or a normal-oil pollinator. Grain obtained using pollen from the high-oil pollinator contained 6–7% oil, while the grain obtained using pollen from the normal-oil pollinator contained only 4.5–5.5% oil. Normal-oil hybrids pollinated by the high-oil pollinators produced grain yields equal to that of hybrids pollinated by the normal-oil pollinators, indicating that at least some hybrids are capable of increasing the resources to provide more energy for high-oil grain (Alexander and Lambert, 1968; Lambert *et al.*, 1998).

To minimize the impact of high oil content on grain yields, DuPont developed a novel topcross system for high-oil grain production. The topcross system used a specially developed high-oil male pollinator line that was mixed and planted with a normal oil cytoplasmic male sterile elite hybrid. The male pollinator usually made up ~10% of the seed, and produced sufficient pollen for pollinating all the ears. The oil content of the high-oil pollinator ranged from 12–15%. Pollen carrying the high-oil genes increased the oil content of the female hybrid seed through what is called a xenia effect. The hybrid seed also showed improved seed growth compared to self-pollinated seed. The average grain oil content ranged from 6–8% due to the xenia effect of the high-oil pollinators. The grain yield from the female parent hybrid was similar to that of the control hybrid that was self-pollinated. Unfortunately, the high-oil male pollinator was not an elite hybrid, and showed a 50% yield reduction compared to the elite hybrid. By using this topcross system, ~5% yield reduction per acre was expected compared to a normal hybrid, mainly due to the 10% portion of non-elite male pollinators in the field. Another potential risk for the topcross system is poor pollination in certain stress conditions, which could further reduce the grain yields. The previously mentioned high-oil breeding program developed a high-oil line based on percentage kernel-oil selection. High kernel-oil percentage can be a result of either larger embryo size and/or higher embryo oil concentration. Although large embryos provide high quality protein with balanced amino acids, large embryos are usually negatively correlated with endosperm size, which could lead to a lower yield potential. In contrast, increased embryo oil concentrations might pose less of an impact on grain yield potential. With a better understanding of oil QTLs that affect component traits, coupled with marker assisted selection, it would be possible to select oil QTLs affecting embryo oil concentration rather than those affecting embryo size. Support for the idea that increased embryo oil concentration could have a minimal yield penalty is supported by transgenic studies with *ZmWRI1* and *ZmDGAT1-2*. Overexpression of *ZmWRI1* and *ZmDGAT1-2* led to an increase in embryo oil concentration without affecting embryo size. Yield trials demonstrated that the grain yields of the high-oil transgenic lines were similar to those of controls (Shen *et al.*, 2010). Thus, embryo oil concentration appears to be a productive target for increased seed oil content.

In conclusion, despite the inherent complexity of developing high-oil maize through breeding and metabolic engineering, the high-oil trait has great potential value for feed, wet milling, and the biofuels industry. Since maize has a much higher grain yield compared to most oil crops, and even with a significant reduction in grain yield, high-oil maize would still produce more oil per acre and more overall value than other oil crops. Marker assisted selection for high embryo oil concentration is a promising strategy to develop high-oil maize with a competitive grain yield.

## References


Alexander, D.E. and Lambert, R.J. (1968) Relationship of kernel oil content to yield in maize. *Crop Science* 8, 273–274.

Alrefai, R., Berke, T.G. and Rocheford, T.R. (1995) Quantitative trait locus analysis of fatty acid concentrations in maize. *Genome* 38, 894–901.

Banas, A., Debski, H., Banas, W., Heneen, W.K., Dahlqvist, A., *et al.* (2007) Lipids in grain tissues of oat (*Avena sativa*): differences in content, time of deposition, and fatty acid composition. *Journal of Experimental Botany* 58, 2463–2470.

Bates, P.D. and Browse, J. (2012) The significance of different diacylglycerol synthesis pathways on plant oil composition and bioengineering. *Frontiers in Plant Science* 3, 147.

Baud, S., Mendoza, M.S., To, A., Harscoët, E., Lepiniec, L. and Dubreucq, B. (2007) WRINKLED1 specifies the regulatory action of LEAFY COTYLEDON2 towards fatty acid metabolism during seed maturation in Arabidopsis. *Plant Journal* 50, 825–838.

Baud, S., Wuillème, S., To, A., Rochat, C. and Lepiniec, L. (2009) Role of WRINKLED1 in the transcriptional regulation of glycolytic and fatty acid biosynthetic genes in Arabidopsis. *Plant Journal* 60, 933–947.

Beló, A., Zheng, P., Luck, S., Shen, B., Meyer, D.J., *et al.* (2008) Whole genome scan detects an allelic variant of *fad2* associated with increased oleic acid levels in maize. *Molecular Genetics and Genomics* 279, 1–10.

Burgal, J., Shockey, J., Lu, C., Dyer, J., Larson, T., Graham, I. and Browse, J. (2008) Metabolic engineering of hydroxy fatty acid production in plants: RcDGAT2 drives dramatic increases in ricinoleate levels in seed oil. *Plant Biotechnology Journal* 6, 819–831.

Cernac, A. and Benning, C. (2004) *WRINKLED1* encodes an AP2/EREB domain protein involved in the control of storage compound biosynthesis in Arabidopsis. *Plant Journal* 40, 575–585.

Chai, Y., Hao, X., Yang, X., Allen, W.B., Li, J., *et al.* (2012) Validation of DGAT1-2 polymorphisms associated with oil content and development of functional markers for molecular breeding of high-oil maize. *Molecular Breeding* 29, 939–949.

Clark, D., Dudley, J.W., Rocheford, T.R. and LeDeaux, J.R. (2006) Genetic analysis of corn kernel chemical composition in the random mated 10th generation of the cross of generations 70 of IHO × ILO. *Crop Science* 46, 807–819.

Cook, J.P., McMullen, M.D., Holland, J.B., Tian, F., Bradbury, P., *et al.* (2012) Genetic architecture of maize kernel composition in the nested association mapping and inbred association panels. *Plant Physiology* 158, 824–834.

Dudley, J.W. and Lambert, R.J. (2004) 100 generations of selection for oil and protein in corn. *Plant Breeding Reviews* 24, 79–110.

Fan, J., Yan, C., Zhang, X. and Xu, C. (2013) Dual role for phospholipid:diacylglycerol acyltransferase: enhancing fatty acid synthesis and diverting fatty acids from membrane lipids to triacylglycerol in Arabidopsis leaves. *Plant Cell* 25, 3506–3518.

Heneen, W.K., Banas, A., Leonova, S., Carlsson, A.S., Marttila, S., Debski, H. and Stymne, S. (2009) The distribution of oil in the oat grain. *Plant Signaling & Behavior* 4, 55–56.

Hsieh, K. and Huang, A.H.C. (2004) Endoplasmic reticulum, oleosins, and oils in seeds and tapetum cells. *Plant Physiology* 136, 3427–3434.

Jako, C., Kumar, A., Wei, Y.D., Zou, J.T., Barton, D.L., *et al.* (2001) Seed-specific over-expression of an Arabidopsis cDNA encoding a diacylglycerol acyltransferase enhances seed oil content and seed weight. *Plant Physiology* 126, 861–874.

Kroon, J., Wei, W., Simon, W. and Slabas, A. (2006) Identification and functional expression of a type 2 acyl-CoA:diacylglycerol acyltransferase (DGAT2) in developing castor bean seeds which has high homology to the major triglyceride biosynthetic enzyme of fungi and animals. *Phytochemistry* 67, 2541–2549.

Lambert, R.J., Alexander, D.E. and Han, Z.J. (1998) A high-oil pollinator enhancement of kernel oil and effects on grain yields of maize hybrids. *Agronomy Journal* 90, 211–215.

Lambert, R.J., Alexander, D.E. and Mejaya, I.J. (2004) Single kernel selection for increased grain oil in maize synthetics and high-oil hybrid development. *Plant Breeding Reviews* 24, 153–175.

Lardizabal, K.D., Mai, J.T., Wagner, N.W., Wyrick, A., Voelker, T. and Hawkins, D.J. (2001) DGAT2 is a new diacylglycerol acyltransferase gene family – purification, cloning, and expression in insect cells of two polypeptides from *Mortierella ramanniana* with diacylglycerol acyltransferase activity. *Journal of Biological Chemistry* 276, 38862–38869.

Laurie, C.C., Chasalow, S.D., LeDeaux, J.R., McCarroll, R., Bush, D., *et al.* (2004) The genetic architecture of response to long-term artificial selection for oil concentration in the maize kernel. *Genetics* 168, 2141–2155.

Li, H., Peng, Z., Yang, X., Wang, W., Fu, J., *et al.* (2013) Genome-wide association study dissects the genetic architecture of oil biosynthesis in maize kernels. *Nature Genetics* 45, 43–50.

Li, Q., Shao, J., Tang, S., Shen, Q., Wang, T., Chen, W. and Hong, Y. (2015) Wrinkled1 accelerates flowering and regulates lipid homeostasis between oil accumulation and membrane lipid anabolism in *Brassica napus. Frontiers in Plant Science* 6, 1015.

Liu, F., Xia, Y., Wu, L., Fu, D., Hayward, A., *et al.* (2015) Enhanced seed oil content by overexpressing genes related to triacylglyceride synthesis. *Gene* 557, 163–171.

Lotan, T., Ohto, M., Yee, K.M., West, M.A., Lo, R., *et al.* (1998) *Arabidopsis* LEAFY COTYLEDON1 is sufficient to induce embryo development in vegetative cells. *Cell* 93, 1195–1205.

Lung, S.C. and Weselake, R.J. (2006) Diacylglycerol acyltransferase: a key mediator of plant triacylglycerol synthesis. *Lipids* 41, 1073–1088.

Ma, W., Kong, Q., Arondel, V., Kilaru, A., Bates, P.D., *et al.* (2013) *WRINKLED1*, a ubiquitous regulator in oil accumulating tissues from *Arabidopsis* embryos to oil palm mesocarp. *PLOS ONE.* 8, e68887.

Ma, W., Kong, Q., Grix, M., Mantyla, J.J., Yang, Y., Benning, C. and Ohlrogge, J.B. (2015) Deletion of a C-terminal intrinsically disordered region of WRINKLED1 affects its stability and enhances oil accumulation in Arabidopsis. *Plant Journal* 83, 864–874.

Maeo, K., Tokuda, T., Ayame, A., Mitsui, N., Kawai, T., *et al.* (2009) An AP2-type transcription factor WRINKLED1 of *Arabidopsis thaliana* binds to AW-box sequence conserved among proximal upstream regions of genes involved in fatty acid synthesis. *Plant Journal* 60, 476–487.

Moose, S.P., Dudley, J.W. and Rocheford, T.R. (2004) Maize selection passes the century mark: a unique resource for 21st century genomics. *Trends in Plant Science* 9, 358–364.

Mu, J., Tan, H., Zheng, Q., Fu, F., Liang, Y., *et al.* (2008) *LEAFY COTYLEDON1* is a key regulator of fatty acid biosynthesis in Arabidopsis. *Plant Physiology* 148, 1042–1054.

Nagasawa, N., Hibara, K., Heppard, E.P., Vander Velden, K.A., Luck, S., *et al.* (2013) *GIANT EMBRYO* encodes CYP78A13, required for proper size balance between embryo and endosperm in rice. *Plant Journal* 75, 592–605.

Oakes, J., Brackenridge, D., Colletti, R., Daley, M., Hawkins, D.J., *et al.* (2011) Expression of fungal *diacylglycerol acyltransferase2* genes to increase kernel oil in maize. *Plant Physiology* 155, 1146–1157.

Ohlrogge, J. and Browse, J. (1995) Lipid biosynthesis. *Plant Cell* 7, 957–970.

Olsen, O.-A. (2001) Endosperm development: cellularization and cell fate specification. *Annual Review of Plant Physiology and Plant Molecular Biology* 52, 233–267.

Perry, T.W. (1988) Corn as a livestock feed. In: Sprague, G.F. and Dudley, J.W. (eds.) *Corn and Corn Improvement*. American Society of Agronomy, Madison, Wisconsin, pp. 941–963.

Peterson, D.M. and Wood, D.F. (1997) Composition and structure of high-oil oat. *Journal of Cereal Science* 26, 121–128.

Pouvreau, B., Baud, S., Vernoud, V., Morin, V., Py, C., *et al.* (2011) Duplicate maize *Wrinkled1* transcription factors activate target genes involved in seed oil biosynthesis. *Plant Physiology* 156, 674–686.

Roesler, K., Meyer, K., Damude, H.G., Shen, B., Li, C., Bermudez, E. and Tarczynski, M. (2012) DGAT genes for increased seed storage lipid production and altered fatty acid profiles in oilseed plants. US Patent 8101818 B2.

Roesler, K., Shen, B., Bermudez, E., Li, C., Hunt, J., *et al.* (2016) An improved variant of soybean type 1 diacylglycerol acyltransferase increases the oil content and decreases the soluble carbohydrate content of soybeans. *Plant Physiology* 171, 878–893.

Saha, S., Enugutti, B., Rajakumari, S. and Rajasekharan, R. (2006) Cytosolic triacylglycerol biosynthetic pathway in oilseeds: molecular cloning and expression of peanut cytosolic diacylglycerol acyltransferase. *Plant Physiology* 141, 1533–1543.

Santos-Mendoza, M., Dubreucq, B., Baud, S., Parcy, F., Caboche, M. and Lepiniec, L. (2008) Deciphering gene regulatory networks that control seed development and maturation in Arabidopsis. *Plant Journal* 54, 608–620.

Shen, B., Sinkevicius, K.W., Selinger, D.A. and Tarczynski, M.C. (2006) The homeobox gene *GLABRA2* affects seed oil content in *Arabidopsis*. *Plant Molecular Biology* 60, 377–387.

Shen, B., Allen, W.B., Zheng, P., Li, C., Glassman, K., *et al.* (2010) Expression of *ZmLEC1* and *ZmWRI1* increases seed oil production in maize. *Plant Physiology* 153, 980–987.

Shockey, J.M., Gidda, S.K., Chapital, D.C., Kuan, J.C., Dhanoa, P.K., *et al.* (2006) Tung tree DGAT1 and DGAT2 have nonredundant functions in triacylglycerol biosynthesis and are localized to different subdomains of the endoplasmic reticulum. *Plant Cell* 18, 2294–2313.

Siloto, R., Truska, M., Brownfield, D., Good, A. and Weselake, R. (2009) Directed evolution of acyl-CoA: *diacylglycerol* acyltransferase: development and characterization of *Brassica napus* DGAT1 mutagenized libraries. *Plant Physiology and Biochemistry* 47, 456–461.

Song, T.M. and Chen, S.J. (2004) Long term selection for oil concentration in five maize populations. *Maydica* 49, 9–14.

Ting, J.T., Lee, K., Ratnayake, C., Platt, K.A., Balsamo, R.A. and Huang, A.H. (1996) Oleosin genes in maize kernels having diverse oil contents are constitutively expressed independent of oil contents: size and shape of intracellular oil bodies are determined by the oleosins/oils ratio. *Planta* 199, 158–165.

To, A., Joubès, J., Barthole, G., Lécureuil, A., Scagnelli, A., *et al.* (2012) WRINKLED transcription factors orchestrate tissue-specific regulation of fatty acid biosynthesis in *Arabidopsis*. *Plant Cell* 24, 5007–5023.

USDA (2017) Oil Crop Yearbook–Table 32. U.S. Department of Agriculture, Washington, D.C. Available at: http://www.ers.usda.gov/data-products/oil-crops-yearbook.aspx (accessed June 5, 2017).

Vanhercke, T., El Tahchy, A., Shrestha, P., Zhou, X.R., Singh, S.P. and Petrie, J.R. (2013) Synergistic effect of WRI1 and DGAT1 coexpression on triacylglycerol biosynthesis in plants. *FEBS Letters* 587, 364–369.

Vanhercke, T., El Tahchy, A., Liu, Q., Zhou, X.R., Shrestha, P., *et al.* (2014) Metabolic engineering of biomass for high energy density: oilseed-like triacylglycerol yields from plant leaves. *Plant Biotechnology Journal* 12, 231–239.

van Erp, H., Kelly, A.A., Menard, G. and Eastmond, P.J. (2014) Multigene engineering of triacylglycerol metabolism boosts seed oil content in Arabidopsis. *Plant Physiology* 165, 30–36.

Voelker, T. and Kinney, A.J. (2001) Variations in the biosynthesis of seed-storage lipids. *Annual Review of Plant Physiology and Plant Molecular Biology* 52, 335–361.

Weber, E.J. (1987) Lipids of the kernel. In: Watson, S.A. and Ramstad, P.E. (eds.) *Corn Chemistry and Technology*. American Association of Cereal Chemists, Inc., St. Paul, Minnesota, pp. 311–349.

Xu, C. and Shanklin, J. (2016) Triacylglycerol metabolism, function, and accumulation in plant vegetative tissues. *Annual Review of Plant Biology* 67, 179–206.

Xu, J., Francis, T., Mietkiewska, E., Giblin, E.M., Barton, D.L., *et al.* (2008) Cloning and characterization of an acyl-CoA-dependent *diacylglycerol acyltransferase 1 (DGAT1)* gene from *Tropaeolum majus*, and a study of the functional motifs of the DGAT protein using site-directed mutagenesis to modify enzyme activity and oil content. *Plant Biotechnology Journal* 6, 799–818.

Yang, X., Ma, H., Zhang, P., Yan, J., Guo, Y., Song, T. and Li, J. (2012) Characterization of QTL for oil content in maize kernel. *Theoretical and Applied Genetics* 125, 1169–1179.

Young, T.E., Geisler-Lee, J. and Gallie, D.R. (2004) Senescence-induced expression of cytokinin reverses pistil abortion during maize flower development. *Plant Journal* 38, 910–922.

Zale, J., Jung, J.H., Kim, J.Y., Pathak, B., Karan, R., *et al.* (2016) Metabolic engineering of sugarcane to accumulate energy-dense triacylglycerols in vegetative biomass. *Plant Biotechnology Journal* 14, 661–669.

Zhang, C., Iskandarov, U., Klotz, E.T., Stevens, R.L., Cahoon, R.E., *et al.* (2013) A thraustochytrid diacylglycerol acyltransferase 2 with broad substrate specificity strongly increases oleic acid content in engineered *Arabidopsis thaliana* seeds. *Journal of Experimental Botany* 64, 3189–3200.

Zhang, M., Fan, J., Taylor, D.C. and Ohlrogge, J.B. (2009) DGAT1 and PDAT1 acyltransferases have overlapping functions in *Arabidopsis* triacylglycerol biosynthesis and are essential for normal pollen and seed development. *Plant Cell* 21, 3885–3901.

Zhang, P., Allen, W.B., Nagasawa, N., Ching, A.S., Heppard, E.P., *et al.* (2012) A transposable element insertion within *ZmGE2* gene is associated with increase in embryo to endosperm ratio in maize. *Theoretical and Applied Genetics* 125, 1463–1471.

Zheng, P., Allen, W.B., Roesler, K., Williams, M.E., Zhang, S., *et al.* (2008) A phenylalanine in DGAT is a key determinant of oil content and composition in maize. *Nature Genetics* 40, 367–372.

Zou, J.T., Wei, Y.D., Jako, C., Kumar, A., Selvaraj, G. and Taylor, D.C. (1999) The *Arabidopsis thaliana* TAG1 mutant has a mutation in a diacylglycerol acyltransferase gene. *Plant Journal* 19, 645–653.

# 14 Maize Seed Storage Proteins

Brian A. Larkins[1,]*, Yongrui Wu[2], Rentao Song[3] and Joachim Messing[4]

[1]*Department of Agronomy and Horticulture, University of Nebraska-Lincoln, USA;*
[2]*Institute of Plant Physiology & Ecology, Shanghai Institutes for Biological Sciences, Chinese
Academy of Sciences, China;* [3]*National Maize Improvement Center of China, China
Agricultural University, China;* [4]*Waksman Institute of Microbiology, Rutgers University, USA*

## 14.1 Introduction

Maize kernels contain several types of storage proteins. By far the most abundant are prolamins, zeins, a unique storage protein found only in cereals. Because of their abundance, zeins have a profound influence on human and livestock nutrition. These proteins also appear to influence the mechanical strength of the kernel, which is important for harvesting and storage, and they affect the functional properties of food products made from corn. While a great deal of research has been devoted to the characterization of genes encoding zeins and the mechanisms by which the proteins are synthesized and stored in endosperm cells, many important questions remain regarding their structure, the regulation of the genes encoding them, and how they influence the formation of the hard, vitreous regions of the mature kernel. This chapter reviews what is known about maize storage proteins, and describes important questions that remain to be answered about their synthesis and functions in the grain. It also considers technical approaches for altering the storage protein content of maize kernels to increase the level of lysine, the most limiting essential amino acid for monogastric animals.

Seeds accumulate proteins to provide a ready source of amino acids and carbon skeletons for the seedling. Arguably, any protein could serve this function, and, to some extent, many do, but for largely unknown reasons evolutionary pressure favored certain types of protein structures and resulted in a few specialized proteins. Among the most common are several water-soluble albumins, salt-soluble globulins, and the alcohol-soluble prolamins (Shewry *et al.*, 1995; Casey and Shewry, 1999). Plants are photoautotrophs and can synthesize all 20 amino acids, but seed storage proteins tend to favor those with extra nitrogen (glutamine and asparagine) and sulfur (cysteine and methionine) at the expense of others, including several essential for humans and other monogastic animals, e.g. lysine, tryptophan, and methionine. Consequently, seeds are often an incomplete source of dietary protein. Plant breeders and chemists have long sought ways to improve the nutritional quality and digestibility of seed proteins, while at the same time eliminating those that are anti-nutritional or allergenic. These goals underlie much research on seed proteins, and they remain important aspirations if we are to feed an ever-expanding human population.

---

*Corresponding author email: blarkins2@unl.edu

 """

## 14.2   Maize Seed Storage Proteins

Maize kernels contain two major types of storage proteins: globulins and prolamins. Globulins-1 and -2 are the primary storage proteins in the embryo, where they account for 10–20% of the total protein. Globulin-1 (63-kD) and globulin-2 (45-kD) are related to the 7S storage globulins common in dicot embryos (Kriz, 1999). The endosperm contains small amounts of two other globulins: legumin-1, a 51-kD protein that resembles an unprocessed form of the 11S storage globulin found in legumes and other dicots (Yamagata *et al.*, 2003); and α-globulin, an 18-kD protein with homologues found in all grasses analyzed (Woo *et al.*, 2001; Gu *et al.*, 2010). Legumin-1 and α-globulin are found in peripheral cell layers of the starchy endosperm, with small amounts in the aleurone (Reyes *et al.*, 2011).

Maize prolamins, of which there are four types, α-, β-, γ-, and δ-zeins, are the most abundant storage proteins in the seed, and they can account for 60–70% of the endosperm protein (Boston and Larkins, 2008). Because of their abundance, zeins are the primary determinants of the amino acid composition of the kernel. Zeins, like most prolamins, are essentially devoid of lysine and tryptophan. Mutations that reduce zein synthesis, like *opaque2* (*o2*) (Fig. 14.1A), can result in kernels with an increased percentage of lysine and tryptophan, which improves their nutritional quality (Mertz *et al.*, 1964).

## 14.3   Zein Protein Structure and Deposition in Protein Bodies

If endosperm flour is extracted with an alkaline buffer containing SDS and 2-mercaptoethanol, which solubilizes all proteins, zeins can be separated by adding sufficient absolute ethanol to create a 70% solution: zeins remain soluble, whereas other proteins precipitate (Wallace *et al.*, 1990). SDS-PAGE analysis of the alcohol-soluble proteins reveals a mixture of polypeptides (Fig. 14.2A), the most abundant of which, the α-zeins, have apparent molecular weights of 22-kD

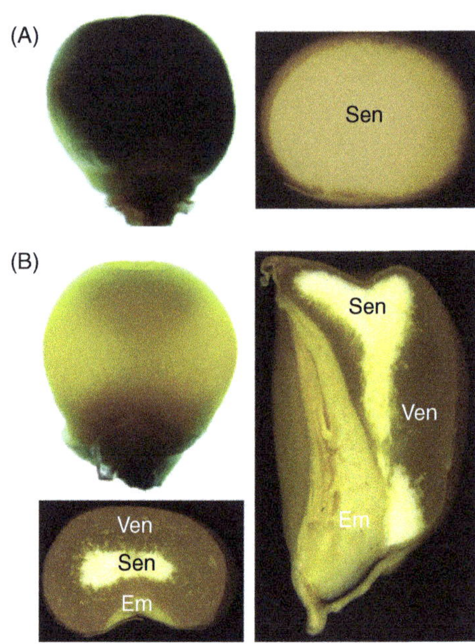



**Fig. 14.1.** Maize kernel phenotypes. (A) Top left, back-illuminated W64A*o2* kernel showing opaque phenotype; right, cross section showing starchy endosperm; (B) illuminated W64A wild-type kernel showing vitreous phenotype; bottom left and right are kernel cross- and longitudinal-sections showing vitreous (Ven) and starchy endosperm (Sen) regions and embryo (Em).

and 19-kD. The second most abundant proteins are the γ-zeins, which have apparent molecular weights of 50-kD, 27-kD, and 16-kD. The β-zein is related to the γ-zeins (Woo *et al.*, 2001) and has an apparent molecular weight of 15-kD. The δ-zeins, which are the least abundant, comprise two related proteins of 18-kD and 10-kD (Swarup *et al.*, 1995).

The primary amino acid sequences of α-, β-, γ-, and δ-zeins show they correspond to three structurally distinct types of proteins that share the common property of hydrophobicity. The α-zeins have an NH2-terminal turn of 36 residues that precedes nine or ten homologous repeating peptides of approximately 18 amino acids, each of which is flanked with glutamine residues (Geraghty *et al.*, 1981; Argos *et al.*, 1982). Circular dichroic measurements of α-zeins in 70% methanol indicate α-helical content of 50–60%, with the remainder turn and

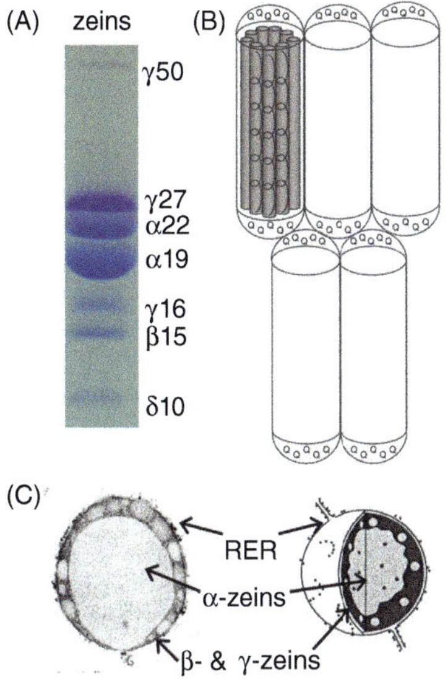



**Fig. 14.2.** Characterization of zein proteins. (A) SDS-PAGE separation showing apparent molecular weights of the α-, β-, γ-, and δ-zeins; (B) proposed 3-dimensional structure of α-zeins based on Argos *et al.* (1982); and (C) transmission electron microscope image (left) of a developing protein body (PB) with diagram illustrating location of zein proteins (right).

random coil configurations. These data suggest a topological model in which the repeats create antiparallel helices that cluster within a distorted cylinder (Fig. 14.2B). The γ-zeins (including β-zein) are sulfur-rich proteins that share a core region related to α-globulin, which in turn belongs to a large protein family that also includes albumin storage proteins (Shewry *et al.*, 1995; Xu and Messing, 2009). This region contains eight conserved cysteine residues that form intra-chain disulfide bonds. Circular dichroic measurements in 50% methanol suggest this region (at least in the β-zein) is composed primarily of β-sheet and turn structures (Pedersen *et al.*, 1986). What distinguishes the 50-kD and the 27-kD γ-zeins from the 15-kD β-zein and 16-kD γ-zein is an extended NH-terminus that contains a large block of multiple tandem repeats. In the 27-kD γ-zein this region contains eight copies of the hexapeptide PPPVHL, and differs from the larger block of polyglutamine repeats at the NH-terminus of the 50-kD γ-zein. Like

the γ-zeins, the two δ-zein proteins are sulfur-rich proteins; however, they contain a higher percentage of methionine (22% and 25%, respectively) than the γ-zeins. The β-zein is also methionine-rich (11%), but it has a high cysteine content like the γ-zeins.

Zein proteins have a signal peptide that directs their synthesis into the lumen of the rough endoplasmic reticulum (RER), where they form insoluble spherical accretions (protein bodies (PBs)) that are generally 1–2 microns in diameter. Immunocytochemical studies (Lending and Larkins, 1989) showed that PBs have an organized structure (Fig. 14.2 B,C). The smallest PB contain principally β- and γ-zeins. As α- and δ-zeins accumulate, they form locules within the β- and γ-zeins at the periphery of the PB and then coalesce within it. There is a subtle layering of α-zeins within the PB; the 19-kD α-zeins concentrate in the center, whereas the 22-kD α-zeins are more peripheral (Holding *et al.*, 2007).

## 14.4 Zein Gene Structure and Expression

The genetic loci encoding α-, β-, γ-, and δ-zeins occur on six of the ten maize chromosomes (Xu and Messing, 2008). The gene duplication that gave rise to the 22-kD and 19-kD α-zeins occurred in the progenitor of maize and sorghum. Following this duplication, there were several chromosomal translocations and many gene duplications and deletions, as well as point mutations, that created the mosaic of α-zein loci that exist among different maize genetic backgrounds. There are (variably) 41–48 α-zein genes organized in complex multigene families on chromosomes 1, 4, and 7 (Miclaus, *et al.*, 2011a). Some of the 19-kD α-zein genes (12 genes in B73 for example) are on the short arm of chromosome 4: these are the so-called z1A1 group (9 genes) and the smaller z1A2 group (3 genes). Additional 19-kD α-zein genes are in three clusters: z1B (9 genes on the short arm of chromosome 7) and z1D (5 genes) on the short arm of chromosome 1 (Song and Messing, 2002). The 22-kD α-zeins, z1C1 and z1C2 (14 genes and one gene, respectively, in B73), are on the short arm of chromosome 4 (Song and Messing, 2003).

In contrast to the highly duplicated nature of α-zein genes, those encoding the β-, γ-, and δ-zeins are typically single copy. The 27-kD γ-zein locus can harbor two nearly identical genes in tandem; however, the locus is somewhat unstable and a recombination can eliminate one (Das *et al.*, 1991). The 50-kD and 27 kD-γ-zein genes are in close proximity on the short arm of chromosome 7 (Yuan *et al.*, 2014), whereas the 16-kD γ-zein is on the long arm of chromosome 2. The 15-kD β-zein is on the short arm of chromosome 6, and the 10-kD and 18-kD δ-zeins are on chromosomes 9 and 6, respectively.

Zein genes are temporally and spatially regulated through the activity of shared cis-acting nucleotide sequences in their promoters and transcription factors (TFs) that recognize them. None of the different types of zein genes contain introns; hence, post-transcriptional processing of mRNAs does not appear to play a major role in gene expression. To date, three different types of zein TFs have been described: bZIP proteins, encoded by *O2* and Opaque2 heterodimerizing protein 1 and 2 (*OHP1/2*); the Prolamin-Box Binding Factor (PBF1), an endosperm-specific zinc finger protein called ZmDOF3; and a MADS box protein, ZmMADS47 (Vicente-Carbajosa *et al.*, 1997; Zhang *et al.*, 2015; Qiao *et al.* 2016). PBF binds the P Box (TGTAAAG), which is a cis-acting regulatory sequence found in the promoters of prolamin genes in most cereals (Wu and Messing, 2012a), while O2 and OHP1/2 recognize the O2 box (TT/CCACGT). ZmMADS47 binds a CATGT motif that flanks the O2 site in α-zein and 50-kD γ-zein promoters. O2 and OHP1 originated from an ancient duplication before the split of maize and rice, while OHP1 and OHP2 are paralogs resulting from allotetraploidization of two maize progenitors. O2 and OHPs appear to have undergone sub-functionalization. O2 is the major and minor TF for α-zein and 27-kD γ-zein genes, respectively; conversely, OHPs are the major and minor TFs for 27-kD γ- and α-zein genes, respectively. In O2, suppression of OHPs does not cause a significant reduction of α-zein gene transcription, but in *o2*, OHPs are critical for expression of a residual level of α-zeins (Zhang *et al.*, 2015; Yang *et al.*, 2016). O2 and PBF mutually promote binding to their promoters, leading to increased transcription. O2 and ZmMADS47 bind near each other in α-zein promoters and form a complex; transactivation by ZmMADS47 relies strictly on O2. The promoter of the 16-kD γ-zein lacks both the P-box and the O2-like boxes, which implies other zein-regulatory TFs remain to be identified. Transcriptional regulation of δ-zeins also remains to be resolved. The δ-zein regulator (*Dzr1*) influences accumulation of this protein via the mRNA noncoding region (Lai *et al.*, 2002).

## 14.5  Mutations Affecting Zein Synthesis

Research on the molecular basis of opaque and floury endosperm mutants has provided insight into the mechanisms that regulate zein gene expression. Four types of naturally occurring mutations affecting zein synthesis have been described: (i) those controlling gene transcription; (ii) those affecting zein protein structure; (iii) those affecting cytoskeletal proteins; and (iv) those influencing metabolic processes. Some of these arose spontaneously and have existed for many years, whereas others were created more recently via chemical mutagenesis and transposon tagging (Thompson and Larkins, 1993). Several mutations have been identified that result from defective signal peptides on α-zein proteins; these include *fl2* (22-kD), *DeB30* (19-kD), and *fl4* (19-kD) (Boston and Larkins, 2008; Wang *et al.*, 2014a). *Mucronate1* (*Mc*) is a frame-shift mutation in the 16-kD γ-zein gene that creates an abnormal protein structure (Kim *et al.*, 2006). All four of these mutations cause physiological stress in the ER, leading to the "unfolded protein response" (UPR), which reduces zein synthesis and elevates the level of chaperones (Kirst *et al.*, 2005). Whether, and how, UPR causes the starchy endosperm phenotype of the mature kernel is unknown (Morton *et al.*, 2015). Two mutations, *o1* (Wang *et al.*, 2012) and *fl1* (Holding *et al.*, 2007), are associated with defects in myosin or myosin-related proteins and may be associated with the cytoskeleton surrounding the RER (Clore *et al.*, 1996). *o1* affects ER streaming and could influence zein translocation into the RER, while *fl1* is associated with the PB membrane and results in a subtle change in zein organization within PB. The *o10* mutation, with similar PB changes as *o1* and *fl1*, encodes a PB internal protein that interacts with Fl1, 22-kD and 16-kD zeins (Yao *et al.*, 2016). Three opaque mutations affect amino acid biosynthesis: *o7* was identified as a defective Acyl-CoA synthetase (Miclaus *et al.*, 2011b). Its function as oxalyl-CoA synthetase was revealed by Arabidopsis *AAE3* (Foster *et al.*, 2012), which creates changes in oxalacetic acid and $\alpha$-ketoglutaric acid, two important precursors of asparagine, lysine and glutamine (Wang *et al.*, 2011). Another mutant was shown to be the consequence of a defective arogenate dehydrogenase, an enzyme involved in tyrosine synthesis that has pleiotropic effects on accumulation of other amino acids (Holding *et al.*, 2010). O6 (*pro1*) encodes a $\Delta^1$-pyrroline-5-carboxylate synthetase; it affects proline synthesis, leading to a general reduction in protein synthesis (Wang *et al.*, 2014b). But not all opaque mutations directly or indirectly affect zein synthesis. The *o5* mutation corresponds to a defective monogalactosyldiacylglyerol synthase (MGD1) and affects the phospholipid composition of amyloplast membranes (Myers *et al.*, 2011).

It is possible to create specific zein gene knockdowns/knockouts by RNA interference (RNAi), which allows reduced expression or silencing of single genes or multigene families (Wu and Messing, 2010; Guo *et al.*, 2013). Also, $\gamma$-irradiation of seeds has been used to create novel kernel mutants, including some with deleted zein loci (Yuan *et al.*, 2014). These approaches make it possible to investigate the role of specific types of zeins in PB formation and kernel phenotypes.

Prior to RNAi mutants, it was hypothesized that $\gamma$-zeins were responsible for nucleating PB and retaining zeins within the lumen of the RER, with the $\alpha$- and $\delta$-zeins acting essentially as filler (Boston and Larkins, 2008). It was also thought that $\gamma$-zeins played an important role in creating vitreous endosperm during kernel maturation. Phenotypes of zein RNAi mutants have largely supported these hypotheses and also provided insight regarding the role of different types of zeins in PB formation. Silencing expression of 22-kD $\alpha$-zeins creates an opaque kernel phenotype, consistent with one of the pleiotropic effects of *o2* (Segal *et al.*, 2003). Silencing both 22-kD and 19-kD $\alpha$-zeins also creates an opaque phenotype, and it dramatically reduces the size of PB without significantly changing their number (Guo *et al.*, 2013). Silencing all the $\gamma$-zeins (50-kD, 27-kD, and 16-kD) and the $\beta$-zein reduces PB size and distorts their structure, similar to what occurs in the *Mc* mutant. However, the most striking effect of reducing all the $\gamma$-zein proteins is a dramatic reduction (~75%) in PB number. This is largely the consequence of eliminating the 27-kD $\gamma$-zein, as PB number is less affected if only the 16-kD and 50-kD $\gamma$-zeins and $\beta$-zein are eliminated. Furthermore, eliminating only the 27-kD $\gamma$-zein had essentially the same phenotypic effect as eliminating all the $\gamma$-zein proteins. In contrast to the above results, elimination of one or both $\delta$-zeins has no significant effect on PB size or kernel phenotype.

The results of the zein knockdown/ knockout experiments are consistent with other studies suggesting that the 27-kD $\gamma$-zein, in particular the NH-terminal domain with its tandem proline-rich repeats and cysteine residues, plays a role in the initiation of PB formation (Mainieri *et al.*, 2014). However, the mechanism by which it does this and interacts with $\alpha$- and $\delta$-zeins is unclear. Experiments based on the yeast two-hybrid system showed strong interactions among the $\gamma$-zein proteins, but only a weak interaction between the 27-kD $\gamma$-zein and the $\alpha$-zeins (Kim *et al.*, 2002). An RNAi transgene that largely eliminated the 16-kD and 27-kD $\gamma$-zeins (Wu and Messing, 2010), or this transgene in combination with a β-zein RNAi (Guo *et al.*, 2013), did not affect α-zein accumulation in PB. This implies that other mechanisms must exist for retention of α-zeins within the RER.

These experiments provide convincing evidence that the 27-kD γ-zein can play an important role in the formation of vitreous endosperm. A 27-kD γ-zein RNAi mutation or a chromosomal deletion mutation (Yuan *et al.*, 2014) both result in a strongly penetrant starchy endosperm phenotype. Furthermore, high-level expression of the 27-kD γ-zein can convert the starchy endosperm of *o2* to a vitreous phenotype (Wu *et al.*, 2010). Gene duplication at the 27-kD γ-zein locus increases the level of gene expression, leading to endosperm modification (Liu *et al.*, 2016). Nevertheless, how the 27-kD γ-zein contributes to the formation of vitreous endosperm is unknown. In summary, these experiments demonstrate that a reduction in the size, number (frequency), and structure of PB in endosperm cells can lead to an opaque/floury endosperm phenotype, although this can also be the consequence of defects in biochemical and metabolic processes, especially those causing ER stress (Morton *et al.*, 2015).

## 14.6 High-lysine Corn and Quality Protein Maize

Knowledge of reduced zeins in *o2* and *fl2* led Mertz and Nelson to investigate them as a possible way to increase kernel lysine content (Mertz *et al.*, 1964; Nelson *et al.*, 1965). They showed *o2* and *fl2* markedly reduce zeins, while nearly doubling lysine content, although this is dependent on genetic background (Moro *et al.*, 1996). Higher lysine is the consequence of three things: (i) zein proteins contain essentially no lysine, so their reduction enhances the percentage of lysine contributed by other endosperm proteins; (ii) *O2* positively regulates expression of lysine ketoglutarate reductase, an enzyme that degrades lysine in the maturing endosperm (Brochetto-Braga, 1992); and (iii) there is a mechanism, proteome rebalancing, that

redistributes nitrogen to other endosperm proteins, some of which are lysine-rich (Jia *et al.*, 2013; Morton *et al.*, 2015).

Unfortunately, one of the many pleotropic effects of *o2* and *fl2* is a kernel with a soft, starchy endosperm that renders it more susceptible to damage during harvesting. Also, flour made from the endosperm has poor functional properties, making it unattractive for food products. Consequently, "high-lysine" corn was abandoned within a few years (Gibbon and Larkins, 2005). However, plant breeders at CIMMYT discovered genetic suppressors of *o2* (*o2* modifiers) that create a vitreous endosperm, restoring the normal kernel phenotype. This led to development of high-lysine varieties called "Quality Protein Maize" (QPM) (Nelson, 2001). Breeding QPM is challenging, as it involves introducing multiple, unlinked *o2* modifier genes into an agronomical adapted *o2* background. QPM varieties are not widely grown, but they are utilized in a few developing counties, where they have value for human and livestock nutrition.

## 14.7 Unanswered Questions

### 14.7.1 Structure of zein proteins and their applications

The evolutionary events leading to zein gene structure and organization have been described (Xu and Messing, 2008, 2009), but the selection pressures responsible for zein protein structures are a mystery; indeed, the tertiary structure of α-, β-, γ-, and δ-zeins are unknown. The near absence of lysine and tryptophan in zeins and their hydrophobic nature are possible consequences of evolutionary selection. The insolubility of zeins, particularly α- and δ-zeins, would facilitate kernel desiccation, an important factor for seed maturation and dormancy. Lysine is a charged amino acid and could be a negative factor in this regard. However, introduction of lysine into α-zein did not alter its ability to form protein bodies (Wallace *et al.*, 1988). Some insects that feed on seeds require lysine in their diet

(Nation, 2016), and the near absence of lysine in zeins, plus their insolubility, could limit their attractiveness as a food source and minimize kernel damage before germination. Also, proteins with extensive disulfide linkages tend to be poorly digested (Hamaker *et al.*, 1987). Consequently, the network of disulfide-linked γ-zeins on the surface of protein bodies might reduce their nutritional value. Some of these features of zeins are shared with prolamins of other cereals; thus, there might have been common selection pressures for prolamin evolution. The ability to downregulate zeins by RNAi and genetically engineer novel proteins provide opportunities to investigate these questions.

The hydrophobicity, abundance, and price of zeins, particularly α-zeins, makes them useful for a variety of applications as plasticizers, coatings, fibers, inks, moldings, and most recently, nanoparticles for delivery of chemotheraputic drugs (Lawton, 2002; Corradini *et al.*, 2014). Nevertheless, we know very little about the structure of zeins, and this information could improve their utility (Xu *et al.*, 2015). Determining α-zein structure is a challenge, as this group of proteins is water-insoluble and structurally heterogeneous. However, a single α-zein can be produced with a yeast protein expression vector (Kim *et al.*, 2002). It should not be difficult to purify γ-zein proteins, as they are water soluble (once intermolecular disulfide bonds are reduced) and single molecular species. Purification of sufficient amounts of γ-zeins for structural analysis is straightforward, although crystallization could be challenging. Cryo-EM might provide a novel approach for structural characterization that does not require protein crystallization (Kühlbrandt, 2014).

Genetically engineered zeins could have features that enhance their functional properties. Higher molecular weight forms of α-zein created by additional α-helical repeats might produce films with greater flexibility, and fusion of α-zein with the NH-terminal cysteine-rich domain of the γ-zeins might lead to polymers that allow slower release of pharmaceuticals. It is possible to produce zeins that contain lysine and tryptophan, although it would require synthesis of large amounts of the engineered proteins in the endosperm to impact the grain's nutritional quality, and the consequences are unknown.

### 14.7.2 How PBs form and are retained within the RER

Zein PB appear to form at the ends of RER cisternae, where they grow to a uniform diameter of 1–2 microns. This begs several questions: (i) Do PB form at unique domains of the RER? (ii) What limits their size? (iii) How are they retained in the RER rather than transported to protein storage vacuoles, as occurs in wheat and other cereals? (iv) Do specialized receptors/chaperones and the actin-myosin cytoskeleton surrounding the RER play a role in these processes? The fact that PBs are spaced within the RER suggests there could be receptors that interact with the 27-kD γ-zein and/or other zein proteins. There is ample evidence of distinct domains in plant ER membranes (Herman, 2008), and receptors for specialized functions are common. Early experiments suggested no difference in polysomes associated with PB and those on vesicular RER (Larkins and Hurkman, 1978); however, the approach may have lacked sufficient resolution. Subsequent experiments with zein and rice storage proteins provided evidence for mRNA targeting to specific domains of the RER (Washida *et al.*, 2009). Targeting mRNAs via the cytoskeleton for translation at unique subcellular locations is well documented (St. Johnston, 2005), and it would be consistent with the actin-myosin network surrounding the RER in endosperm cells. Nevertheless, *in situ* hybridization of 27-kD γ-zein and 22-kD α-zein mRNAs did not show spatial differences in their location on RER membranes (Kim *et al.*, 2002), and evidence of rice prolamin and glutelin mRNA targeting is not necessarily applicable to zeins, since the rice storage proteins are deposited at different subcellular locations. The facts that *o1* and *fl1* are defects in myosin and myosin-related proteins associated with the RER, and *fl1* alters zein organization within the PB, are consistent with the cytoskeleton playing a role in PB formation.

If there are receptors for zeins at discrete locations in the RER, it could account for zein spatial distribution and perhaps the limited growth of PBs. Proteomic analysis of intact PB with the associated RER/cytoskeleton could provide insight into this question. It is also possible that hydrophobic–hydrophilic interactions between γ-zeins and α- and δ-zeins are sufficient to explain PB organization, although it appears there is a mechanism(s) independent of γ-zeins that retains α- and δ-zeins within the RER. The unique NH-terminal structures of γ-zeins suggest they have functions beyond sulfur storage, but these roles are less clear for the 50- and 16-kD γ-zeins compared to the 27-kD γ-zein. It is possible that the proline-rich domains interact with membrane lipids and influence RER retention or PB membrane curvature (Kogan *et al.*, 2004). One might also wonder what distinguishes the function of δ-zeins from γ-zeins for sulfur storage; perhaps there is a special need for methionine during germination.

### 14.7.3   Formation of vitreous endosperm

Midway through kernel development, it is not possible to visualize vitreous endosperm: wild-type and opaque/floury mutants are indistinguishable. But by the late dough stage (35–40 days after pollination (DAP)), yellow, vitreous endosperm begins to form at the periphery of a normal, wild-type kernel (Fig. 14.1B). The reason for concentration of carotenoids in this region is not understood, but it might result from hydrophobic interaction with protein bodies. (Note, vitreous endosperm is not yellow in white corn varieties.) Much earlier, beginning in the center of the endosperm and progressing outward, cells undergo programmed cell death; by the late dough stage, most cells are dead (Woo *et al.*, 2001). Consequently, vitreous endosperm formation does not appear to require metabolic energy. It could form simply as a consequence of

cellular contents condensing onto starch granules (Gayral *et al.*, 2016), and it might also involve a chemical reaction, such as intramolecular disulfide bond formation between PBs and other proteins. Electron micrographs of vitreous endosperm show compressed starch grains and protein bodies embedded in desiccated cytoplasm, while there are air spaces between starch grains in opaque/starchy endosperm. Vitreous endosperm is thought not to form in opaque/floury mutants because they have fewer or perhaps abnormal PBs and less cytoplasmic material.

The mechanism that creates vitreous endosperm is unknown, but it has agronomic importance. Vitreous endosperm strengthens the kernel, making it less fragile (Chandrashekar and Mazhar, 1999). It is also important for food processing: vitreous endosperm is the origin of grits, corn chips and corn flakes. Some dent corn hybrids are unsuitable for food processing because they contain too little vitreous endosperm. Flint corn and popcorn have a large proportion of vitreous endosperm, and they are harder than dent corn. Humans likely selected for kernels with vitreous endosperm. Teosinte kernels are small, like those of other cereals, and they have a hard outer shell, the glume, that protects them (see Chapter 1). Through selection, the glume was lost and modern maize developed into a large flat kernel (the largest cereal caryopsis) that is much more subject to fracturing.

The 27-kD γ-zein appears to play an important role in vitreous endosperm formation. Mutants lacking this protein are soft and starchy, like *o2*, and high levels of the protein create vitreous endosperm in starchy *o2* mutants. This could be the consequence of high levels of 27-kD γ-zein creating more PBs, which would provide more surface area and the potential for a larger number of disulfide bonds. It is not clear whether teosinte develops vitreous endosperm; however, it does contain γ-zeins, although in smaller amounts than modern corn varieties (Flint-Garcia *et al.*, 2009; Chapter 1, this volume).

Besides PB number and their ability to form disulfide bonds, starch granule structure also appears to influence vitreous endosperm formation (Gibbon *et al.*, 2003; Wu *et al.*, 2015). The *o5* effect on amyloplast membrane structure supports this hypothesis. During kernel desiccation, starch grains are freed of amyloplast membranes and become embedded in a matrix of protein bodies and dried cytoplasmic contents. This creates a glue-like intercellular substance that will transmit light; consequently it is vitreous. Starch mutants have few starch grains and protein bodies and their kernels are vitreous, but their glassy appearance appears to be a consequence of factors different from those that create the typical vitreous endosperm.

Insight about reactions that create vitreous endosperm could come from characterization of additional opaque/floury mutants, although it is clear the non-vitreous phenotype can arise from a variety of mutations. There have been few biochemical studies of endosperm that is becoming vitreous, and this approach might provide insight into the process. For example, it could be related to the redox potential of the cells as they become desiccated.

### 14.7.4 Regulation of storage protein gene expression

High-level expression, strict tissue specificity and temporal regulation are among features of storage protein gene transcriptional regulation. Although several major TFs have been characterized, several questions are unanswered: (i) Which TFs are sufficient for zein gene expression and what epigenetic processes regulate their activation and suppression? (ii) What upstream TFs activate O2 and PBF expression? (iii) What TFs besides VP1 regulate storage globulins?

O2, PBF and OHPs influence zein gene expression by additive and synergistic interaction, but they do not explain activation of all zein genes. Some genes containing a P-box, like the 15-kD β-zein, are not affected by *PbfRNAi*, suggesting another DOF is redundant with PBF. Since only 46 DOFs are predicted for the B73 genome, it would be possible to test the remaining 45 to find one that activates the 15-kD β-zein

gene. Additional TFs could be identified by yeast two-hybrid screens using a known TF as bait, as was done with ZMADS47, a TF that increases activation of O2 through protein–protein interactions (Qiao *et al.*, 2016).

O2 and PBF are endosperm-specific TFs with expression beginning 8–10 DAP. But not all zein TFs, e.g. *Ohps* and *ZmMADS47*, are endosperm-specific. Ectopic expression of O2 and PBF with the 35S promoter failed to activate zein gene expression in leaf tissue, suggesting that epigenetic factors are important for tissue specificity (Wu and Messing, 2012a). Indeed, most α-zein promoters are more highly methylated in leaf than endosperm tissue, where their degree of methylation is variable (Xu *et al.*, 2016). To investigate whether demethylation and ectopic expression of O2 and PBF (OHPs are not endosperm specific) are sufficient to activate zein gene expression, one could screen mutants that reduce DNA methylation of zein promoters and examine if transcription occurs in the presence of a constitutive promotor, like 35S-O2 and 35S-PBF. The 16-kD γ-zein gene is useful to study temporal and tissue-specific TFs because its promoter lacks the P-box and O2 box, but its expression pattern is exactly the same as its paralog, the 27-kD γ-zein gene. Expression of the 16-kD γ-zein gene is not affected by *o2*, *PbfRNAi*, and *OhpRNAi*, indicating the existence of additional, unknown TFs. Defects in these TFs might not create a visible phenotype, but one might find suppression of the *Mc1* opaque phenotype among progeny of an EMS-treated *Mc1* population.

Transcriptional regulation of storage protein genes in the embryo appears to be conserved in monocots and dicots. Furthermore, some evidence suggests these regulatory pathways are conserved in fern spores and the seeds of gymnosperms (Schallau *et al.*, 2008). Genes encoding 2S albumins and 12S cruciferins in *Arabidopsis* are mainly regulated by B3s (AtABI3, AtFUS3 and AtLEC2) and bZIPs (AtbZIP10 and AtbZIP25), the latter of which are homologs of maize O2. Maize storage globulins are dramatically reduced by mutation of VP1, a homolog of AtABI3, indicating that VP1 and AtABI3 are functionally conserved. As is true of the 2S albumin and 12S cruciferin promoters, a G box is found in the *Glob* promoter and is probably recognized by a bZIP TF. One could use the yeast two-hybrid system to identify the interacting bZIP.

A reduced form of sulfur, which is important for seed germination, is stored in the seed in two amino acids: cysteine and methionine. γ-zeins are the main sink for cysteine, while δ-zeins are the main sink for methionine; β-zein is a reservoir for both. What determines utilization of each sink is unclear, but a reduction in the synthesis of one is generally compensated by increased synthesis of the other. For example, methionine content can be increased by overexpressing the 10-kD δ-zein, but there is a compensatory decrease in β- and γ-zeins (Wu *et al.*, 2012). The capacity for sulfate absorption and reduction is limited, which restricts synthesis of all three proteins. One can utilize several strategies to increase the content of methionine, which, like lysine, is an essential amino acid: (i) overexpression of APS reductase (APR) and serine acetyl-transferase (SAT) could increase sulfur reduction and cysteine synthesis, respectively, from sulfate and create a larger pool of sulfur-containing amino acids for protein synthesis; and (ii) methionine/10-kD δ-zein contents are variable among maize inbred lines and range from sufficient to insufficient levels to support normal animal growth. Using a GWAS approach, one could identify quantitative trait loci (QTLs) associated with superior alleles to create a high methionine phenotype.

### 14.7.5   Proteome rebalancing

It appears that seeds have a mechanism that monitors protein accumulation, resulting in a consistent level in the endosperm and/or embryo at maturity. In soybeans, proteome rebalancing occurs between the two major storage proteins, conglycinin and glycinin, or other proteins (Herman, 2014). Rebalancing is not observed among α-zeins, but it is among β-, γ-, and δ-zeins, suggesting that sulfur availability levers the balance between these proteins (Wu *et al.*, 2012).

The increased lysine content in *o2* and α-zein RNAi kernels occurs through proteome redistribution from zeins to other proteins (non-zeins) (Wu and Messing, 2012b), some of which are high in lysine. Two patterns of proteome rebalancing have been observed in maize: one suggests a general, global increase in non-zein proteins, while the other shows significant increases in specific proteins (Morton *et al.*, 2015). Theoretically, significant accumulation of proteins with more than 4% lysine would enhance the lysine content. The identity of such proteins and their contribution to total lysine can be established by LC-MS/MS. Several proteins, including elongation factor 1A (eEF1A) and glyceraldehyde-3-phosphate dehydrogenase (GAPDH), have been shown to contribute to the high-lysine phenotype (Morton *et al.*, 2015). It has not been demonstrated these proteins correspond to QTLs predictive of lysine content, although eEF1A is highly correlated (Habben *et al.*, 1995). A GWAS study could show whether or not this is the case.

In soybean, proteome rebalancing involves minor changes in transcription, with post-transcriptional and translational regulation playing a critical role; the role of these processes in maize proteome rebalancing has not been investigated. Transcriptome and proteome analysis of developing endosperm and embryo in a 22- and 19-kD α-zein RNAi background could identify genes and proteins that respond to proteome rebalancing, and modeling the affected pathways would provide insight into the processes involved. EMS mutagenesis of an appropriate inbred expressing a reporter gene (e.g. GFP attached to the promoter of a rebalancing-responsive gene) could be used to identify genes regulating this process, the corresponding genetic defects mapped, and responsible gene(s) characterized.

### 14.7.6   Quality protein maize

Like most cereals, maize is an important source of food and feed, particularly in developing countries, where its deficiency of essential amino acids can negatively impact human and livestock growth and development. Efforts to improve the lysine content of maize led to the development of QPM (Vasal *et al.*, 1980), which addresses many deficiencies of *o2*. However, as previously noted, QPM breeding is a complex process involving selection of multiple, unlinked *o2* modifier loci, while maintaining a homozygous recessive *o2* background. Recent research identified DNA markers linked to modifier loci (Holding *et al.*, 2011). These markers can accelerate the breeding process, but more could be achieved if the nature of the modifier genes and the mechanism(s) by which they convert starchy endosperm to a hard, vitreous phenotype were understood. There is good evidence at least part of the mechanism involves increased synthesis of the 27-kD γ-zein (Liu *et al.*, 2016), but whether or not this is necessary and sufficient is unknown.

There are more effective means than *o2* to reduce zein content and bring about compensatory increases in lysine-containing proteins. RNAi creates a dominant trait that silences/suppresses zein gene expression, and it does not have the pleiotropic effects of *o2* that reduce starch synthesis and yield (Zhang *et al.*, 2016). The *o2* modifiers can create vitreous endosperm in α-zein RNAi mutants, making it possible to identify *o2* modifier alleles and create novel QPMs (Wu and Messing, 2011, 2012b). More than 1000 maize inbreds have been genotyped by next-generation sequencing, and with GWAS one could identify *o2* modifiers by crossing these lines with pollen from an α-zein RNAi mutant. The resulting $F_1$ progeny can be classified by the extent to which vitreous endosperm is created, and then different genetic backgrounds can be screened for the most effective alleles. Coupling this approach with optimal proteome rebalancing should make it possible to create maize kernels that have a suitable phenotype and meet the nutritional requirements for monogastric animals. This would provide a source of vegetable protein that meets human needs and has far less environmental impact than feeding the grain to livestock.

Agricultural biotechnology companies have successfully used genetic engineering to improve the protein nutritional quality of maize; however, these varieties were not released for production because they lack a successful path to commercialization. Nevertheless, the value of a high-yielding, quality protein maize to the ever-growing human population is clear, and it could eventually be a commercial success.

## References


Argos, P., Pedersen, K., Marks, M.D. and Larkins, B.A. (1982) A structural model for maize zein proteins. *Journal of Biological Chemistry* 257, 9984–9990.

Boston, R.S. and Larkins, B.A. (2008) The genetics and biochemistry of maize zein proteins. In: Bennetzen, J.L. and Hake, S.C. (eds.) *The Maize Handbook. Volume II: History and Practice of Genetics, Genomics and Improvement.* Springer, New York, pp. 715–730.

Brochetto-Braga, M.R., Leite, A. and Arruda, P. (1992) Partial purification and characterization of lysine-ketoglutarate reductase in normal and opaque-2 maize endosperms. *Plant Physiology* 98, 1139–1147.

Casey, R. and Shewry, P.R. (1999) *Seed Proteins.* Kluwer Academic Publishers, Dordrecht, The Netherlands.

Chandrashekar, A. and Mazhar, H. (1999) The biochemical basis and implications of grain strength in sorghum and maize. *Journal of Cereal Science* 30, 193–207.

Clore, A.M., Dannenhoffer, J.M. and Larkins, B.A. (1996) EF-1α is associated with a cytoskeletal network surrounding protein bodies in maize endosperm cells. *Plant Cell* 11, 2003–2014.

Corradini, E., Curti, P.S., Meniqueti, A.B., Martins, A.F., Rubira, A.F. and Muniz, E.C. (2014) Recent advances in food-packing, pharmaceutical and biomedical applications of zein and zein-based materials. *International Journal of Molecular Science* 15, 22438–22470. DOI:10.3390/ijms151222438

Das, O.P., Ward, K., Ray, S. and Messing, J. (1991) Sequence variation between alleles reveals two types of copy correction at the 27-kDa zein locus of maize. *Genomics* 11, 849–856.

Flint-Garcia, S.A., Bodnar, A.L. and Scott, M.P. (2009) Wide variability in kernel composition, seed characteristics, and zein profiles among diverse maize inbreds, landraces and teosite. *Theoretical and Applied Genetics* 119, 1129–1142.

Foster, J., Kim, H., Nakata, P. and Browse, J. (2012) A previously unknown oxalyl-CoA synthetase is important for oxalate catabolism in *Arabidopsis*. *Plant Cell* 24, 1217–1229.

Gayral, M., Gaillard, C., Bakan, B., Dalgalarrondo, M., Klmorjani, K., *et al.* (2016) Transition from vitreous to floury endosperm in maize (*Zea mays* L.) kernels is related to protein and starch. *Journal of Cereal Science* 68, 148–154.

Geraghty, D., Peifer, M.A., Rubenstein, I. and Messing, J. (1981) The primary structure of a plant storage protein: zein. *Nucleic Acids Research* 9, 5163–5173.

Gibbon, B.C. and Larkins, B.A. (2005) Molecular genetic approaches to developing quality protein maize. *Trends in Genetics* 21, 227–233.

Gibbon, B.C., Wang, X. and Larkins, B.A. (2003) Altered starch structure is associated with endosperm modification in quality protein maize. *Proceedings of the National Academy of Sciences of the United States of America* 100, 15329–15334.

Gu, Y.Q., Wanjugi, H., Coleman-Derr, D., Kong, X. and Anderson, O.A. (2010) Conserved globulin gene across eight grass genomes identify fundamental units of the loci encoding seed storage proteins. *Functional Integrated Genomics* 10, 111–122.

Guo, X., Yuan, L., Chen, H., Sato, S.J., Clemente, T.E. and Holding, D.R. (2013) Nonredundant function of zeins and their correct stoichiometric ratio drive protein body formation in maize endosperm. *Plant Physiology* 162, 1359–1369.

Habben, J.E., Moro, G.L., Hunter, B.G., Hamaker, B. and Larkins, B.A. (1995) Elongation factor 1α concentration is highly correlated with the lysine content of maize endosperm. *Proceedings of the National Academy of Sciences of the United States of America* 92, 8640–8644.

Hamaker, B.R., Kirlies, A.W., Butler, L.G., Axtell, J.D. and Mertz, E.T. (1987) Improving the *in vitro* protein digestibility of sorghum with reducing agents. *Proceedings of the National Academy of Sciences of the United States of America* 84, 626–628.

Herman, E.M. (2008) Endoplasmic reticulum bodies: solving the insoluble. *Current Opinion in Plant Biology* 11, 672–679.

Herman, E.M. (2014) Soybean seed proteome rebalancing. *Frontiers in Plant Science* 5, 437. Available at: http://journal.frontiersin.org/article/10.3389/fpls.2014.00437/full (accessed June 4, 2017).

Holding, D.R., Otegui, M., Li, B.-L., Meeley, R.B., Dam, T., *et al.* (2007) The maize *floury1* gene encodes a novel ER protein involved in zein protein body formation. *Plant Cell* 19, 2569–2582.

Holding, D.R., Meeley, R.B., Hazebroek, J., Selinger, D., Gruis, F., Jung, R. and Larkins, B.A. (2010) Identification and characterization of the maize arogenate dehydrogenase gene family. *Journal of Experimental Botany* 61, 3663–3673.

Holding, D.R., Hunter, B.G., Klingler, J.P., Wu, S., Guo, X., *et al.* (2011) Characterization of *opaque2* modifier QTLs and candidate genes in recombinant inbred lines derived from the K0326Y quality protein maize inbred. *Theoretical and Applied Genetics* 122, 783–794.

Jia, M., Wu, H., Clay, K.L., Jung, R., Larkins, B.A. and Gibbon, B.C. (2013) Identification and characterization of lysine-rich proteins and starch biosynthesis genes in the *opaque2* mutant by transcriptional and proteomic analysis. *BMC Plant Biology* 13, 60. Available at: http:/www.biomedcentral.com/1471-2229/13/60 (accessed June 4, 2017).

Kim, C.S., Woo, Y.-M., Clore, A.M., Burnett, R.J., Carneiro, N.P. and Larkins, B.A. (2002) Zein protein interactions, rather than the asymmetric distribution of zein mRNAs on endoplasmic reticulum membranes, influence protein body formation in maize endosperm. *Plant Cell* 14, 655–672.

Kim, C.S., Gibbon, B.C., Gilikin, J.W., Larkins, B.A., Boston, R.S. and Jung, R. (2006) The maize *Mucronate* mutation is a deletion in the 16-kD γ-zein gene that induces the unfolded protein response. *Plant Journal* 48, 440–451.

Kirst, M.E., Meyer, D.J., Gibbon, B.C., Jung, R. and Boston, R.S. (2005) Identification and characterization of endoplasmic reticulum-associated degradation proteins differentially affected by endoplasmic reticulum stress. *Plant Physiology* 138, 218–231.

Kogan, M.J., López, O., Cocera, M., López-Iglesias, C., De la Maza, A. and Giralt, E. (2004) Exploring the interaction of the surfactant N-terminal domain of γ-Zein with soybean phosphatidylcholine liposomes. *Biopolymers* 73, 258–268.

Kriz, A.L. (1999) 7S globulins of cereals. In: Casey, R. and Shewry, P.R. (eds.) *Seed Proteins*. Kluwer Academic Publishers, Dordrecht, The Netherlands, pp. 477–498.

Kühlbrandt, W. (2014) Cryo-EM enters a new era. *eLife* 3, e03678. DOI:10.7554/eLife.03678

Lai, J. and Messing, J. (2002) Increasing maize seed methionine by mRNA stability. *Plant Journal* 30, 395–402.

Larkins, B.A. and Hurkman, W.J. (1978) Synthesis and deposition of zein in protein bodies of maize endosperm. *Plant Physiology* 62, 256–263.

Lawton, J.W. (2002) Zein: a history of processing and use. *Cereal Chemistry* 79, 1–18.

Lending, C.R. and Larkins, B.A. (1989) Changes in the zein composition of protein bodies during maize endosperm development. *Plant Cell* 23, 1011–1023.

Liu, H., Shi, J., Sun, C., Gong, H., Fan, X., *et al.* (2016) Gene duplication confers enhanced expression of 27-kDa γ-zein for endosperm modification in quality protein maize. *Proceedings of the National Academy of Sciences of the United States of America* 113, 4964–4969.

Mainieri, D., Morandini, F., Maîtrejean, M., Saccani, A., Pedrazzini, E. and Vitale, A. (2014) Protein body formation in the endoplasmic reticulum as an evolution of storage protein sorting to vacuoles: insights from maize γ-zein. *Frontiers in Plant Science* 5, 331. Available at: http://dx.doi.org/10.3389/fpls.2014.00331 (accessed June 4, 2017).

Mertz, E.T., Bates, L.S. and Nelson, O.E. (1964) Mutant gene that changes protein composition and increases lysine content of maize endosperm. *Science* 145, 279–280.

Miclaus, M., Xu, J.-H. and Messing, J. (2011a) Differential gene expression and epiregulation of alpha zein gene copies in maize haplotypes. *PLOS Genetics* 7, e1002131. Available at: https://doi.org/10.1371/journal.pgen.1002131 (accessed June 4, 2017).

Miclaus, M., Wu, Y., Xu, J-.H., Dooner, H.K. and Messing, J. (2011b) The maize high-lysine mutant *opaque7* is defective in an acyl-coA synthetase-like protein. *Genetics* 189, 1271–1280.

Moro, G.L., Habben, J.E., Hamaker, B.R. and Larkins, B.A. (1996) Characterization of the variability in lysine content for normal and *opaque2* maize endosperm. *Crop Science* 36, 1651–1659.

Morton, K.J., Jia, S., Zhang, C. and Holding, D.R. (2015) Proteomic profiling of maize opaque endosperm mutants reveals selective accumulation of lysine-enriched proteins. *Journal of Experimental Botany* 67, 1381–1396.

Myers, A.M., James, M.G., Lin, Q., Yi, G., Stinard, P.S., Hennen-Bierwagen, T.A. and Becraft, P.W. (2011) Maize *opaque5* encodes monogalactosyldiacylglycerol synthase and specifically affects C18:3/C18:2 galactolipids necessary for amyloplast and chloroplast function. *Plant Cell* 23, 2331–2347.

Nation, J.L., Sr. (2016) *Insect Physiology and Biochemistry* (3rd edn.). CRC Press, Boca Raton, Florida.

Nelson, O.E. (2001) Maize: the long trail to QPM. In: Reeve, E.C.R. and Black, I. (eds.) *Encyclopedia of Genetics*. Fitzroy Dearborn, London and Chigaco, Illinois, pp. 657–660.

Nelson, O.E., Mertz, E.T. and Bates, L.S. (1965) Second mutant gene affecting the amino acid pattern of maize endosperm proteins. *Science* 150, 1469–1470.

Pedersen, K., Argos, P., Naravana, S.V.L. and Larkins, B.A. (1986) Sequence analysis and characterization of a maize gene encoding a high-sulfur zein protein of Mr 15,000. *Journal of Biological Chemistry* 261, 6279–6284.

Qiao, Z., Qi, W., Wang, Q., Feng, Y., Yang, Q., *et al.* (2016) ZmMADS47 regulates zein gene transcription through interaction with Opaque2. *PLOS Genetics* 12, e1005991.

Reyes, F., Chung, T., Holding, D., Jung, R., Vierstra, R. and Otegui, M.S (2011) Delivery of prolamins to the protein storage vacuole in maize aleurone cells. *Plant Cell* 23, 769–784.

Schallau, A., Kakhovskaya, I., Tewes, A., Czihal, A., Tiedemann, J., *et al.* (2008) Phylogenetic footprints in fern spore- and seed-specific gene promoters. *Plant Journal* 53, 414–424.

Segal, G., Song, R. and Messing, J. (2003) A new opaque variant of maize by a single dominant RNA-interference-inducing transgene. *Genetics* 165, 387–397.

Shewry, P.R., Napier, J.A. and Tatham, A.S. (1995) Seed storage proteins: structures and biosynthesis. *Plant Cell* 7, 945–956.

Song, R. and Messing, J. (2002) Contiguous genomic DNA sequence comprising the 19-kDa-zein gene family from *Zea mays*. *Plant Physiology* 130, 1626–1635.

Song, R. and Messing, J. (2003) Gene expression of a gene family in maize based on noncollinear haplotypes. *Proceedings of the National Academy of Sciences of the United States of America* 100, 9055–9060.

St. Johnston, D. (2005) Moving messages: the intracellular localization of mRNAs. *Nature Reviews Molecular Cell Biology* 6, 363–375.

Swarup, S., Timmermans, M.C.P., Chaudhuri, S. and Messing, J. (1995) Determinants of the high-methionine trait in wild and exotic germplasm may have escaped selection during early cultivation of maize. *Plant Journal* 8, 359–368.

Thompson, G.A. and Larkins, B.A. (1993) Characterization of the zein genes and their regulation in maize endosperm. In: Walbot, V. and Freeling, M. (eds.) *The Maize Handbook*. Springer, Heidelberg, Germany, pp. 639–647.

Vasal, S.K., Villegas, E., Bjarnason, M., Gelaw, B. and Gortz, P. (1980) Genetic modifiers and breeding strategies in developing hard endosperm *opaque2* materials. In: Pollmer, W.G. and Philips, R.H. (eds.) *Improvement of Quality Traits of Maize for Grain and Silage Use*. Martinus Nijhoff, London, pp. 37–73.

Vicente-Carbajosa, J., Moose, S.P., Parsons, R.L. and Schmidt, R.J. (1997) A maize zinc-finger protein binds the prolamin box in zein promoters and interacts with the basic leucine zipper transcriptional activator Opaque2. *Proceedings of the National Academy of Sciences of the United States of America* 94, 7685–7690.

Wallace, J.C., Galili, G., Kawata, E.E., Cuellar, C.E., Shotwell, M.A. and Larkins, B.A. (1988) Aggregation of lysine-containing zeins into protein bodies in Xenopus oocytes. *Science* 240, 662–664.

Wallace, J.C., Lopes, M.A., Paiva, E. and Larkins, B.A. (1990) New methods for extraction and quantitation of zeins reveal a high content of gamma-zein in modified *opaque2* maize. *Plant Physiology* 92, 191–196.

Wang, G., Sun, X., Wang, G., Wang, F., Gao, Q., *et al.* (2011) *Opaque7* encodes an acyl activating enzyme-like protein that affects storage protein synthesis in maize endosperm. *Genetics* 189, 1281–1295.

Wang, G., Wang, F., Wang, G., Wang, F., Zhang, X., *et al.* (2012) *Opaque1* encodes a myosin XI motor protein that is required for endoplasmic reticulum motility and protein body formation in maize endosperm. *Plant Cell* 24, 3447–3462.

Wang, G., Qi, W.W., Wu, Q., Yao, D.S., Zhang, J.S., *et al.* (2014a) Identification and characterization of maize *floury4* as a novel semi-dominant opaque mutant that disrupts protein body assembly. *Plant Physiology* 165, 582–594.

Wang, G., Zhang, J., Wang, G., Fan, X., Sun, X., *et al.* (2014b) *Proline responding1* plays a critical role in regulating general protein synthesis and cell cycle in maize. *Plant Cell* 26, 2582–2600.

Washida, H., Kaneko, S., Crofts, N., Sugino, A., Wang, C. and Okita, T.W. (2009) Identification of *cis*-localization elements that target glutelin RNAs to a specific subdomain of the cortical endoplasmic reticulum in rice endosperm cells. *Plant and Cell Physiology* 50, 1710–1714.

Woo, Y.-M., Hu, D.W.-N., Larkins, B.A. and Jung, R. (2001) Genomics analysis of genes expressed in maize endosperm identifies novel seed proteins and clarifies patterns of zein gene expression. *Plant Cell* 13, 2297–2317.

Wu, H., Clay, K., Thompson, S.S., Hennen-Bierwagen, T.A., Andrews, B.J., Zechmann, B. and Gibbon, B.C. (2015) Pullulanase and starch synthase III are associated with formation of vitreous endosperm in Quality Protein Maize. *PLOS ONE* 10, e0130856. DOI:10.1371/journal.pone.0130856

Wu, Y. and Messing, J. (2010) RNA interference-mediated change in protein body morphology and seed opacity through loss of different zein proteins. *Plant Physiology* 153, 337–347.

Wu, Y. and Messing, J. (2011) Novel genetic selection system for quantitative trait loci of Quality Protein Maize. *Genetics* 188, 1019–1022.

Wu, Y. and Messing, J. (2012a) Rapid divergence of prolamin gene promoters of maize after gene amplification and dispersal. *Genetics* 192, 507–519.

Wu, Y. and Messing, J. (2012b) RNA interference can rebalance the nitrogen sink of maize seeds without losing hard endosperm. *PLOS ONE* 7, e32850.

Wu, Y., Holding, D.R. and Messing, J. (2010) Gamma-zein is essential for endosperm modification in quality protein maize. *Proceedings of the National Academy of Sciences of the United States of America* 107, 12810–12815.

Wu, Y., Wang, W. and Messing, J. (2012) Balancing of sulfur in maize seed. *BMC Plant Biology* 12, 77. DOI:10.1186/1471-2229-12-77

Xu, H., Shen, L. and Yang, Y. (2015) Controlled delivery of hollow corn protein nanoparticles via non-toxic crosslinking: *in vivo* and drug loading study. *Biomedical Microdevices* 17, 8. DOI:10.1007/s10544-014-9926-5

Xu, J.-H. and Messing, J. (2008) Organization of the prolamin gene family provides insight into the evolution of the maize genome and gene duplications in grass species. *Proceedings of the National Academy of Sciences of the United States of America* 105, 14330–14335.

Xu, J.-H. and Messing, J. (2009) Amplification of prolamin storage protein genes in different subfamilies of the Poaceae. *Theoretical and Applied Genetics* 119, 1397–1412.

Xu, J.-H., Wang, R., Xinxin, L., Miclaus, M. and Messing, J. (2016) Locus- and site-specific DNA methylation of 19 kDa zein genes in maize. *PLOS ONE* 11, e0146416. DOI:10.1371/journal.pone.0146416

Yamagata, T., Kato, H., Kuroda, S., Abe, S. and Davies, E. (2003) Uncleaved legumin in developing maize endosperm: identification, accumulation and putative subcellular localization. *Journal of Experimental Botany* 54, 913–922.

Yang, J., Ji, C. and Wu, Y. (2016) Divergent transactivation of maize storage protein zein genes by the transcription factors Opaque2 and OHPs. *Genetics* 204, 581–591.

Yao, D., Qi, W., Li, X., Yang, Q., Yan, S., *et al.* (2016) Maize *opaque10* encodes a cereal specific protein that is essential for the proper distribution of zeins in endosperm protein bodies. *PLOS Genetics* 12, e1006270.

Yuan, L., Dou, Y., Kianian, S.F., Zhang, C. and Holding, D.R. (2014) Deletion mutagenesis identifies a haploinsufficient role for γ-zein in *opaque2* endosperm modification. *Plant Physiology* 164, 119–130.

Zhang, Z., Yang, J. and Wu, Y. (2015) Transcriptional regulation of zein gene expression in maize through the additive and synergistic action of opaque2, prolamine-box binding factor and O2 heterodimerizing proteins. *Plant Cell* 27, 1162–1172.

Zhang, Z., Zheng, X., Yang, J., Messing, J. and Wu, Y. (2016) Maize endosperm-specific transcription factors O2 and PBF network the regulation of protein and starch synthesis. *Proceedings of the National Academy of Sciences of the United States of America* 113, 10842–10847.

# 15 Determinants of Kernel Sink Strength


**Karen E. Koch* and Fangfang Ma**
*Department of Horticultural Science, University of Florida, Gainesville, Florida, USA*


## 15.1  Introduction

As a large, $C_4$-photosynthetic plant, maize provides an abundant amount of photosynthate for sinks, compared to the more "source-limited" small-grain species (rice, wheat, barley) with $C_3$ photosynthesis. In a general sense, "sink strength" refers to factors responsible for transport of metabolites from one plant part to another. This is typically phloem-borne sucrose from leaves, but other nutrients, plant organs, and non-phloem paths can be involved, e.g. endosperm-to-embryo transfer. Sinks acquire a spectrum of assimilates containing C, N, S, as well as other vital resources. By the time the maize kernel reaches maturity, its cumulative sink strength accounts for its composition. Consequently, yield is highly dependent on sink strength.

Some determinants of sink strength in maize kernels are common among other plant species, but differences exist. While we have learned much about how sink strength is generated, many aspects of this process are poorly understood. Our overall understanding is as follows: (i) Sucrose moves down a turgor gradient through phloem tissue toward the kernel, which has diverse mechanisms, hence "sink strength," to attract it. (ii) After leaving the phloem, sugars move toward starchy endosperm cells across one of the longest non-vascular distances of any crop species. The path includes maternal as well as endosperm tissues, with both contributing to cleavage and resynthesis of sucrose in transit. Little oxygen is available in the endosperm portion of this path, or at sites of starch deposition, implying that oxidative phosphorylation (ATP synthesis) might not significantly contribute to this process. (iii) Transporters play a central role in sink strength, initially by compartmentalizing sugars in the vacuoles of expanding maternal cells, and later by mediating sugar transfer into the endosperm. (iv) Nutrient sensing systems modulate kernel sink strength, but much is unknown about how they operate. (v) All of the above are changing during kernel development. This chapter explains these five topics and identifies many associated questions that remain to be answered.

## 15.2  Roles of Phloem and Post-phloem Water Flow in Sink Strength

Low turgor at the sink end of the phloem path is essential to generate a strong sink for

---


*Corresponding author e-mail: kekoch@ufl.edu



"

sucrose transport via the Munch Pressure Flow mechanism (Fig. 15.1A). This mechanism and its variations (Turgeon, 2010) depend on a source-to-sink turgor gradient in the phloem that directs solution flow to where the gradient is steepest. The pressure-flow mechanism is an unusual instance where water (phloem sap in this

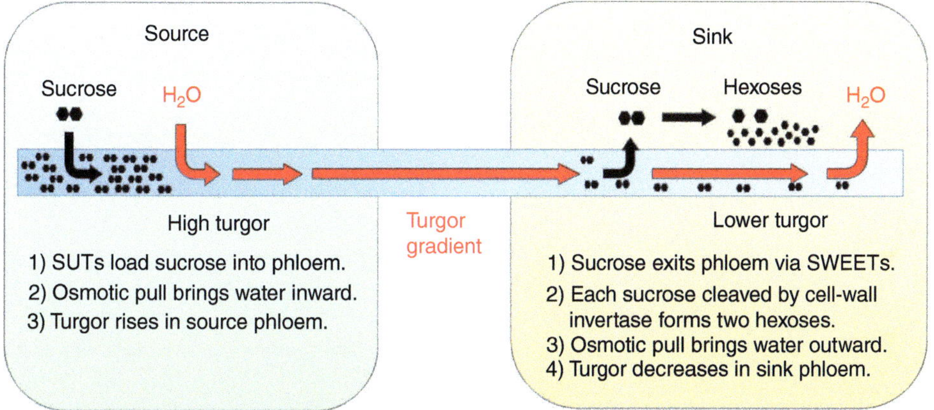



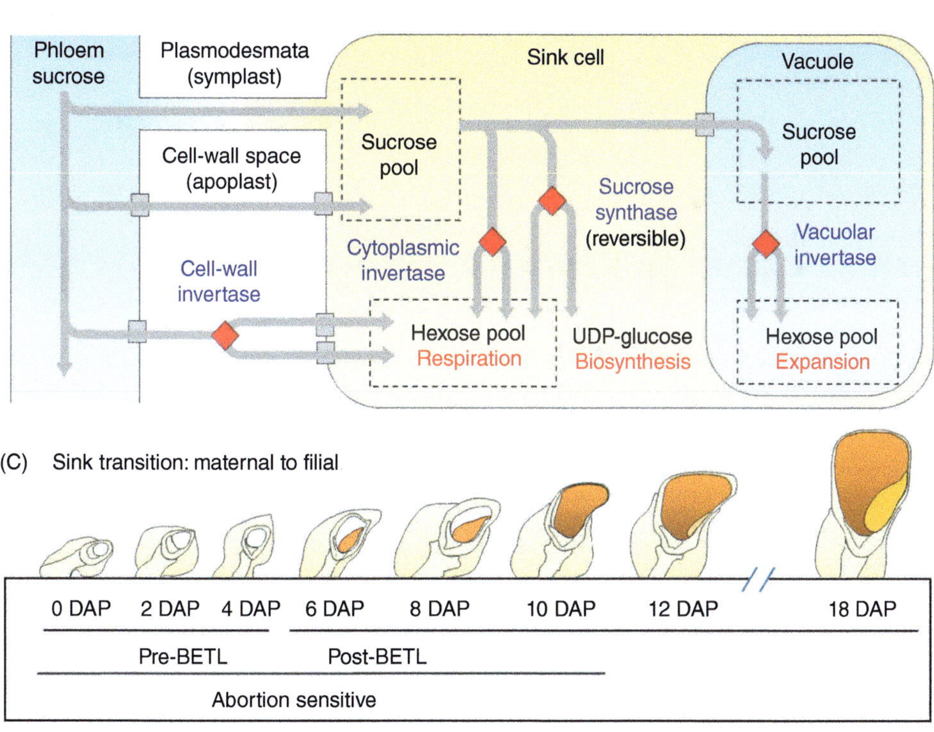



Fig. 15.1. Determinants of sink strength in maize. (A) Phloem translocation from source leaves to sink kernels is driven by a turgor gradient. The essential water entry and exit from phloem is shown in red. Water is drawn into phloem of source leaves by the osmotic pull of sucrose loaded into sieve tubes. High turgor

case) flows along a turgor gradient ($\Delta\Psi p$) rather than the gradient in total water potential ($\Delta\Psi w$). At least two aspects of kernel sink strength relate to the Munch pressure-flow mechanism. The first is low turgor inside the phloem sieve tube elements of a strong sink. This is enhanced by sucrose cleavage in the apoplast (cell-wall space) surrounding phloem of an importing structure. Since hydrolysis of sucrose yields two hexoses (thus doubling osmotic potential), its cleavage immediately outside the sieve tubes draws water from these cells. This in turn reduces turgor at the sink end of the pressure-flow continuum, with stronger sinks generating a steeper gradient. The importance of sucrose cleavage to this hydraulic component of sink strength is distinct from its role in metabolism and signaling and is often underappreciated.

The second aspect of Munch pressure flow relevant to kernel sink strength is the fate of phloem water, especially on a diurnal basis. This fluid is readily accommodated by young kernels during their early expansion phase, and can also be lost via transpiration. Later in development, even small amounts of water loss from kernels could influence the influx of phloem fluid. Diurnal pulses, such as those reported for barley and wheat

(Rolletschek *et al.*, 2015), could result. Like many fruits, maize kernels undergo most of their expansion relatively early in development, leaving little space later to accommodate phloem water. Water movement into and out of the kernel interior is limited to basal regions by the presence of a suberized layer enveloping much of the seed (see Chapter 11). Nonetheless, entry and exit of phloem water is essential. One path for potential loss is via xylem back-flow to adjacent structures with lower water potentials, e.g. from fruit to nearby transpiring leaves (Huang *et al.*, 1992; Tilbrook and Tyerman, 2009). Still another mechanism for possible ebb and flow of kernel fluid is the diurnal cycles of root pressure that deliver pre-dawn water to maize leaves and ears (Tang and Boyer, 2008; Boyer and Koch, unpublished). Root pressure in maize exceeds that typical of small herbaceous plants and can send kernels a regular pre-dawn pulse of water carrying not only xylem contents, but also any assimilates in extracellular portions of the post-phloem transport path. Collectively, these tidal movements of kernel water could have important effects on assimilate uptake and sink strength.

Another hydraulic aspect of kernel sink strength lies in the movement of water and

---

Fig. 15.1. Continued.

results, and phloem sap moves to sink sites with lowest turgor. Strong sinks can enhance the exit of phloem water by compartmentalizing sucrose in vacuoles or by doubling the osmotic pull of unloaded sucrose through its cleavage by extracellular invertases. Much remains to be learned about fates of phloem water in maize kernels. (B) The fates of unloaded sucrose differ with their subcellular path. Phloem sucrose can move into a sink cell through plasmodesmata (symplastic path) or across the cell-wall space (apoplastic path) (grey arrows). For sugars crossing the apoplast, transporters control both their efflux and uptake (small grey squares). Note that sucrose can move across the apoplast unaltered. Four enzymes (red diamonds) can metabolize imported sucrose. Cell-wall invertase can cleave extracellular sucrose. Cytoplasmic sucrose can be hydrolyzed by cytoplasmic invertases, which are typically less abundant. Both sources of hexoses can fuel respiration and stimulate sugar sensing systems. The reversible sucrose synthase reaction mediates a different sucrose cleavage in the cytoplasm, which produces UDP-glucose for starch and cell-wall biosynthesis as well as for a sensing system. Vacuolar invertase can enhance expansion by increasing vacuolar osmotic content, especially during initial phases of maternal-driven kernel growth. (C) Maize kernel sinks are initially maternal, then shift to filial. Maternal tissues are shown in light yellow, filial tissues in dark orange. Mechanisms of sucrose import change dramatically, and operate differently within each of the tissue types shown. Note the predominance of maternal tissues during establishment and initial development of kernel sinks. Even after development of the basal endosperm transfer layer (BETL), kernels remain sensitive to abortion until endosperm becomes the physically predominant structure and displaces the nucellus inside the pericarp (between 10 and 12 DAP under Florida conditions). Tissue outlines are from images of fresh sagittal sections of maize kernels (W22, Florida).

solutes through the extensive post-phloem transport path. In maize, the distance between phloem termini and sites of assimilate deposition in the endosperm is longer than in other crop species. Rates of diffusion typically applied to short-distance transfer, even when combined with cytoplasmic streaming, have limited capacity for long, non-vascular paths (Koch and Avigne, 1990; Slewinski, 2011). A greater speed and volume of non-vascular transport can be achieved by micro mass-flow along the cell-wall apoplast (Koch and Avigne, 1990). A small, but effective means of mass-flow in the apoplast can thus be invoked for maternal portions of the kernel and could also be important in filial tissues. This mechanism offers a sufficient means of post-phloem movement, but it is sensitive to cycles of water flow and metabolism by extracellular enzymes, including cell-wall invertases where present.

Key questions about phloem water flow and sink strength include the following:

**1.** What osmotic processes operate at phloem termini near the base of maize kernels to establish the source-to-sink turgor gradient driving phloem translocation? Models suggest turgor gradients inside phloem can be enhanced by invertases immediately outside points of sucrose unloading. If invertase cleaves a molecule of sucrose exiting phloem, then the localized contribution to turgor reduction will be two-fold greater (as noted above). Are cell-wall invertases in the basal endosperm transfer layer (BETL) close enough to phloem termini to reduce phloem turgor? How does this function early in development prior to BETL formation? Do invertases in maternal cells near sites of phloem-unloading influence turgor-driven sink strength? Consistent with this hypothesis, invertase mRNAs localize not only to the BETL, but also to maternal phloem (Sosso *et al.*, 2015). Are BETL invertases only part of the story? Are spatial and temporal (including diurnal) patterns of invertase expression more important than previously envisioned?

**2.** To what extent does xylem back-flow contribute to water removal from developing kernels? What happens when the influx of phloem water exceeds potential for kernel expansion, and when pre-dawn root pressure pushes xylem and other apoplastic water into ears? Estimates of sucrose requirements by developing kernels (Koch, unpublished) indicate a surplus of phloem water is likely during grain filling. Is suberization of the inner seed coat an impediment to water movement? Could fluid arrival via phloem and exit via xylem occur in tidal cycles during the day? Support for this hypothesis comes from magnetic resonance imaging (MRI), indicating diurnal patterns of assimilate movement into grains of barley and wheat (Rolletschek *et al.*, 2015). Is transpiration pull from nearby leaves strong enough to draw water from kernels in preference to distant roots (Tang and Boyer, 2008; Tilbrook and Tyerman, 2009)? Might this contribute to regulating turgor aspects of kernel sink strength?

**3.** What effects would tidal ebb and flow of kernel fluid have on sink strength? Might it enhance or inhibit sugar uptake, and what environmental factors could influence this process? The xylem back-flow hypothesis predicts a need for sucrose retrieval from xylem. Such a mechanism has been proposed for the SUT1 sucrose transporter in maternal tissues (Baker *et al.*, 2016). What other kernel functions are adapted for ebb and flow of water? Do real-time measurements of fluid movement reflect transfer and metabolism of assimilates over the diurnal cycle? What is the response to perturbation? How does movement of water into kernels grown in culture (without transpiration) compare with that of kernels borne on tassel-seed plants (with potentially unrestricted transpiration)?

**4.** Does mass-flow through the cell wall apoplast of the post-phloem transport path affect sink strength? As noted above, the capacity and speed of transport would be greater for apoplastic movement along the extracellular cell wall matrix. Again, potential effects of phloem water and root pressure could extend well into the endosperm, where tidal movement of kernel fluid could occur. Under these circumstances, effects of even slight changes in water movement could have a greater impact than activity of transporters or metabolism alone. Given this possibility, could the BETL be adapted for yet-to-be explored roles related to water flux and/or pressure sensing? In addition to increasing membrane surface area for metabolite transport, BETL cell structure could also aid water flow through the cell wall matrix. The vertical orientation of wall ingrowths favors fluid flow along this cell wall path. What is the relationship between apoplastic and symplastic transfer across the BETL? Does it change during development and/or the diurnal cycle of water movement?

**5.** What structural features of tissues affect fluid and assimilate movement through the phloem and post-phloem path? Maternal portions of the path are complex, and we know little of the physical features that influence their function. At least three tissue layers (two layers of seed coat and one of nucellus) separate phloem termini from the BETL. Each of these layers has properties that could differentially affect apoplastic versus symplastic water movement. Programmed cell death (PCD), for example, begins in tissues nearest the BETL and spreads toward the vascular bundles. PCD increases the volume of the apoplastic space available for transport and extracellular metabolism. To what degree does PCD affect sink strength? Does the fluid pool adjacent to the endosperm influence water flux and assimilate import/retention? How does its role compare to that of the liquid endosperm of the Pooideae grasses (rice, wheat, barley), and the post-phloem fluid sac in sorghum (Jain *et al.*, 2008). Can we build a 3-D model for analyzing fluid transfer capacity, surface areas, volumes, and their potential bottlenecks? Did these features change during maize domestication? In what ways could hydraulic roles of fruit-case tissues (outer glume and rachid) differ for teosinte? Is there trade-off between the constraint of a protective exterior and the potential for kernel expansion? Does variation in maize pericarp thickness affect the fluidics of sink strength? Does diversion of water and resources from expanding silks within hours

of pollination play a role in establishing early sink strength (Xu *et al.*, 1996)?

## 15.3 Metabolic Microenvironments and Assimilate Movement

### 15.3.1 Roles of the pedicel, pericarp, and maternal transfer zone

Prior to pollination, the sink strength of ovaries is entirely determined by maternal tissues (Fig. 15.1C), and varying degrees of this influence the first third of kernel development. Kernel size is primarily a maternal trait (Zhang *et al.*, 2016; Chapter 16, this volume), and maternal tissues also regulate kernel abortion during early periods of growth (Andersen *et al.*, 2002; Guan and Koch, 2015; Nuccio *et al.*, 2015). Kernel expansion is a key component of maternal sink strength, and the caryopsis, as for other fruits, can expand even when embryos are aborted. Thus, mutants with dysfunctional seeds manifest a characteristic "empty-pericarp" phenotype analogous to a seedless fruit. Maternal contributions to sink strength are clearly evident in these instances, since the caryopsis itself is not aborted. The seed-based effectors of sink strength are readily apparent in such mutants. For example, a deficiency in cell wall invertase results in a miniature kernel, but not an aborted one (see Chapter 5).

A key process underlying maternal tissue sink strength is the import and cleavage of sucrose in vacuoles of expanding cells (Fig. 15.1B). Vacuolar invertases, like those of the cell wall, catalyze a 2-for-1 increase in osmotic potential, but in this instance it translates into the turgor needed for rapid cell expansion (Koch, 2004; Slewinski, 2011; Bihmidine *et al.*, 2013). Interestingly, as long as hexoses remain in the vacuole, sink strength is temporarily uncoupled from the metabolic costs of these sugars and their consequent impact on sugar sensing.

Maternal portions of the post-phloem path that supply assimilates to filial tissues are complex (Bihmidine *et al.*, 2013) (Fig. 15.2A). As noted above, maternal phloem is separated

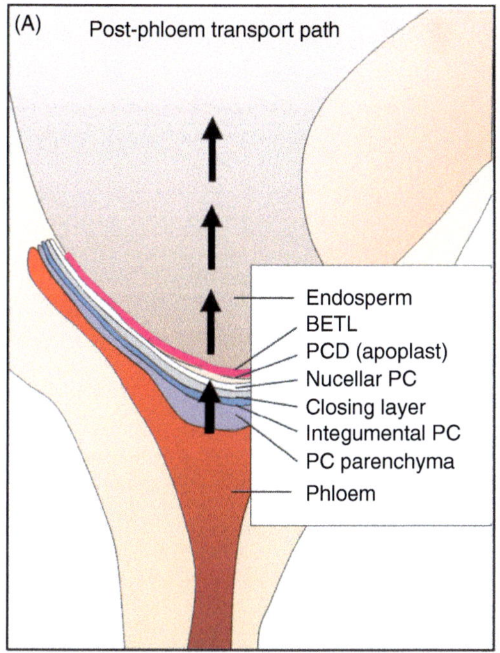


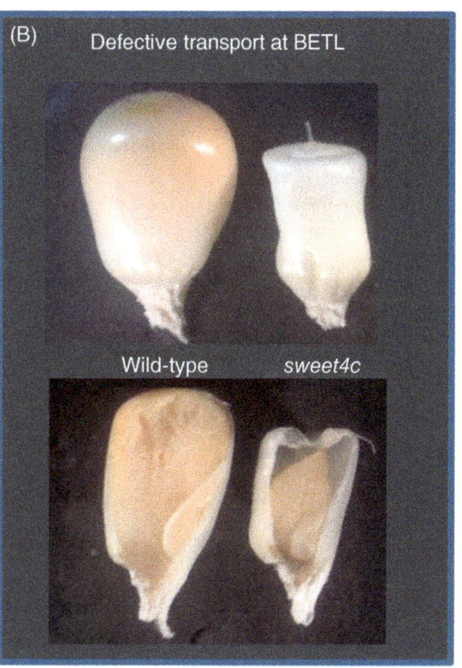


**Fig. 15.2.** Determinants of sink strength in maize. (A) The post-phloem transport path in kernels is one of the longest of any grain species and includes a complex series of maternal tissues, a requisite apoplastic step, and a long non-vascular distance through the low-oxygen endosperm. Phloem is shown in red, with multiple layers of the PC (placental chalazal) transfer tissues immediately above it. Maternal portions of the post-phloem path are separated from filial tissues by an apoplastic space that progressively enlarges by PCD (programmed cell death). The endosperm portion of the post-phloem path begins with the BETL (basal endosperm transfer layer) shown as a thin rose-colored line, and continues through the lengthy distance of low-oxygen endosperm cells. (B) Defective transport at the BETL has a striking effect on kernel development. The *sweet4c* mutant shown here is lacking a functional hexose transporter normally expressed in the BETL (see also Chapter 5).

from the BETL by three cell layers of maternal tissues, plus an apoplastic region free of plasmodesmatal connections. Much attention has been directed to the apoplastic step of the post-phloem path (see below). However, terminal portions of the maternal path remain virtually unexplored in terms of assimilate movement, metabolism, and metabolic microenvironments. Specific mechanisms of movement through each of these different layers are unknown.

Ample evidence supports Shannon's observation (1972) that a portion of the sucrose initially cleaved by invertases during import into the maize kernel is resynthesized along the post-phloem transport path (Bihmidine *et al.*, 2013). This seemingly futile resynthesis presumably occurs via the

sucrose phosphate synthase + sucrose phosphate phosphatase (SPS-SPP) pathway. Final sucrose cleavage is envisioned to occur via invertase or reversible sucrose synthase reactions, depending on where the sucrose is ultimately metabolized. However, the significance of resynthesis to sink strength is unclear. Also unknown is the extent of resynthesis, where it occurs, what factors affect it, and what underlying mechanisms are involved.

Sucrose resynthesis would plausibly impact water flow, sugar metabolism, signaling, and the hydraulic implications discussed above. At the metabolic level, sucrose resynthesis could decrease hexose levels in the low-oxygen central endosperm and reduce the ATP demand associated with entry into glycolysis (Bihmidine *et al.*, 2013; Chapter 11, this volume). In addition, sucrose cycling could provide a mechanism for balancing sucrose and hexose sensing (Guan and Koch, 2015; see below).

Key questions regarding metabolic microenvironments include the following:

**1.** By what means do maternal tissues influence sink strength during early kernel development? Maternal effects on kernel size (Zhang *et al.*, 2016) and abortion (Hanft and Jones, 1986) are well known, but much remains to be learned about the underlying mechanisms. What regulates the dramatic initial expansion of the ovary? What sustains expansion of the caryopsis in "seedless," empty-pericarp mutants? Vacuolar invertases are strong candidates for this role and are sensitive to pre-pollination stresses (Andersen *et al.*, 2002). In addition, early enhancement of seed set, especially under stress conditions, can be induced at pollination by overexpression of a signaling system linked to sucrose metabolism in ovaries (Guan and Koch, 2015; Nuccio *et al.*, 2015). Is this indeed a key player in maternal control of sink strength? Does the size, carbohydrate status, and/or developmental state of a given ovary at pollination affect its subsequent sink strength (Schussler and Westgate, 1995)?

**2.** What relationship, if any, exists between the sink strength of an expanding silk and import by its subtending ovary? A sudden change is triggered by pollination, as silk expansion and the gradient of invertase expression along its length drops within hours of pollination (Xu *et al.*, 1996). Do similar amounts of water and sugar continue to enter ovaries?

**3.** Do maternal tissues limit sink strength later in development by constraining physical expansion of the kernel? Do protective features of the pericarp ultimately impact maximum kernel size? Is this one reason why seed size is a maternal trait (Chapter 16)? Did domestication release kernel expansion from constraints of the protective fruit case of teosinte (Chapter 1)?

**4.** How do structures in the maternal portion of the post-phloem path affect kernel sink strength? PCD progresses from the BETL toward the vascular tissues, physically enlarging the apoplastic space. Does this affect movement and metabolism of assimilates en route, and what changes occur at each point? Are metabolic microenvironments encountered and do they impact transfer? How does this process operate in conjunction with filial portions of the post-phloem path (see below)?

**5.** To what degree does cleavage and resynthesis of sucrose occur in maternal tissues? Does this process, including resynthesis, take place throughout the transport path? Shannon's original model (1972) is well supported, but how is it regulated and how is it advantageous? Is this process linked to endogenous water flow? To what extent does sucrose cycling occur and where? Does it impact sugar-sensing systems?

**6.** What metabolite levels, gradients, and compartmentalization exist within maternal tissues? Are these storage pools, osmotic constituents, and/or signaling effectors? Do they change? Each of the metabolic microenvironments along the post-phloem path could affect the movement and metabolism of entering assimilates. Assimilates include amino acids, and previous work indicates their interconversions could be particularly important in maternal portions of the post-phloem path.

**7.** Do metabolic microenvironments within maternal portions of the post-phloem path impact gene expression? Nutrient sensing systems affect diverse aspects of sink strength, from the cell cycle to genes influencing sucrose metabolism and storage (Koch, 1996, 2004; Bihmidine *et al.*, 2013).

### 15.3.2 Assimilate movement in low-oxygen regions of the endosperm

Sucrose and hexoses crossing the BETL into the endosperm must still traverse the longest portion of an already-long post-phloem path. This distance, one of the longest in any crop species, extends most of the kernel length, which can exceed 1.5 cm in gourd-seed type kernels. By contrast, in small cereal grains, like rice, wheat, and barley, assimilates are delivered by a vascular bundle running the entire length of the endosperm (Chapter 11). Endosperm sink strength thus involves long-distance, non-vascular transfer, with ample opportunities for metabolism along the way.

Central to our knowledge of endosperm sink strength is an appreciation of spatial divisions with respect to function, micro-environments, and metabolic processes (Chapters 3 and 11) (Fig. 15.2A). In addition to starch deposition, two important considerations are: (i) the role of the BETL; and (ii) the low-oxygen state of endosperm. The BETL lies at a key position immediately following a mandatory apoplastic step in post-phloem transport. Its distinctive cell-wall ingrowths and elaborated membrane surface are typical of transfer cells, and there are many unknown activities in this cell layer (Chapter 5). Two roles in sink strength are highlighted by phenotypes of *mn1* and *SWEET4c* mutants (Fig. 15.2B), which are deficient in cell wall invertase and hexose transport, respectively (Sosso *et al.*, 2015; Chapter 5, this volume).

The low oxygen microenvironment of starchy endosperm undoubtedly affects sucrose cleavage and resynthesis by negatively impacting energetics. However, experimental evidence indicates sugar recycling occurs nonetheless (Alonso *et al.*, 2011). If the BETL is better oxygenated than the interior endosperm, it could be an important site for sucrose resynthesis. Although sucrose and hexoses move into the endosperm interior, the hexoses could present a greater metabolic challenge in a low-oxygen microenvironment, where ATP demand for their metabolism could be limited.

The low-oxygen state of the starchy endosperm could also affect transport, metabolism, and assimilate storage. Central metabolism is adapted to low oxygen in the endosperm, yet it remains responsive to changes in external oxygen. Among metabolic adjustments are increased glycolytic activities, reduced production of reactive oxygen species, and changes in redox regulation. In wheat, when grains are exposed to sub-ambient levels of external oxygen, phloem transport and fluxes to seeds decline (van Dongen *et al.*, 2004). In endosperm of barley (see Chapter 11) and rice (Yang *et al.*, 2015), hypoxia activates interconversion of pyruvate and alanine by alanine aminotransferase (AlaAT). This could be essential for restraining pyruvate fermentation, while oxalacetic acid (OAA) provides reducing equivalents for the TCA cycle via glutamate dehydrogenase (GDH). Consequently, neither lactate nor ethanol accumulate during grain fill. Even with these metabolic modifications to increase ATP production, localized limitations remain (see Chapter 11).

Key questions related to effects of low oxygen in starchy endosperm include the following:

**1.** How do assimilates move into and within low-oxygen regions of the endosperm? We know that both sucrose and hexoses cross the BETL, but do they do so via the apoplast, symplast, or both; what is the balance? How much metabolism and/or resynthesis occurs en route, and is it affected by specific states of energy charge, redox, and ATP level? Can more targeted, quantitative measurements address these questions? (see question 5 below).

**2.** What contributions to sink strength by the BETL remain to be discovered? Transcriptomic and metabolomic studies reveal a wide range of sink-related functions possible for the BETL, in addition to the importance of its cell wall invertase and SWEET4c hexose transporters (see Chapter 5). Oxygen concentration could be critically important in the BETL. What is its oxygen status? One might predict that it differs based on the extent of invertase expression, since genes for this enzyme are typically down-regulated by hypoxia (Zeng *et al.*, 1999). In addition, the energetic demands of active transport would be difficult to support in a low-oxygen environment. Does sucrose resynthesis occur in the BETL? Transcript and proteomic analysis of the BETL will help identify genes and enzymes potentially involved in sucrose resynthesis and substrate recycling.

**3.** What happens to plasmodesmatal function in endosperm regions with low oxygen? This activity is essential for symplastic transport.

**4.** What creates the low-oxygen state inside the endosperm? This could result from oxygen draw-down, diffusion barriers, or both (Chapter 11). How does this compare in the aleurone? How is the embryo better oxygenated than the endosperm (Rolletschek *et al.*, 2005)?

**5.** Can technical challenges be overcome to investigate movement and metabolism of assimilates traversing the post-phloem path? Approaches used thus far include isotope labeling, mutant analyses, metabolite and transcript profiling, flux balance analysis (FBA), and noninvasive imaging. A current unknown is the extent of metabolism occurring along the post-phloem path and how it affects the movement of assimilates. To date, *in vivo* fluxes have been estimated by metabolic flux analysis (MFA) based on long-term, stationary-state labeling with $^{13}C$ (Alonso *et al.*, 2011). The results are consistent with extensive metabolic recycling over time, but spatial and temporal resolution are unknown. Can we determine the fate of assimilates as they move through regions of the maize kernel? Can a more rapid, pulse-based time course labeling be devised to appraise sink-strength determinants? Recently, transient labeling and non-stationary modeling (INST-MFA and KFP) have been combined to address questions in multicellular systems such as Arabidopsis, where fluxes were resolvable at cellular and sub-cellular levels (Ma *et al.*, 2014; Allen, 2016). In related work, transient labeling of $^{13}C$-sucrose also allowed determination of fluxes through key paths in linseed embryos (Troufflard *et al.*, 2007); these results have been inaccessible using steady-state approaches.

**6.** Is cellular activity in low-oxygen micro-environments aided by NADH-dependent reduction of fructose to sorbitol via sorbitol dehydrogenase (SDH)? The function of sorbitol in maize kernels is a long-standing question; it forms rapidly from imported $^{14}C$-sucrose (Shaw and Dickinson, 1984). The gene encoding SDH is strongly upregulated by low oxygen and is abundantly expressed in maize kernels (de Sousa *et al.*, 2008). Is sorbitol transported from site to site within the kernel and is it related to sink strength?

## 15.4 Transporters

The important role of sugar transporters in sink strength is highlighted by the phenotype of defective-kernel mutants lacking a functional SWEET4c hexose transporter in the BETL (Sosso *et al.*, 2015) (Fig. 15.2B). Unlike active transporters, SWEETs are thought to facilitate equilibration of solutes (Chen *et al.*, 2010; Slewinski, 2011); SWEET4c is proposed to enhance sugar movement from the hexose-rich apoplast to the BETL. Although highly expressed in the BETL, SWEET4c transcripts appear in the young kernel prior to BETL formation, and thus the impact of this transporter could extend beyond current models.

Other transporters implicated in kernel resource import include a diversity of SWEETs (some of which secrete sucrose) and SUTs (sucrose transporters). These can function sequentially in processes such as phloem loading (Braun, 2012; Chen *et al.*, 2010, 2012) and sucrose uptake into sink tissues. Additional players, designated MSTs (monosaccharide transporters) (Slewinski, 2011) and TSTs (tonoplast sugar transporters) (Hedrich *et al.*, 2015), handle hexoses and compartmentalization of sugars in vacuoles, respectively. Although their transcripts are abundant in developing maize kernels (Slewinski, 2011; Baker *et al.*, 2016), virtually nothing is known about their individual roles. Some could be central to sucrose transport, since sucrose and hexoses can potentially move in and out of the apoplast (and/or vacuoles) along the entire post-phloem pathway.

Tonoplast transporters (TSTs) are likely crucial for the expansion-driven sink strength that supports ovary formation and early grain development (Slewinski, 2011; Hedrich *et al.*, 2015). TSTs import sucrose into vacuoles, where its cleavage drives expansion through hexose effects on osmotic potential. Vacuolar transport also contributes to sink strength throughout development by controlling transient storage of sucrose and hexoses along the post-phloem path.

Key questions on the role of these transporters include the following:

**1.** How are the secretory and equilibration functions of SWEETs balanced with active uptake by SUT, MST, and TST transporters in the maize kernel? Do secretory functions predominate in the maternal transfer zone, with active uptake in the BETL? The story may not be this simple, since multiple SWEETS, SUTs, MSTs, and TSTs are expressed in both locations. What processes are mediated by each of the transporters and where? How might interpretation of their roles be altered by ebb and flow of kernel fluids? Specifically, does the proposed function of SUT1 in retrieval of sucrose from xylem contribute to sink strength of maize kernels? Much attention has been directed to the BETL due to its prominent structural and functional role. However, each site of membrane crossing along the post-phloem path is a point of potential control for assimilates en route to sites of deposition. Are the diversity of functions and compartments in endosperm cells reflected in the diversity of transporter genes expressed at each locale?

**2.** Which TSTs are most important for vacuole-driven expansion? To what degree do sugars along the post-phloem path constitute transient reserves stored in vacuoles?

## 15.5  Nutrient Sensing and Kernel Sink Strength

Microenvironments within the kernel have specialized metabolism, so it is not surprising that exquisitely sensitive systems respond to availability of sugars and nitrogen assimilates and their flux. Sink strength is an integrated feed-forward response to these signals, which, along with the sensing systems, likely vary in different parts of the kernel (from import to deposition).

The "feast-and-famine" hypothesis proposed over a decade ago (Koch, 1996) continues to provide a framework for interpreting nutrient-responsive gene expression in a source–sink context (Bihmidine *et al.*, 2014). Carbohydrate abundance upregulates genes favoring sink strength, while carbohydrate deprivation enhances expression of those for acquiring this limiting resource (e.g. photosynthesis). Notably, sugars have the capacity to upregulate genes for their use in sink tissues. This feed-forward mechanism regulates sink capacity in relation to available resources. Also, photosynthetic rates increase when sugar-repression of "famine genes" for carbohydrate acquisition is relieved. Gene expression is thus responsible for much of the feed-back inhibition previously attributed to build-up of starch and metabolites alone.

A key mechanism for hexose signaling is mediated by hexokinase, which has dual roles as the first enzyme of glycolysis (converting hexose to hexose-P) and as a hexose sensor (Harrington and Bush, 2003; Bihmidine *et al.*, 2013). The two roles are related, because hexose signals are generated only when active hexokinase enzymes have initiated but not completed their catalytic reaction. The resulting conformational changes are associated with incorporation of hexokinase into a transcription-factor complex (Cho *et al.*, 2006; Sheen 2014). Since this process is initiated in the cytoplasm, hexokinase does not sense hexoses while they are in the apoplast or inside vacuoles. For maize kernels, this means hexose compartmentalization, as well as sites and times of its formation, can be central to hexokinase signaling of nutrient abundance. An additional consideration is the one-ATP cost of the hexokinase reaction. Since ATP supplies can be limiting in low-oxygen endosperm and these cells make minimal use of hexokinase (see Chapter 11), one could expect little output from the hexokinase-based sensing system in these cells. The same would *not* be true elsewhere in the kernel, which could have widespread effects on nutrient-responsive gene expression.

Another key player is the TOR (target of rapamycin)-mediated sensing system that is proposed to operate in concert with hexokinase (Lastdrager *et al.*, 2014; Sheen, 2014; Xiong and Sheen, 2014; Dobrenel *et al.*, 2016). Both of these respond to sugar abundance and upregulate diverse aspects of sink strength, including central energy metabolism. However their action is counterbalanced by the starvation-inducible Snf1-related kinases (SnRKs). Downstream impacts of this ying/yang system extend from transcription and translation to carbohydrate metabolism and cell proliferation, and from carbohydrate partitioning and cell growth to senescence and autophagy (Lastdrager *et al.*, 2014; Xiong and Sheen, 2014; Dobrenel *et al.*, 2016). TOR kinase can also interact with inositol polyphosphate signaling (Couso *et al.*, 2016), and in rice can regulate membrane biogenesis (Sun *et al.*, 2016). Perturbation of this system in maize kernels could clearly impact sink strength.

A third nutrient sensing system with pronounced effects on sink strength involves synthesis and metabolism of trehalose-6-P (T6P). Dramatic changes result from perturbation of T6P biosynthesis by TPS (trehalose-6P synthase) or breakdown by TPP (trehalose-6P phosphatase). A striking example is the amelioration of drought-stress effects on young kernels and ovaries by overexpression of TPP (Guan and Koch, 2015; Nuccio *et al.*, 2015). Reactions initiating T6P-related signals are closely associated with sucrose formation and cleavage, so proposed roles include sensing of sucrose versus hexoses and/or sucrose cycling (Guan and Koch, 2015).

Thus far, none of these systems appears capable of sensing extracellular sugars in plants, leaving this a centrally important unknown mechanism for maize. One candidate for this role in Arabidopsis is a pair of interacting factors, the first being a G protein-coupled receptor1 (GCR1) and the second a regulator of G protein signaling1 (RGS1). Chen and Jones (2004) propose that, together, these could function in sensing exogenous glucose.

Finally, recent work indicates that G4 quadruplex structures in DNA can potentially mediate responses to low-energy states, such as those in maize endosperm (Andorf *et al.*, 2014). These four-stranded DNA "kinks" are well-known components of telomers, and appear periodically throughout the genome. Their roles have been unclear, but evidence from maize shows that G4s are associated with genes expressed under "energy emergencies," such as ATP depletion (Andorf *et al.*, 2014) observed in endosperm. Examples include *Sh1* (sucrose synthase) and *Sdh1* (sorbitol dehydrogenase), both of which are upregulated by oxygen deprivation, abundant in endosperm, and central to sucrose metabolism in the maize kernel.

Key questions on nutrient sensing include the following:

**1.** To what degree are the hexose products of sucrose cleavage accessible to the hexokinase-based signaling system? Compartmentalization of sugars along the transport path could greatly impact this mechanism.

**2.** In what ways could energy metabolism in different compartments of the maize kernel be affected by signals of nutrient availability? How does the TOR system function in maize and other cereal grains? How and when are its signals of abundance countered by those of starvation-induced Snf1-related kinase (SnRK)?

**3.** In what way does trehalose-6-P influence maize kernel sink strength? Does more rapid sucrose cleavage and cycling allow for T6P-related signaling? Does this system facilitate sensing of glucose versus sucrose and delineate "feast and famine" conditions in immediately adjacent cells (Guan and Koch, 2015; Nuccio *et al.*, 2015)?

**4.** If sucrose-specific sensing is possible (via T6P-related signaling or other mechanisms), might it provide a mechanistic basis for different effects of sucrose and hexoses on development? Wobus and Webber (1999) suggested hexoses enhance cell division, while sucrose favors differentiation. This is consistent with effects of glucose-sensing systems on the cell cycle, but we do not yet have a mechanism for the sucrose side of this story.

## 15.6 Developmental Changes in Sink-strength Determinants

Determinants of sink strength change throughout maize kernel development, with at least three phases: (i) early sink establishment prior to BETL formation; (ii) a post-BETL, abortion-sensitive phase; and (iii) an abortion-resistant grain-filling phase (Fig. 15.1C). Early events establishing sink strength are detectable within hours of pollination, even before fertilization (Xu *et al.*, 1996). Recent studies suggest that trehalose phosphate and sugar signaling by maternal tissues both contribute to ovule sink strength prior to fertilization (Andersen *et al.*, 2002; Guan and Koch, 2015; Nuccio *et al.*, 2015). Early induction of soluble invertase genes in ovaries is followed by upregulation of SWEET4C and cell wall invertase that amplify sugar signals implicated in BETL differentiation. This differentiation 6–8 days after pollination (DAP) is critical for establishment of the endosperm sink strength. However, the developing kernel remains susceptible to stress-induced abortion until about 12 DAP (Hanft and Jones, 1986; Chapter 17, this volume). Endosperm cell number and volume changes could be centrally important during this time, since both are targets of early heat stress (Hanft and Jones, 1986).

Key questions on developmental changes in sink strength determinants include the following:

**1.** What role do soluble and insoluble invertases play in progression of sugar signals leading to BETL differentiation?
**2.** What molecular and physiological events distinguish the abortion-sensitive and abortion-resistant phases of grain filling?
**3.** What roles do the relative volumes of maternal and filial tissues play in the transition to grain filling? How do kernel volume changes influence the fluid-flow aspects of sink strength?
**4.** How do environmental stresses influence sink strength during the three different developmental phases of kernel development?

In conclusion, important gaps remain in our understanding of kernel sink strength and how it mediates sucrose import during development. One of these unknowns is the role of phloem water and fluid flux in kernels. Another is the unresolved mechanism of assimilate transfer through the long non-vascular distance from phloem termini to the starchy endosperm. Key questions about this movement involve the nature and function of maternal tissues and the effects of the low-oxygen microenvironment in the endosperm. Also, we know little about the nature and contributions of transporters in maternal and filial tissues. There are important but poorly understood effects of nutrient-sensing systems on kernel sink strength. Finally, each of these aspects changes during kernel development; how is this coordinated? Advances in our understanding of the above areas have aided efforts to improve yield and resistance to biotic and abiotic stresses; however, much remains to be learned.

# References


Allen, D.K. (2016) Quantifying plant phenotypes with isotopic labeling and metabolic flux analysis. *Current Opinion in Biotechnology* 37, 45–52.

Alonso, A.P., Val, D.L. and Shachar-Hill, Y. (2011) Central metabolic fluxes in the endosperm of developing maize seeds and their implications for metabolic engineering. *Metabolic Engineering* 13, 96–107.

Andersen, M.N., Asch, F., Wu, Y., Jensen, C.R., Naested, H., Mogensen, V.O. and Koch, K.E. (2002) Soluble invertase expression is an early target of drought stress during the critical, abortion-sensitive phase of young ovary development in maize. *Plant Physiology* 130, 591–604.

Andorf, C.M., Kopylov, M., Dobbs, D., Koch, K.E., Stroupe, M.E., Lawrence, C.J. and Bass, H.W. (2014) G-quadruplex (G4) motifs in the maize (*Zea mays* L.) genome are enriched at specific locations in thousands of genes coupled to energy status, hypoxia, low sugar, and nutrient deprivation. *Journal of Genetics and Genomics* 41, 627–647.

Baker, R.F., Leach, K.A., Boyer, N.R., Swyers, M.J., Benitez-Alfonso, Y., *et al.* (2016) Sucrose transporter *ZmSut1* expression and localization uncover new insights into sucrose phloem loading. *Plant Physiology* 172, 1876–1898.

Bihmidine, S., Hunter III, C.T., Johns, C.E., Koch, K.E. and Braun, D.M. (2013) Regulation of assimilate import into sink organs: update on molecular drivers of sink strength. *Frontiers in Plant Science* 4, 177.

Braun, D.M. (2012) SWEET! The pathway is complete. *Science* 335, 173–174.

Chen, J.G. and Jones, A.M. (2004) AtRGS1 function in *Arabidopsis thaliana*. *Methods in Enzymology* 389, 338–350.

Chen, L.-Q., Hou, B.H., Lalonde, S., Takanaga, H., Hartung, M.L., *et al.* (2010) Sugar transporters for intercellular exchange and nutrition of pathogens. *Nature* 468, 527–532.

Chen, L.-Q., Qu, X.-Q., Hou, B.-H., Sosso, D., Osorio, S., Fernie, A.R. and Frommer, W.B. (2012) Sucrose efflux mediated by SWEET proteins as a key step for phloem transport. *Science* 335, 207–211.

Cho, Y.-H., Yoo, S.-D. and Sheen, J. (2006) Regulatory functions of nuclear hexokinase1 complex in glucose signaling. *Cell* 127, 579–589.

Couso, I., Evans, B., Li, J., Liu, Y., Ma, F., Diamond, S., Allen, D.K. and Umen, J.G. (2016) Synergism between inositol polyphosphates and TOR kinase signaling in nutrient sensing, growth control and lipid metabolism in Chlamydomonas. *Plant Cell* 28, 2026–2042. DOI:10.1105/tpc.16.00351

de Sousa, S.M., Paniago Mdel, G., Arruda, P. and Yunes, J.A. (2008) Sugar levels modulate sorbitol dehydrogenase expression in maize. *Plant Molecular Biology* 68, 203–213.

Dobrenel, T., Caldana, C., Hanson, J., Robaglia, C., Vincentz, M., Veit, B. and Meyer, C. (2016) TOR signaling and nutrient sensing. *Annual Review of Plant Biology* 67, 261–285.

Guan, J.C. and Koch, K.E. (2015) A time and a place for sugar in your ears. *Nature Biotechnology* 33, 827–828.

Hanft, J.M. and Jones, R.J. (1986) Kernel abortion in maize II. Distribution of [14]C among kernel carbohydrates. *Plant Physiology* 81, 511–515.

Harrington, G.N. and Bush, D.R. (2003) The bifunctional role of hexokinase in metabolism and glucose signaling. *Plant Cell* 15, 2493–2496.

Hedrich, R., Sauer, N. and Neuhaus, H.E. (2015) Sugar transport across the plant vacuolar membrane: nature and regulation of carrier proteins. *Current Opinion in Plant Biology* 25, 63–70.

Huang, T., Darnell, R. and Koch, K. (1992) Water and carbon budgets of developing citrus fruit. *Journal of the American Society for Horticultural Science* 117, 287–293.

Jain, M., Chourey, P.S., Li, Q.-B. and Pring, D.R. (2008) Expression of cell wall invertase and several other genes of sugar metabolism in relation to seed development in sorghum (*Sorghum bicolor*). *Journal of Plant Physiology* 165, 331–344.

Koch, K.E. (1996) Carbohydrate-modulated gene expression in plants. *Annual Review of Plant Physiology and Plant Molecular Biology* 47, 509–540.

Koch, K.E. (2004) Sucrose metabolism: regulatory mechanisms and pivotal roles in sugar sensing and plant development. *Current Opinion in Plant Biology* 7, 235–246.

Koch, K.E. and Avigne, W.T. (1990) Postphloem, nonvascular transfer in citrus: kinetics, metabolism, and sugar gradients. *Plant Physiology* 93, 1405–1416.

Lastdrager, J., Hanson, J. and Smeekens, S. (2014) Sugar signals and the control of plant growth and development. *Journal of Experimental Botany* 65, 799–807.

Ma, F., Jazmin, L.J., Young, J.D. and Allen, D.K. (2014) Isotopically nonstationary [13]C flux analysis of changes in *Arabidopsis thaliana* leaf metabolism due to high light acclimation. *Proceedings of the National Academy of Sciences of the United States of America* 111, 16967–16972.

Nuccio, M.L., Wu, J., Mowers, R., Zhou, H.P., Meghji, M., *et al.* (2015) Expression of trehalose-6-phosphate phosphatase in maize ears improves yield in well-watered and drought conditions. *Nature Biotechnology* 33, 862–869.

Rolletschek, H., Koch, K., Wobus, U. and Borisjuk, L. (2005) Positional cues for the starch/lipid balance in maize kernels and resource partitioning to the embryo. *Plant Journal* 42, 69–83.

Rolletschek, H., Grafahrend-Belau, E., Munz, E., Radchuk, V., Kartausch, R., *et al.* (2015) Metabolic architecture of the cereal grain and its relevance to maximize carbon use efficiency. *Plant Physiology* 169, 1698–1713.

Schussler, J. and Westgate, M. (1995) Assimilate flux determines kernel set at low water potential in maize. *Crop Science* 35, 1074–1080.

Shannon, J.C. (1972) Movement of [14]C-labeled assimilates into kernels of *Zea mays* L. I. Pattern and rate of sugar movement. *Plant Physiology* 49, 198–202.

Shaw, J.R. and Dickinson, D.B. (1984) Studies of sugars and sorbitol in developing corn kernels. *Plant Physiology* 75, 207–211.

Sheen, J. (2014) Master regulators in plant glucose signaling networks. *Journal of Plant Biology* 57, 67–79.

Slewinski, T.L. (2011) Diverse functional roles of monosaccharide transporters and their homologs in vascular plants: a physiological perspective. *Molecular Plant* 4, 641–662.

Sosso, D., Luo, D., Li, Q.-B., Sasse, J., Yang, J., *et al.* (2015) Seed filling in domesticated maize and rice depends on SWEET-mediated hexose transport. *Nature Genetics* 47, 1489–1493.

Sun, L., Yu, Y., Hu, W., Min, Q., Kang, H., *et al.* (2016) Ribosomal protein S6 kinase1 coordinates with TOR-Raptor2 to regulate thylakoid membrane biosynthesis in rice. *Biochimica et Biophysica Acta* 1861, 639–649.

Tang, A.C. and Boyer, J.S. (2008) Xylem tension affects growth-induced water potential and daily elongation of maize leaves. *Journal of Experimental Botany* 59, 753–764.

Tilbrook, J. and Tyerman, S.D. (2009) Hydraulic connection of grape berries to the vine: varietal differences in water conductance into and out of berries, and potential for backflow. *Functional Plant Biology* 36, 541–550.

Troufflard, S., Roscher, A., Thomasset, B., Barbotin, J.-N., Rawsthorne, S. and Portais, J.-C. (2007) *In vivo* $^{13}$C NMR determines metabolic fluxes and steady state in linseed embryos. *Phytochemistry* 68, 2341–2350.

Turgeon, R. (2010) The puzzle of phloem pressure. *Plant Physiology* 154, 578–581.

van Dongen, J.T., Roeb, G.W., Dautzenberg, M., Froehlich, A., Vigeolas, H., Minchin, P.E. and Geigenberger, P. (2004) Phloem import and storage metabolism are highly coordinated by the low oxygen concentrations within developing wheat seeds. *Plant Physiology* 135, 1809–1821.

Wobus, U. and Weber, H. (1999) Sugars as signal molecules in plant seed development. *Biological Chemistry* 380, 937–944.

Xiong, Y. and Sheen, J. (2014) The role of target of rapamycin signaling networks in plant growth and metabolism. *Plant Physiology* 164, 499–512.

Xu, J., Avigne, W.T., McCarty, D.R. and Koch, K.E. (1996) A similar dichotomy of sugar modulation and developmental expression affects both paths of sucrose metabolism: evidence from a maize invertase gene family. *Plant Cell* 8, 1209–1220.

Yang, J., Kim, S.R., Lee, S.K., Choi, H., Jeon, J.S. and An, G. (2015) Alanine aminotransferase 1 (OsAlaAT1) plays an essential role in the regulation of starch storage in rice endosperm. *Plant Science* 240, 79–89.

Zeng, Y., Wu, Y., Avigne, W.T. and Koch, K.E. (1999) Rapid repression of maize invertases by low oxygen: invertase/sucrose synthase balance, sugar signaling potential, and seedling survival. *Plant Physiology* 121, 599–608.

Zhang, X., Hirsch, C.N., Sekhon, R.S., de Leon, N. and Kaeppler, S.M. (2016) Evidence for maternal control of seed size in maize from phenotypic and transcriptional analysis. *Journal of Experimental Botany* 67, 1907–1917.

# 16 Natural Variations in Maize Kernel Size: A Resource for Discovering Biological Mechanisms

Xia Zhang[1] and Shawn K. Kaeppler[2,*]

[1]Biotechnology Research Institute, Chinese Academy of Agricultural Sciences, Beijing, China; [2]Department of Agronomy and Great Lakes Bioenergy Research Center, University of Wisconsin, Madison, Wisconsin, USA

## 16.1 Introduction

Seed size is a trait that has been selected during the domestication and improvement of multiple crop species. In maize, the caryopsis or kernel is a fruit composed of the maternal pericarp surrounding the zygotic seed tissues. Kernel size increased dramatically during the domestication of maize (*Zea mays* ssp. *mays*) from its wild progenitor, teosinte (*Zea mays* ssp. *parviglumis*) (Doebley and Stec, 1993). Today, maize is one of the most important crops worldwide, providing food for human consumption, feed for livestock, and raw materials for industrial products. Given the increasing size of the human population and concomitant demand for food and renewable resources, crop scientists are striving to increase the productivity and sustainability of maize and other primary agricultural crops. Yield components, such as kernel size, are among targets to increase yield potential in maize. In industrialized countries, maize kernel size and shape are of great consequence to growers because of their relevance to mechanized cultivation, harvesting, and processing.

This chapter reviews current knowledge about a variety of developmental, molecular, and genetic factors involved in kernel size determination. Information covered includes: (i) background and practical considerations of kernel size in maize production and utilization; (ii) kernel development and its relation to the inherent establishment of kernel size; (iii) maternal effects; and (iv) phytohormone regulation and (v) genes/QTL controlling kernel size. Finally, the chapter reviews future prospects for research on kernel size, and opportunities these studies could provide to enhance grain yield.

## 16.2 Background Considerations

For thousands of years following its domestication, farmers planted and harvested maize manually and used the resulting crop for multiple purposes. This resulted in tremendous diversity in kernel size, shape, color, and composition (Fig. 16.1). In industrialized countries, mechanization and transition to large farms within the last 100 years required greater uniformity in the size

---

*Corresponding author e-mail: smkaeppl@wisc.edu

**Fig. 16.1.** Maize kernel diversity for color, size, and shape.

and shape of kernels. In addition, since millions of seed units are produced and distributed each year, the volume and weight of seed has an impact on distribution. Therefore, maize breeders have had to consider several seed characteristics of female inbreds and commercial hybrids, including kernel size and shape (Fig. 16.2).

The female parent largely determines many kernel characteristics that affect seed sold for planting. However, xenia effects— the genetic influence of pollen on kernel development—can be important for some traits, including kernel size and composition. The extreme example shown in Fig. 16.3 involves material from Krug Large Seed (KLS) and Krug Small Seed (KSS) populations, where mixed pollen from the two populations was applied to a KSS ear. Kernels that developed from KLS pollination are clearly larger than those from KSS pollination. In commercial hybrids, xenia effects have been shown for kernel oil concentration (see Chapter 13). Pollen from a high-oil genotype was found to increase embryo oil concentration of hybrids with normal and low oil contents; this is a consequence of increased embryo size and weight (Tanaka and Maddonni, 2008). Xenia is not generally

observed for traits such as starch, protein, and mineral content (Letchworth and Lambert, 1998; Pletsch-Rivera and Kaeppler, 2007). Evidence also shows that kernel weight, and therefore yield, of commercial hybrids can be increased after cross-pollination with an unrelated pollinator (Tsai and Tsai, 1990). Higher yield is often associated with changes in grain growth rate and grain filling duration (Seka and Cross, 1995; Bulant and Gallais, 1998). These observations indicate that untapped yield potential in modern hybrids can be harnessed by increasing kernel sink strength. However, the degree of the xenia effect is dependent on genotype and environmental conditions.

Kernel size exhibits tremendous phenotypic variation in maize. Examples range from small-kernel popcorn to the very large types sold as a snack under the brand Corn Nuts. In a survey of diverse maize inbred lines, including corn-belt dent, pop, sweet, flint, and tropical lines, we observed nearly threefold variation in various measures of kernel size (Miller *et al.*, 2016). Broader collections of maize from across the world, including floury and flint corn used for food, provide an even greater range of kernel size, color, and composition. The tremendous

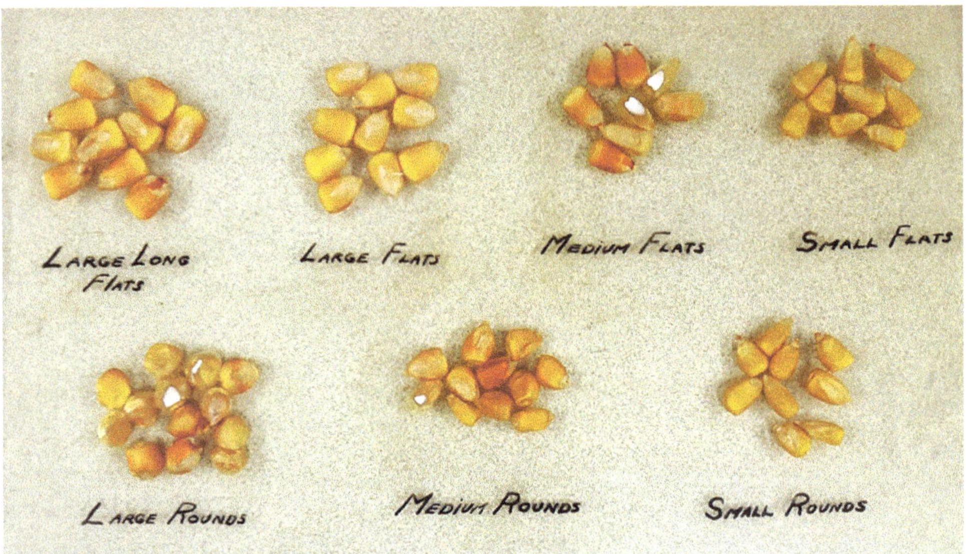



**Fig. 16.2.** Example seed grades based on size and shape.

**Fig. 16.3.** Xenia effect on kernel size. A plant from the Krug Small Seed C30 population was pollinated with a mixture of Krug Large Seed C30 (KLS30) and Krug Small Seed C30 (KSS30) pollen. Kernels resulting from pollination with KLS30 pollen are noticeably larger than those resulting from pollination with the KSS30 pollen. Kernel size in these populations is quantitative; this example is not the result of complementing a major gene qualitative mutation.

phenotypic diversity of maize kernel characteristics has been maintained over thousands of years by many generations of smallholder farmers.

The wide phenotypic diversity of maize kernels, coupled with observations regarding a strong maternal effect on kernel characteristics and indications of untapped source potential revealed as a xenia effect, support the hypothesis that natural variation in maize is a powerful resource to dissect mechanisms controlling kernel size and composition. In this context, the following discussion provides an overview of biological discoveries relevant to maize kernel size.

## 16.3 Overview of Kernel Size Determination

Kernel development in maize begins with a double fertilization event that leads to the formation of three genetically distinct compartments: the embryo, endosperm, and pericarp. The triploid endosperm, with two maternal and one paternal genome, constitutes approximately 83% of the dry weight of the mature kernel and as such plays a central role in determining kernel size. As the endosperm develops, it mediates transfer of maternal nutrients to the embryo and stores most of the metabolic reserves. The pericarp is a completely maternal tissue that not only provides protection for the kernel, but also functions in transmitting physical, environmental, and developmental cues to the seed, thus affecting kernel size (Dante *et al.*, 2014). Consequently, kernel size is coordinately determined by the growth of the embryo, endosperm, and pericarp, and it is influenced by a wide range of genetic and developmental factors (Sundaresan, 2005). Although kernel size has been extensively studied and utilized by biologists and plant breeders over the past century, it is only in the past decade that we have made big steps towards understanding the molecular/genetic/developmental basis of its regulation. This is mainly a consequence of mutant analysis, genome synteny across cereal species, and advances in next-generation sequencing technologies.

## 16.4 Kernel Development Programs and Kernel Size

Kernel development and the inherent establishment of its final size is a multi-step processes controlled by cellular and developmental factors. Kernel growth and development is partitioned into three partially overlapping phases. Phase I, beginning within a few hours after ovary fertilization, is an initial lag period characterized by cell proliferation and rapid increase in kernel water content, but with minimal dry matter accumulation. Following double fertilization, the coenocytic endosperm sequentially undergoes syncytium formation, cellularization, and differentiation into distinct domains, coupled with cell proliferation. During the lag phase, all cells comprising the mature endosperm are generated through cell division (Sabelli and Larkins, 2009). Eventually, cells cease dividing and switch to a period of DNA endoreduplication, which is proposed to drive cell expansion. Therefore, while Phase I contributes little directly to kernel biomass, active cell proliferation and cell expansion provide two key driving forces for endosperm growth and development and thus attainment of final kernel

size. The cell population established during the lag phase sets the demand of sink tissue for subsequent accumulation of nutrient reserves. An abundance of maize mutants with defective endosperm development (*de* and *dek* mutants) have been described (Neuffer and Sheridan, 1980). Many *dek* mutants are defective in mitotic activity and/or endoreduplication and appear to have pleiotropic effects on seed growth, i.e. aborted or small seeds (Kowles *et al.*, 1992). Maize *Rough Endosperm 3* (*Rgh3*) encodes a predicted RNA splicing factor and is required for endosperm cells to switch from a proliferative to a differentiation stage. Mutant *rgh3* plants are defective in endosperm differentiation and have a reduced kernel size (Fouquet *et al.*, 2011).

Following the lag phase, there is a period with significant dry matter gain involving rapid storage metabolite deposition and continued cell enlargement. This period is generally referred to as the "effective grain-filling phase" (Phase II). Finally, kernels approach maturation and drying (Phase III), which is associated with maximum dry matter accumulation and desiccation tolerance. Kernels eventually reach "physiological maturity," a point at which nutrients are no longer translocated to or from the kernel. After this, kernels begin to dry down and enter a quiescent state.

Measurement across diverse inbred lines of the transition through this developmental program is challenging due to the difficulty in non-destructively accessing developing kernels in field-grown plants. Two grain-filling traits, kernel growth rate (KGR) and grain-filling duration (GFD), based on the rate and duration of storage compound accumulation (primarily starch and storage proteins), have been applied to predict final kernel size and weight (Alvarez Prado *et al.*, 2013). Research on the maize Krug populations previously described demonstrated the causal relationship between kernel size and grain filling characteristics. The linear phase of grain filling in the large-seeded KLS30 initiates later, but progresses at a faster rate and terminates later, suggesting the important role of the onset, rate, and duration of grain filling in determining kernel size. The prolonged lag phase observed in large kernels is associated with a greater endosperm cell number, which contributes to the nutrient sink strength and rapid growth of the kernel (Sekhon *et al.*, 2014; Zhang *et al.*, 2016).

Factors that control translocation of carbohydrates to the developing kernel are important in determining whether the kernel will reach its genetic potential for size. Cell wall invertases in the basal endosperm transfer layer (BETL) drive sucrose unloading from the mother plant to the sink organ (kernel) by converting sucrose to hexoses. The BETL, adjacent to the pedicel tissue, is located at the entry of the endosperm and has a postulated function of nutrient acquisition from the mother plant. *Miniature1* (*Mn1*) encodes a cell wall invertase (INCW2) in the BETL, and loss of *Mn1* function in the *miniature-1* (*mn1*) mutant renders a phenotype with up to 70% reduction in kernel mass at maturity (Miller and Chourey, 1992; Cheng *et al.*, 1996). Kernel size in *mn1* is reduced because there are fewer and smaller endosperm cells and because carbohydrate allocation to developing kernels is restricted (Vilhar *et al.*, 2002). Manipulating *Mn1* expression improves grain filling by promoting carbohydrate transport from source to sink organs and brings about an obvious increase in kernel size and kernel weight (Li *et al.*, 2013). One defective kernel mutant, *rgf1* (*reduced grain filling*), is morphologically similar to *mn1* seed and shows a substantial reduction in the rate of starch accumulation and fresh weight increase. Distinct from *mn1*, the reduced grain filling conditioned by *rgf1* mutation is associated with alterations in both basal endosperm and pedicel development. Therefore, the *Rgf1* gene, which has not been characterized, may be involved in sugar uptake or perception during early endosperm development (Maitz *et al.*, 2000). The broad importance of metabolic reserve deposition as it applies to kernel size is also revealed by starch-synthesizing enzymes. The endosperm-specific *shrunken2* (*Sh2*) and *brittle2* (*Bt2*) genes encode the large and small subunits, respectively, of the starch-synthetic enzyme ADP-glucose pyrophosphorylase (AGP). Overexpression of *Bt2* and *Sh2* enhances starch accumulation during grain filling and exerts pleiotropic effects on overall plant and kernel weight (Li *et al.*, 2011a). Maize *Sh2-Rev6*, a variant of the *Sh2* locus created by transposon mutagenesis, contains two amino acid insertions and conditions a substantial increase in kernel weight. This increase is not only associated with a greater starch content, but also with a stronger carbon sink leading to the enhanced synthesis of kernel components other than starch (Giroux *et al.*, 1996) (Chapter 12, this volume).

## 16.5 Maternal Control of Kernel Size

Variation in an individual's phenotype is determined not only by its genotype and the environment it experiences, but also by maternal effects, i.e. the contribution of the environment and genotype of the maternal parent to the phenotype of its offspring beyond the equal chromosomal contribution expected from each parent. Kernel size is a trait for which large maternal effects have been demonstrated through differences between reciprocal crosses. The maternal plant exerts effects on seed size via (i) the pericarp; (ii) maternal provisioning during seed development, with nutrient resources being provisioned to seed by the maternal parent; (iii) maternal determination of progeny plasticity in response to developmental signals and environmental cues; and (iv) the genetic contribution to the endosperm where gene imprinting occurs most often (Chapter 9). Pericarp is derived from the ovule integuments after pollination and is entirely maternal in origin. It nourishes the seed and imposes mechanical constraints on embryo and endosperm development, thereby setting an upper limit to final kernel size (Borisjuk and Radchuk, 2014). Maternal provisioning of resources during kernel growth and development can control the allocation of carbohydrates and other storage compounds to the developing kernel, which in turn influences the final size and weight of kernels. Partitioning of carbohydrates is not only important for the supply of storage compounds to the kernel but also provides key developmental cues that eventually determine kernel size. Pericarp and kernel provisioning commonly exhibit phenotypic plasticity in response to environmental factors during kernel development; these maternal effects, along with maternal determination of progeny plasticity to developmental signals and environmental cues, are important transmission mechanisms of maternal environmental effects. In cereals, including maize, the triploid endosperm is where gene imprinting most often occurs. Imprinted genes, identified by differential expression of alleles in a parent-of-origin specific manner, are thought to influence endosperm growth and nutrient transfer to the seed, consequently affecting seed size (Chapter 9).

Our knowledge regarding maternal control of seed size is enlightened by studies of Arabidopsis and rice that support regulation of cell proliferation and/or cell expansion in integuments as one causal basis for seed size (Huang *et al.*, 2013; Li and Li, 2015). Maize *Mn1* and *Rgf1,* as discussed previously, also influence the maternal tissue that controls nutrient flow to the endosperm. The *mn1* and *rgf1* mutants show degeneration of the maternal chalaza region of pericarp, which presumably could interrupt nutrient supply and metabolite storage in the endosperm and ultimately reduce seed size (Chaudhury and Berger, 2001). In addition, maternal effect mutations that affect development and function of the male or female gametophytes, e.g. *maternal effect lethal1* (*mel1*) (Evans and Kermicle, 2001), *baseless1* (*bsl1*) (Gutiérrez-Marcos *et al.*, 2006), and *stunter1* (*stt1*) (Phillips and Evans, 2011), have been reported to affect seed size. These mutants have defective endosperm and embryo development and display miniature kernels. Recently, maternal effects on kernel size were investigated on maize inbred lines and their reciprocal $F_1$ crosses with contrasting kernel phenotypes (Zhang *et al.*, 2016). The reciprocal crosses largely mirrored the maternal parent in kernel characteristics, including endosperm size, grain filling pattern, and gene transcriptional profiles, supporting a role for the maternal parent in determining kernel size. Further, differentially co-expressed genes in parental lines and reciprocal crosses highlighted underlying developmental pathways and identified candidate genes involved in the process of kernel development.

Gene imprinting in plants is most prevalent in endosperm (Chapter 9) and has been proposed to regulate seed size (Springer and Gutiérrez-Marcos, 2009; Bai and Settles, 2014). Basic knowledge regarding imprinting as a mechanism to influence seed phenotype is mainly from research on Arabidopsis. Mutations in the imprinted Arabidopsis Polycomb group genes FERTILIZATION-INDEPENDENT ENDOSPERM (FIE), MULTICOPY SUPPRESSOR OF IRA1, MEDEA (MEA), and FERTILIZATION-INDEPENDENT SEED2 (FIS2) induce proliferation defects during endosperm development and pleiotropically affect seed growth and development (Luo *et al.*, 2000; Chapter 3, this volume). *FIE* orthologs have been identified in rice (*OsFIE1* and *OsFIE2*), but only the imprinted gene *OsFIE1* shows a direct impact on seed growth, as plants overexpressing *OsFIE1* have reduced seed size and exhibit precocious cellularization (Folsom *et al.*, 2014).

Although gene imprinting was first observed in maize, so far only a few allele-specific imprinted genes have been reported, including *Meg1, Nrp1, Mez1, ZmFie1,* and *ZmFie2* (Danilevskaya *et al.*, 2003; Guo *et al.*, 2003; Haun *et al.*, 2007). Among them, *maternally expressed gene1* (*Meg1*) was shown to impact kernel size. The *Meg1* locus encodes a small, secreted polypeptide exclusively localized in the BETL of the developing endosperm, and the maternal allele is preferentially expressed during early endosperm development (Gutiérrez-Marcos *et al.*, 2004). *Meg1* positively regulates transfer tissue development and function, thereby promoting nutrient uptake and allocation to the seed and ultimately increasing biomass/yield. RNAi knockdown of *Meg1* results in reduced transfer cell differentiation and smaller seeds than non-transgenic controls (Costa *et al.*, 2012). These results support the functional relevance of genomic imprinting in regulating seed traits via maternal nutrient provisioning.

RNA-seq analysis of the endosperm transcriptome revealed many imprinted genes, including hundreds of paternally and maternally expressed genes (Waters *et al.*, 2013; Xin *et al.*, 2013). These genes have a wide range of functional annotations in the processes of nutrient transport, hormone signaling, and transcriptional regulation of endosperm development, but the biological function of most of them in maize is unknown. Future efforts to functionally characterize these imprinted candidate genes will be important in determining the role that epigenetic regulation plays on endosperm development and seed size.

## 16.6 Phytohormones and Kernel Size

Phytohormones, including auxins, cytokinins, and brassinosteroids, are thought to play an important role in the regulation of seed size. Auxin in its most common *in planta* form, indole-3-acetic acid (IAA), controls numerous aspects of plant growth and development including coordinated communication between seed compartments (primarily embryo and endosperm), influencing the final seed size. Several mutants defective in auxin biosynthesis were identified in maize, including *orange pericarp*, *vanishing tassel2*, *sparse inflorescence1*, and various single-locus recessive mutants (*defective kernels* [*dek*]) (Locascio *et al.*, 2014). Of these, defective endosperm-B18 (*de18*) is a viable auxin-related mutant showing a small kernel size phenotype (Bernardi *et al.*, 2012). Throughout kernel development, *de18* kernels are reduced in size and accumulate less dry matter. Application of synthetic auxin to developing kernels largely rescues the *de18* phenotype, indicating impairment in auxin biosynthesis as the cause of the kernel size changes. Molecular characterization identified an endosperm-specific YUCCA1 protein in maize, *ZmYuc1*, which encodes a flavin monooxygenase that catalyzes auxin biosynthesis. *ZmYuc1* is tightly linked to the De18 locus and is the causal basis for IAA deficiency and the small kernel phenotype in *de18* (Bernardi *et al.*, 2012). The role exerted

by auxin in the regulation of plant growth and development strongly depends on numerous auxin transporters and signaling factors. A large number of candidate genes involved in auxin responses have been identified in plants ranging from Arabidopsis to various crop plants. Among these are auxin response factors (ARFs), a superfamily of transcription factors that bind to auxin-responsive elements (AuxREs) in promoters of auxin-regulated genes and heterodimerize with Auxin/Indole-3-acetic (Aux/IAA) proteins, promoting an auxin signaling transduction cascade. It has been suggested that these ARFs play an important role in gene expression in response to auxin and potential regulation of seed development. One example supporting a role for ARFs in seed size regulation is *AtARF2*. Mutation in *ARF2* dramatically increases seed size and seed weight in "megaintegumenta" (mnt)/*arf2* mutants, because extra cell division in the integuments causes an enlarged seed coat (Schruff *et al.*, 2006). Maize genome-wide analysis has successfully identified 31 ARFs and 34 Aux/IAA gene family members (Ludwig *et al.*, 2013). Maize Aux/IAA proteins BARREN INFLORESCENCE1 and BARREN INFLORESCENCE4 (BIF1 and BIF4), along with ARF transcriptional regulators, have recently been reported to be involved in the formation of axillary meristems in maize inflorescences (Galli *et al.*, 2015), but there is currently no evidence for a functional role of these auxin-related factors in maize kernel development.

Cytokinins (CKs) promote cell proliferation and differentiation and control many facets of plant growth and development, including seed development. In fact, the major naturally occurring CK, zeatin, was first isolated from immature maize kernels (Letham, 1963). A tight, positive correlation between cytokinin levels and the phase of cell division has been shown in developing kernels (Dietrich *et al.*, 1995). CKs are active throughout kernel development, with the highest levels found 6–8 days after pollination in the BETL and embryo-surrounding regions (ESR); there are very low levels in the embryo (Chen *et al.*, 2014). CKs accumulate during cell division and active growth at the very early phase of endosperm development. The limited existing evidence for CKs controlling maize kernel size comes from studies of the *mn1* mutant (Rijavec *et al.*, 2009, 2012). Based on CK levels in *Mn1* and *mn1* during kernel development, Rijavec *et al.* (2009) proposed that crosstalk among CKs, the cell cycle, and cell wall invertase is causal to increased cell number and sink strength in *Mn1*, which accounts for a larger endosperm and kernel size. A more recent study on functional analysis of cytokinin receptors identified seven maize histidine kinases genes (*ZmHK*). Ectopic expression of *ZmHK* genes in *Arabidopsis* resulted in a reduction in seed size, suggesting a repressor role of *ZmHKs* in seed development (Wang *et al.*, 2014). Similarly, a triple mutant of the *Arabidopsis HK* genes (*AHK2, AHK3*, and *AHK4*) produced seeds more than twice as large as wild-type due to cytokinin-dependent endospermal and/or maternal control of embryo size (Riefler *et al.*, 2006).

In contrast to auxin and cytokinins, information about brassinosteroid (BR) control of maize kernel size is extremely limited. Studies performed in Arabidopsis and rice indicated a role for BR on seed size. Brassinosteroid deficient/insensitive *dwarf* mutants (i.e. *dwarf5, dwf5; dwf1; shrink1-D; BR-Insensitive1, bri1*) can produce small seeds in Arabidopsis (Locascio *et al.*, 2014), but the mechanisms underlying seed size regulation remain largely unclear. In maize, a few BR biosynthesis and signaling genes, such as *Zmdwf1, Zmdwf4*, and *Zmbrd1,* have been isolated. However, no data are available about the specific effect of BR on maize kernel development and kernel size, so this remains to be explored.

## 16.7 Genetic and Genomic Basis of Kernel Size

As described above, our understanding of the regulation of kernel size is predominantly based on mutations of genes in key pathways affecting kernel development. Complementary to this approach, quantitative trait locus (QTL) mapping has been widely applied to uncover genomic regions containing genes affecting seed size. Researchers have resolved hundreds of QTLs associated with grain size and grain yield traits (Liu *et al.*, 2014; Zhang *et al.*, 2014). However, few studies have validated causal genes in maize. In contrast, several rice genes controlling QTLs for grain size and grain weight have been cloned and characterized, including grain width and weight 2 (*GW2*) (Song *et al.*, 2007), and grain size 3 and grain size 5 (*GS3* and *GS5*) (Fan *et al.*, 2006; Li *et al.*, 2011b).

Identifying and cloning maize orthologs of rice seed size genes has proven helpful in understanding mechanisms controlling maize kernel size. *GW2* is a major QTL for rice grain width and weight (Song *et al.*, 2007). Two maize homologs of *GW2*, *ZmGW2-CHR4* and *ZmGW2-CHR5*, have been found (Li *et al.*, 2010a). Linkage, association, and expression analyses show that the two genes could have conserved functions in controlling kernel size and weight. *GS3*, a major negative regulator of grain length, encodes four putative domains functioning differently in rice grain size regulation (Fan *et al.*, 2006). The *GS3* ortholog in maize, *ZmGS3*, is thought to play a role in kernel development. Association mapping revealed polymorphisms in the exon and promoter of *ZmGS3* are associated with kernel length and weight (Li *et al.*, 2010b). *GS5*, a minor rice grain weight QTL, encodes a putative serine carboxypeptidase and functions as a positive regulator of grain size (Li *et al.*, 2011b). The orthologous gene in maize, *ZmGS5*, has a conserved function in seed development. *ZmGS5* and its trans-regulator *ZmBAK1-7,* identified by expression QTL (eQTL) analysis, are located in kernel-related QTL intervals and associated with kernel length (KL) and width (KW) (Liu *et al.*, 2015). The heterologous expression of *ZmGS5* in Arabidopsis caused a significant increase in seed size and cell number, suggesting a possible role in kernel development (Liu *et al.*, 2015). Collectively, gene identification based on orthologs of the well-annotated genome in other cereal species provides the opportunity to explore genetic resources related to kernel size control in maize.

Conventional QTL mapping performed with bi-parental recombinant populations generally offers relatively coarse linkage resolution due to low allele numbers and insufficient recombination events. To overcome these limitations, new types of maize mapping populations that enable more abundant genetic diversity and higher mapping resolution have been developed by crossing more than two inbred founder lines, such as Multi-parent Advanced Generation Integrated Cross (MAGIC) (Dell'Acqua *et al.*, 2015). QTL mapping coupled with expression analysis in a maize MAGIC population identified three suggestive QTL loci for grain yield, and two candidate causal genes (*GRMZM2G054651* and *GRMZM2G101875*) were claimed to affect kernel size (Dell'Acqua *et al.*, 2015). More recently—using a four-way cross mapping population derived from four maize inbred lines with varied kernel size—Chen *et al.* (2016) reported ten QTLs associated with kernel size, with two for kernel length (*qKL3-1* and *qKL7-1*) and one for kernel width (*qKW5-1*). These QTLs show stable expression across different genetic backgrounds and consistently co-localize with known QTLs conferring kernel size, such as *ZmGW2-Chr5*, *ZmGW2-Chr5*, and glutamine synthetase isoenzyme (*Gln1-4*) (Martin *et al.*, 2006; Li *et al.*, 2010a). These fine-mapped QTL regions support the value of natural variation to study maize kernel size.

With the advent of next-generation sequencing (NGS) technologies, large amounts of sequencing data have been generated that facilitate dissection of the genetic and molecular basis for phenotypic variation in crops. Various types of maize NGS data have been released, such as transcriptome sequencing of 503 diverse maize inbred lines representative of the major U.S. grain heterotic groups, sweet maize, and popcorn, as well as exotic maize lines (Hansey *et al.*, 2011); genotyping-by-sequencing (GBS) of 2815 USA national maize inbred accessions (Romay *et al.*, 2013); and genotyping of maize Nested Association Mapping (NAM) population (McMullen *et al.*, 2009). NGS-enabled genome-wide association studies (GWAS), as an alternative to conventional QTL linkage mapping, has been successfully used to identify genetic loci associated with important agronomic traits in maize, including flowering time, leaf architecture, kernel composition, and vegetative phase change (Tian *et al.*, 2011; Hirsch *et al.*, 2014b). In the case of kernel size and kernel number, GWAS performed in an association panel of 513 inbred lines identified significant association signals on chromosome 7 for kernel width, and chromosome 1 for kernel number per row. The associated single nucleotide polymorphism (SNP) positions co-localized with previously identified kernel trait QTLs. Six candidate genes for kernel width (GRMZM2G354539, GRMZM2G052893, GRMZM2G052817, GRMZM2G354525, GRMZM2G052610, and GRMZM2G052509) and four candidate genes for kernel number per row (GRMZM2G088524, GRMZM2G022822, GRMZM2G108180, and GRMZM2G052666) were predicted, with GRMZM2G052509 coinciding with an eQTL locus for kernel width (Yang *et al.*, 2014).

Phenotype-based pooling of diverse accessions from natural populations and subsequent sequence-based determination of allele frequency are providing a powerful approach for identifying genes/genomic regions in crop species. Regions of the maize genome under selection for kernel size were explored by sequencing pools of individuals from pairs of extreme populations derived from a long-term divergent selection program for small and large kernel size (Hirsch *et al.*, 2014a). This genome-wide scan resulted in the identification of 94 divergent regions with a median of six genes per region contributing to control of seed size in maize. Along with this work, extensive transcriptome analysis aimed at characterizing the genetic and developmental basis involved in kernel size control was performed (Sekhon *et al.*, 2014; Zhang *et al.*, 2016). In their studies, gene expression patterns, co-expressed gene modules and differentially expressed genes between two contrasting kernel size types were classified and clustered into subgroups, and several interesting regulatory networks were proposed. A similar method designated "extreme-phenotype GWAS" (XP-GWAS) was recently described that relies on measurement of allele frequencies in pools of individuals from a diversity panel that has extreme phenotypes. By using the kernel row number trait as an example, several linked QTLs were resolved and trait-associated variants within a single gene under a QTL peak were detected (Yang *et al.*, 2015). These rich resources from phenotype-based sequencing studies are providing valuable candidates for the follow-up analysis, for allele mining to identify functional variation, and for marker development in kernel size regulation.

## 16.8 Future Prospects

Kernel size in maize is an important and complex life-history trait for farmers, breeders, and geneticists. Extensive studies have achieved significant progress in understanding some of the mechanisms underlying kernel size variation, which is mainly attributed to the identification of genes and elements that are involved in key pathways of kernel development. Genome-wide scans for natural variation in kernel size and genetic mapping are now possible in maize. The process will be accelerated by high-throughput and high-resolution platforms for both phenotyping and genotyping, and will be facilitated by continuing to collect diverse maize germplasm worldwide and constructing association panels, as well as developing new statistical models. NGS-enabled gene expression profiles and GWAS have enriched genetic resources linked to kernel size, but follow-up gene functional analyses will be required to characterize the causal genes. Gene annotation and expression profiles may provide useful clues for the post-GWAS/QTL analysis. Additional experimental studies, including transposon mutagenesis, chemical mutagenesis, and transgenic analysis of candidate genes, will be necessary to conclusively identify causal genes and causal variants.

A prime target of crop breeding is to improve grain yield. Kernel size and weight, as a primary yield component, has a direct effect on final grain yield as well as indirect effects through other yield-contributing components such as ear length, kernel row number, and kernel number per ear. Due to the interrelated correlations and compensatory effects among yield components, e.g. the trade-off between kernel size and kernel number, improving grain yield therefore requires simultaneous consideration of their genetic correlations. Dissecting the molecular and genetic basis of seed size and its interaction with other yield components are critical steps to effective ideotype breeding for increasing grain yield in maize and other cereal crops.

## References


Alvarez Prado, S., López, C.G., Gambín, B.L., Abertondo, V.J. and Borrás, L. (2013) Dissecting the genetic basis of physiological processes determining maize kernel weight using the IBM (B73×Mo17) Syn4 population. *Field Crops Research* 145, 33–43.

Bai, F. and Settles, A.M. (2014) Imprinting in plants as a mechanism to generate seed phenotypic diversity. *Frontiers in Plant Science* 5, 780.

Bernardi, J., Lanubile, A., Li, Q.-B., Kumar, D., Kladnik, A., *et al.* (2012) Impaired auxin biosynthesis in the *defective endosperm18* mutant is due to mutational loss of expression in the *ZmYuc1* gene encoding endosperm-specific YUCCA1 protein in maize. *Plant Physiology* 160, 1318–1328.

Borisjuk, L. and Radchuk, V. (2014) Physical, metabolic and developmental functions of the seed coat. *Frontiers in Plant Science* 5, 510. DOI:10.3389/fpls.2014.00510

Bulant, C. and Gallais, A. (1998) Xenia effects in maize with normal endosperm. I. Importance and stability. *Crop Science* 38, 1517–1525.

Chaudhury, A.M. and Berger, F. (2001) Maternal control of seed development. *Seminars in Cell and Developmental Biology* 12, 381–386.

Chen, J., Lausser, A. and Dresselhaus, T. (2014) Hormonal responses during early embryogenesis in maize. *Biochemical Society Transactions* 42, 325–331.

Chen, J., Zhang, L., Liu, S., Li, Z., Huang, R., *et al.* (2016) The genetic basis of natural variation in kernel size and related traits using a four-way cross population in maize. *PLOS ONE* 11, e0153428.

Cheng, W.H., Taliercio, E.W. and Chourey, P.S. (1996) The *Miniature1* seed locus of maize encodes a cell wall invertase required for normal development of endosperm and maternal cells in the pedicel. *Plant Cell* 8, 971–983.

Costa, L.M., Yuan, J., Rouster, J., Paul, W., Dickinson, H. and Gutiérrez-Marcos, J.F. (2012) Maternal control of nutrient allocation in plant seeds by genomic imprinting. *Current Biology* 22, 160–165.

Danilevskaya, O.N., Hermon, P., Hantke, S., Muszynski, M.G., Kollipara, K. and Ananiev, E.V. (2003) Duplicated fie genes in maize: expression pattern and imprinting suggest distinct functions. *Plant Cell* 15, 425–438.

Dante, R.A., Larkins, B.A. and Sabelli, P.A. (2014) Cell cycle control and seed development. *Frontiers in Plant Science* 5, 493.

Dell'Acqua, M., Gatti, D.M., Pea, G., Cattonaro, F., Coppens, F., *et al.* (2015) Genetic properties of the MAGIC maize population: a new platform for high definition QTL mapping in *Zea mays*. *Genome Biology* 16, 1–23.

Dietrich, J., Kaminek, M., Blevins, D., Reinbott, T. and Morris, R. (1995) Changes in cytokinins and cytokinin oxidase activity in developing maize kernels and the effects of exogenous cytokinin on kernel development. *Plant Physiology and Biochemistry* 33, 327–336.

Doebley, J. and Stec, A. (1993) Inheritance of the morphological differences between maize and teosinte: comparison of results for two $F_2$ populations. *Genetics* 134, 559–570.

Evans, M.M. and Kermicle, J.L. (2001) Interaction between maternal effect and zygotic effect mutations during maize seed development. *Genetics* 159, 303–315.

Fan, C., Xing, Y., Mao, H., Lu, T., Han, B., *et al.* (2006) GS3, a major QTL for grain length and weight and minor QTL for grain width and thickness in rice, encodes a putative transmembrane protein. *Theoretical and Applied Genetics* 112, 1164–1171.

Folsom, J.J., Begcy, K., Hao, X., Wang, D. and Walia, H. (2014) Rice fertilization-independent endosperm1 regulates seed size under heat stress by controlling early endosperm development. *Plant Physiology* 165, 238–248.

Fouquet, R., Martin, F., Fajardo, D.S., Gault, C.M., Gómez, E., *et al.* (2011) Maize rough endosperm3 encodes an RNA splicing factor required for endosperm cell differentiation and has a nonautonomous effect on embryo development. *Plant Cell* 23, 4280–4297.

Galli, M., Liu, Q., Moss, B.L., Malcomber, S., Li, W., *et al.* (2015) Auxin signaling modules regulate maize inflorescence architecture. *Proceedings of the National Academy of Sciences of the United States of America* 112, 13372–13377.

Giroux, M.J., Shaw, J., Barry, G., Cobb, B.G., Greene, T., Okita, T. and Hannah, L.C. (1996) A single mutation that increases maize seed weight. *Proceedings of the National Academy of Sciences of the United States of America* 93, 5824–5829.

Guo, M., Rupe, M.A., Danilevskaya, O.N., Yang, X. and Hu, Z. (2003) Genome-wide mRNA profiling reveals heterochronic allelic variation and a new imprinted gene in hybrid maize endosperm. *The Plant Journal* 36, 30–44.

Gutiérrez-Marcos, J.F., Costa, L.M., Biderre-Petit, C., Khbaya, B., O'Sullivan, D.M., *et al.* (2004) *Maternally expressed gene1* is a novel maize endosperm transfer cell-specific gene with a maternal parent-of-origin pattern of expression. *Plant Cell* 16, 1288–1301.

Gutiérrez-Marcos, J.F., Costa, L.M. and Evans, M.M.S. (2006) Maternal gametophytic *baseless1* is required for development of the central cell and early endosperm patterning in maize (*Zea mays*). *Genetics* 174, 317–329.

Hansey, C.N., Johnson, J.M., Sekhon, R.S., Kaeppler, S.M. and de Leon, N. (2011) Genetic diversity of a maize association population with restricted phenology. *Crop Science* 51, 704–715.

Haun, W.J., Laoueillé-Duprat, S., O'Connell, M.J., Spillane, C., Grossniklaus, U., *et al.* (2007) Genomic imprinting, methylation and molecular evolution of maize enhancer of zeste (Mez) homologs. *The Plant Journal* 49, 325–337.

Hirsch, C.N., Flint-Garcia, S.A., Beissinger, T.M., Eichten, S.R., Deshpande, S., *et al.* (2014a) Insights into the effects of long-term artificial selection on seed size in maize. *Genetics* 198, 409–421.

Hirsch, C.N., Foerster, J.M., Johnson, J.M., Sekhon, R.S., Muttoni, G., *et al.* (2014b) Insights into the maize pan-genome and pan-transcriptome. *Plant Cell* 26, 121–135. DOI:10.1105/tpc.113.119982

Huang, R., Jiang, L., Zheng, J., Wang, T., Wang, H., Huang, Y. and Hong, Z. (2013) Genetic bases of rice grain shape: so many genes, so little known. *Trends in Plant Science* 18, 218–226.

Kowles, R.V., McMullen, M.D., Yerk, G., Phillips, R.L., Kraemer, S. and Srienc, F. (1992) Endosperm mitotic activity and endoreduplication in maize affected by defective kernel mutations. *Genome* 35, 68–77.

Letchworth, M.B. and Lambert, R.J. (1998) Pollen parent effects on oil, protein, and starch concentration in maize kernels. *Crop Science* 38, 363–367.

Letham, D.S. (1963) Zeatin, a factor inducing cell division from *Zea mays*. *Life Sciences* 8, 569–573.

Li, B., Liu, H., Zhang, Y., Kang, T., Zhang, L., *et al.* (2013) Constitutive expression of cell wall invertase genes increases grain yield and starch content in maize. *Plant Biotechnology Journal* 11, 1080–1091.

Li, N. and Li, Y. (2015) Maternal control of seed size in plants. *Journal of Experimental Botany* 66, 1087–1097. DOI:10.1093/jxb/eru549

Li, N., Zhang, S., Zhao, Y., Li, B. and Zhan, J. (2011a) Over-expression of AGPase genes enhances seed weight and starch content in transgenic maize. *Planta* 233, 241–250.

Li, Q., Li, L., Yang, X., Warburton, M.L., Bai, G., *et al.* (2010a) Relationship, evolutionary fate and function of two maize co-orthologs of rice GW2 associated with kernel size and weight. *BMC Plant Biology* 10, 1–15.

Li, Q., Yang, X., Bai, G., Warburton, M.L., Mahuku, G., *et al.* (2010b) Cloning and characterization of a putative GS3 ortholog involved in maize kernel development. *Theoretical and Applied Genetics* 120, 753–763.

Li, Y., Fan, C., Xing, Y., Jiang, Y., Luo, L., *et al.* (2011b) Natural variation in GS5 plays an important role in regulating grain size and yield in rice. *Nature Genetics* 43, 1266–1269.

Liu, J., Deng, M., Guo, H., Raihan, S., Luo, J., *et al.* (2015) Maize orthologs of rice GS5 and their trans-regulator are associated with kernel development. *Journal of Integrative Plant Biology* 57, 943–953.

Liu, Y., Wang, L., Sun, C., Zhang, Z., Zheng, Y. and Qiu, F. (2014) Genetic analysis and major QTL detection for maize kernel size and weight in multi-environments. *Theoretical and Applied Genetics* 127, 1019–1037.

Locascio, A., Roig-Villanova, I., Bernardi, J. and Varotto, S. (2014) Current perspectives on the hormonal control of seed development in Arabidopsis and maize: a focus on auxin. *Frontiers in Plant Science* 5, 412.

Ludwig, Y., Zhang, Y. and Hochholdinger, F. (2013) The maize (*Zea mays* L.) *AUXIN/INDOLE-3-ACETIC ACID* gene family: phylogeny, synteny, and unique root-type and tissue-specific expression patterns during development. *PLOS ONE* 8, e78859.

Luo, M., Bilodeau, P., Dennis, E.S., Peacock, W.J. and Chaudhury, A. (2000) Expression and parent-of-origin effects for FIS2, MEA, and FIE in the endosperm and embryo of developing Arabidopsis seeds. *Proceedings of the National Academy of Sciences of the United States of America* 97, 10637–10642.

Maitz, M., Santandrea, G., Zhang, Z., Lal, S., Hannah, L.C., Salamini, F. and Thompson, R.D. (2000) rgf1, a mutation reducing grain filling in maize through effects on basal endosperm and pedicel development. *The Plant Journal* 23, 29–42.

Martin, A., Lee, J., Kichey, T., Gerentes, D., Zivy, M., *et al.* (2006) Two cytosolic glutamine synthetase isoforms of maize are specifically involved in the control of grain production. *Plant Cell* 18, 3252–3274.

McMullen, M.D., Kresovich, S., Villeda, H.S., Bradbury, P., Li, H., *et al.* (2009) Genetic properties of the maize nested association mapping population. *Science* 325, 737–740.

Miller, M.E. and Chourey, P.S. (1992) The maize invertase-deficient miniature-1 seed mutation is associated with aberrant pedicel and endosperm development. *Plant Cell* 4, 297–305.

Miller, N.D., Haase, N.J., Lee, J., Kaeppler, S.M., de Leon, N. and Spalding, E.P. (2016) A robust, high-throughput method for computing maize ear, cob, and kernel attributes automatically from images. *The Plant Journal* 89, 169–178. DOI:10.1111/tpj.13320

Neuffer, M.G. and Sheridan, W.F. (1980) Defective kernel mutants of maize. I. Genetic and lethality studies. *Genetics* 95, 929–944.

Phillips, A.R. and Evans, M.M.S. (2011) Analysis of stunter1, a maize mutant with reduced gametophyte size and maternal effects on seed development. *Genetics* 187, 1085–1097.

Pletsch-Rivera, L.A. and Kaeppler, S.M. (2007) Phosphorus accumulation in maize kernel is not influenced by xenia (*Zea mays* L.). *Maydica* 52, 151–157.

Riefler, M., Novak, O., Strnad, M. and Schmülling, T. (2006) *Arabidopsis* cytokinin receptor mutants reveal functions in shoot growth, leaf senescence, seed size, germination, root development, and cytokinin metabolism. *Plant Cell* 18, 40–54.

Rijavec, T., Kovač, M., Kladnik, A., Chourey, P.S. and Dermastia, M. (2009) A comparative study on the role of cytokinins in caryopsis development in the maize *miniature1* seed mutant and its wild type. *Journal of Integrative Plant Biology* 51, 840–849.

Rijavec, T., Li, Q.-B., Dermastia, M. and Chourey, P.S. (2012) Cytokinins and their possible role in seed size and seed mass determination in maize. In: Montanaro, G. and Dichio, B. (eds.) *Advances in Selected Plant Physiology Aspects.* InTech, Rijeka, Croatia, pp. 293–308. DOI:10.5772/33076

Romay, M.C., Millard, M.J., Glaubitz, J.C., Peiffer, J.A., Swarts, K.L., *et al.* (2013) Comprehensive genotyping of the USA national maize inbred seed bank. *Genome Biology* 14, 1–18.

Sabelli, P.A. and Larkins, B.A. (2009) The contribution of cell cycle regulation to endosperm development. *Sexual Plant Reproduction* 22, 207–219.

Schruff, M.C., Spielman, M., Tiwari, S., Adams, S., Fenby, N. and Scott, R.J. (2006) The *AUXIN RESPONSE FACTOR 2* gene of *Arabidopsis* links auxin signalling, cell division, and the size of seeds and other organs. *Development* 133, 251–261.

Seka, D. and Cross, H.Z. (1995) Xenia and maternal effects on maize kernel development. *Crop Science* 35, 80–85.

Sekhon, R.S., Hirsch, C.N., Childs, K.L., Breitzman, M.W., Kell, P., *et al.* (2014) Phenotypic and transcriptional analysis of divergently selected maize populations reveals the role of developmental timing in seed size determination. *Plant Physiology* 165, 658–669.

Song, X.-J., Huang, W., Shi, M., Zhu, M.-Z. and Lin, H.-X. (2007) A QTL for rice grain width and weight encodes a previously unknown RING-type E3 ubiquitin ligase. *Nature Genetics* 39, 623–630.

Springer, N.M. and Gutiérrez-Marcos, J.F. (2009) Imprinting in maize. In: Bennetzen, J.L. and Hake, S. (eds.) *Handbook of Maize: Genetics and Genomics.* Springer, New York, pp. 429–440.

Sundaresan, V. (2005) Control of seed size in plants. *Proceedings of the National Academy of Sciences of the United States of America* 102, 17887–17888.

Tanaka, W. and Maddonni, G.A. (2008) Pollen source and post-flowering source/sink ratio effects on maize kernel weight and oil concentration. *Crop Science* 48, 666–677.

Tian, F., Bradbury, P.J., Brown, P.J., Hung, H., Sun, Q., *et al.* (2011) Genome-wide association study of leaf architecture in the maize nested association mapping population. *Nature Genetics* 43, 159–162.

Tsai, C.L. and Tsai, C.Y. (1990) Endosperm modified by cross-pollinating maize to induce changes in dry matter and nitrogen accumulation. *Crop Science* 30, 804–808.

Vilhar, B., Kladnik, A., Blejec, A., Chourey, P.S. and Dermastia, M. (2002) Cytometrical evidence that the loss of seed weight in the *miniature1* seed mutant of maize is associated with reduced mitotic activity in the developing endosperm. *Plant Physiology* 129, 23–30.

Wang, B., Chen, Y., Guo, B., Kabir, M.R., Yao, Y., *et al.* (2014) Expression and functional analysis of genes encoding cytokinin receptor-like histidine kinase in maize (*Zea mays* L.). *Molecular Genetics and Genomics* 289, 501–512.

Waters, A.J., Bilinski, P., Eichten, S.R., Vaughn, M.W., Ross-Ibarra, J., Gehring, M. and Springer, N.M. (2013) Comprehensive analysis of imprinted genes in maize reveals allelic variation for imprinting and limited conservation with other species. *Proceedings of the National Academy of Sciences of the United States of America* 110, 19639–19644.

Xin, M., Yang, R., Li, G., Chen, H., Laurie, J., *et al.* (2013) Dynamic expression of imprinted genes associates with maternally controlled nutrient allocation during maize endosperm development. *Plant Cell* 25, 3212–3227.

Yang, J., Jiang, H., Yeh, C.-T., Yu, J., Jeddeloh, J.A., Nettleton, D. and Schnable, P.S. (2015) Extreme-phenotype genome-wide association study (XP-GWAS): a method for identifying trait-associated variants by sequencing pools of individuals selected from a diversity panel. *The Plant Journal* 84, 587–596.

Yang, N., Lu, Y., Yang, X., Huang, J., Zhou, Y., *et al.* (2014) Genome wide association studies using a new nonparametric model reveal the genetic architecture of 17 agronomic traits in an enlarged maize association panel. *PLOS Genetics* 10, e1004573.

Zhang, X., Hirsch, C.N., Sekhon, R.S., de Leon, N. and Kaeppler, S.M. (2016) Evidence for maternal control of seed size in maize from phenotypic and transcriptional analysis. *Journal of Experimental Botany* 67, 1907–1917.

Zhang, Z., Liu, Z., Hu, Y., Li, W., Fu, Z., *et al.* (2014) QTL analysis of kernel-related traits in maize using an immortalized $F_2$ population. *PLOS ONE* 9, e89645.

# 17 Effects of Drought Stress on Maize Kernel Set


Jeffrey E. Habben* and Jeffrey R. Schussler

*Research and Development, DuPont-Pioneer, Johnston, Iowa, USA*


## 17.1 Introduction

For centuries, humans have depended on maize for sustenance, and thus have endeavored to increase its productivity. During most of this time grain yield was determined by the maize grower, who selected the most "attractive" ears after each harvest as a source of seed for the following year (Crabb, 1947). Unfortunately, even though this process provided a ready source of seed, yields were not significantly improved (Duvick, 2005). In the early 1900s, this scheme changed when hybrid maize was developed by professional geneticists and plant breeders; thereafter, grain yields began to increase steadily. Selection for improved reproductive resilience, combined with better agronomics, resulted in maize germplasm with higher yield potential and greater yield stability. In 2014, for the first time in recorded history, the winner of the U.S. National Corn Growers Association contest exceeded 500 bushels acre$^{-1}$ (31 Mg ha$^{-1}$) (Jeschke, 2016). This noteworthy achievement was bettered in 2015 with a certified yield of 532 bushels acre$^{-1}$ (33 Mg ha$^{-1}$). In addition to these record yields, in the years 2013–2015, 15 other entries exceeded 400 bushels acre$^{-1}$

(24 Mg ha$^{-1}$). Interestingly, these record yields defied the predictions of some academics who posited that given the modeled limits of maize physiology, the maximum grain yield would peak around 320 bushels acre$^{-1}$ (20 Mg ha$^{-1}$) (e.g. Sinclair, 2011). Despite the record grain yields produced by contest winners, the average maize yield in the United States is approximately threefold less. In 2015, for example, the mean yield was only 169 bushels acre$^{-1}$ (11 Mg ha$^{-1}$). Why is there such a large gap between the current maximum grain yield and that realized by the average maize grower in the U.S.? Why is the average U.S. maize yield not more in line with record yields? There are numerous explanations for this, and a comprehensive examination of them extends beyond the scope of this chapter. Therefore, we focus on one of the major constraints— drought stress—and confront the key questions that persist around water limitations to grain yield in maize.

Drought stress is considered the most consistent environmental factor that negatively impacts the grain yield of maize (Boyer, 1982). In a classic study by Shaw (1983), it was determined that the greatest decrease in yield in the U.S. Corn Belt occurred when


*Corresponding author e-mail: jeffrey.habben@pioneer.com

maize was drought-stressed around anthesis (flowering), followed by drought stress during the grain fill phase, with yield being least affected by a drought during early vegetative development. This result was reinforced by Campos *et al.* (2004), who demonstrated more sensitivity of maize to water deficits occurring at flowering and early grain fill, compared to a terminal grain fill stress. The drought-stress treatments bracketing flowering and early grain fill significantly impacted kernel set and tip kernel abortion, while the terminal drought-stress impacted primarily kernel mass after establishment of final kernel number per ear. From a temporal perspective, we can conclude that the greatest impact of drought stress on grain yield is during reproductive growth and this phase provides the most logical target for improvements to drought tolerance of maize (Araus *et al.*, 2012).

Final maize grain yield is the product of three female reproductive factors: the number of ears per plant × kernels per ear × mass per kernel. In the early years of hybrid maize use, a crop exposed to drought stress was more likely to exhibit female (ear) barrenness (Duvick *et al.*, 2004). However, because of strong selection pressure applied by maize breeders over the decades, fertile ear number is more stable and one ear per plant is the norm for most hybrids grown across a diversity of environments. Significant barrenness in modern germplasm only occurs under extreme abiotic stress conditions. Multiple studies have shown that of the two remaining yield components, kernels per ear is the key parameter to improve grain yield under drought stress. For example, Otegui *et al.* (1995) showed that approximately 85% of the variability in grain yield under drought stress is related to the number of kernels produced per acre, while the remaining 15% is related to kernel mass. Similarly, when drought was applied prior to or during the maize flowering period, kernel number was most closely related to final grain yield (Schussler and Westgate, 1994). Thus, from a spatial perspective a priority target for increasing grain yield is enhancing kernel set on plants exposed to drought stress.

Before exploring control points of kernel set under drought stress, it is important to understand how ovaries form on ears under favorable environmental conditions. The female inflorescence (ear) develops from lateral meristems in leaf axils. The basic unit of the ear inflorescence is the pistillate spikelet, composed of a floret subtended by a pair of glumes. Development of the ear is defined by several meristem transitions starting with spikelet pair meristems (SPMs) arising from the flanks of each inflorescence meristem (IM). Each SPM gives rise to two spikelet meristems (SMs) that terminate as a floral meristem (FM) (Cheng *et al.*, 1983; Tanaka *et al.*, 2013). The FM produces different organ primordia, including the gynoecium (female reproductive structure) which terminates in formation of an ovary composed of the embryo sac and the silk. After silks exert from the husk, the female florets are fertilized by pollen shed from the tassel, and kernel development commences (Chapter 2).

The ontogeny of ovary and kernel development is such that there is a strong temporal gradient along the ear (rachis), starting at the base and progressing to the tip (apical end). This acropetal development of reproductive structures is not unique to maize, and has been observed in many other species (Bangerth, 1989). It is likely this gradient provides flexibility in female fecundity, such that resources (minerals, water, and assimilates) are prioritized for development of a minimum number of basal kernels under resource-limited conditions; but in highly favorable environments kernels on the remainder of the ear continue to develop and mature to full size. In modern, intensively managed maize, reduction of the temporal gradient in floret development along the ear is desirable, so that kernel set in water-limiting environments might be improved. The extent of kernel set during flowering establishes the upper limit of ear yield potential. Of course, ongoing resource supply is required to fill kernels for the ear to express its maximum yield potential. But one cannot expect to attain maximum grain yield potential without starting with strong sink (ear) strength at flowering (Chapter 15).

Drought stress reduces the resources available to support normal ear establishment and kernel development in maize. These resources include both carbon and nitrogen assimilates, as well as the water and minerals necessary to support cell division and expansion of the ear. The integrated response of maize to these resource limitations is a consequence of many regulatory decision points at the molecular, biochemical, tissue, and organ levels. Minimizing rate-limiting processes at any one of these levels could potentially reduce the impact of drought stress on grain yield.

In this chapter we pose critical questions that persist regarding factors that limit maize kernel set during exposure to drought stress conditions, and we augment these questions with specific insights.

## 17.2 Question 1: Is there Opportunity for Improving Floral Synchrony to Enhance Kernel Set under Drought Stress?

The anthesis-to-silking interval (ASI) is the difference in time between pollen shed and silk exertion from the husk. If pollen shed (at least one anther dehiscing pollen) and silk exertion (at least one silk exerting from the husk) of 50% of the plants in a population (or plot) occur on the same day, the ASI = 0. If silking occurs prior to pollen shed, then the ASI is negative, while silking that occurs after shed results in a positive ASI. ASI can be expressed in chronological days, or in growing degree day units. Under water-limiting conditions, there is a predisposition for maize to have an increased ASI, as there is typically a delay in silk exertion relative to pollen shed. This differential response is likely a reflection of the tassel forming from the apical meristem and the ear originating from an axillary meristem. In other words, there appears to be enhanced apical dominance in maize exposed to drought-stress conditions prior to and during pollination. Through selection, breeders have done an excellent job in decreasing the overall apical dominance of the modern maize plant (so-called feminizing the plant). Interestingly, this reduction in apical dominance is a result of intensive selection for increased grain yield, not direct selection for diminished tassel size (Duvick and Cassman, 1999).

The inherent dominance of tassel (apical) over ear (axillary) development limits the sink strength of the ear during the critical flowering window, when the number of kernels per ear is determined (Loussaert et al., 2017). Sink strength is the product of sink size × sink activity (Ho, 1988). Compared to its size at maturity, the ear is relatively small at the time of first silk exertion (commonly, 1 to 4 g dry weight on the day of silk exertion versus >100 grams dry weight at maturity (Schussler and Westgate, 1994). Sink activity is difficult to measure at this early stage, but there is evidence that indicates utilization of incoming assimilates can be limited by reductions in ovary water potential $(\psi_w)$ (Schussler and Westgate, 1991; Zinselmeier, 1991). Thus, the product of a small sink size and reduced sink activity under water-limited conditions may be substantially due to decreased sink strength, and improvement of sink strength, either through increasing sink size or activity, would be appropriate targets for enhanced kernel set under abiotic stress conditions. Thus, for Question 1 we would strongly argue that additional improvements can be made to enhance kernel set in elite commercial maize hybrids via improved floral synchrony. What follows are insights to help underpin this question.

### 17.2.1 Insight 1: continue selection for negative ASI

Selection for a reduced ASI has been accomplished across many breeding programs over a number of decades (Edmeades et al., 2000; Duvick et al., 2004), and is associated with more rapid plant and ear growth rates (Borras et al., 2009). The improved synchrony of silk exertion with pollen shed has been associated with improved yield and yield stability across many environments. Silk exertion prior to pollen shed (protogyny) facilitates a naturally synchronous pollination of a higher proportion of ovary cohorts, and presumably would minimize the negative impact of basal ovary pollination on acropetal ovary cohorts. This creates the opportunity for greater sink strength and more kernels per ear.

Currently, drought tolerant hybrids may express an ASI of 0 to –1 under favorable conditions and an ASI of +1 to +3 days under water-limiting conditions. In contrast, drought susceptible hybrids can exhibit ASI values of +4 to +8 days under drought stress, thus severely limiting kernel set. Undoubtedly this was also the case with early generation hybrids, prior to targeted selection for improved drought tolerance and reduced ASI (Barker *et al.*, 2005).

When moderate drought stress occurs prior to or during silk exertion, drought tolerant hybrids with a naturally reduced or even negative ASI are better able to manage the inherent delay in silking that occurs with plant water deficits (Oury *et al.*, 2016a,b). Thus, if hybrids were selected to have a –2 day ASI in favorable conditions, they would be expected to shift to a 0 to +2 day ASI under moderate drought-stress conditions. This modest shift in ASI under stress could then minimize the loss in kernels per ear, since silk exertion is adequate to enable the majority of ovaries to become fertilized.

Ongoing selection for a negative ASI should have positive results for improving tolerance to drought stress at flowering (Westgate, 1997). Drought stress occurring prior to flowering impacts female reproductive development, but has relatively little impact on timing of tassel development and pollen shed. In some cases, pollen shed may actually occur slightly earlier under drought stress. Selection of hybrids that silk up to 4 days prior to pollen shed under favorable conditions should pose very little downside risk, since under favorable conditions silks do not begin to senesce until 7 to 8 days after exertion from the husk (Bassetti and Westgate, 1993a). However, it is likely that selection for a negative ASI should not proceed beyond –4 days, due to the increased risk of some of the earliest exerting silks losing their viability. In the presence of moderate drought stress, it is unlikely that the ASI in selected hybrids would be more negative than –1 day, and silks with reduced water potential should not senesce for at least 2–3 days after initial exertion (Bassetti and Westgate, 1993b), thus providing an overlap of more viable silks with optimum pollen shed. Initially, selection for a negative ASI in germplasm can be conducted under favorable environments (Westgate, 1997), since they are more easily managed and allow for practical high-throughput phenotyping. However, validation of the best negative ASI selections should then be conducted in drought-stress environments, preferably at managed stress environment locations (Cooper *et al.*, 2014; Barker *et al.*, 2005), where the probability of a discriminating drought stress at flowering is enhanced.

### 17.2.2 Insight 2: exploit selection for enhanced silking kinetics

ASI represents one direct visual method of assessing silking activity, and significant genetic variation in this trait allows for effective selection in corn breeding populations (Edmeades *et al.*, 2000). The silking component of ASI is determined by identifying the date when silks first exert from husks, but does not account for the rate of silk exertion of subsequent silk cohorts on the ear. Thus, the rate of silk exertion across all cohorts on an ear represents another level of genetic variation to select for silking efficiency. Quantification of exerted silk number during the effective pollination window is a fine-tuned method to select for improved kernel set under drought. This is a direct measure of the ability of a genotype to overcome the first limitation of kernel set, i.e. silk exertion. High-throughput collection of exposed silks from a diversity of hybrids exposed to defined levels of plant water deficits, followed by rapid quantification of silks via digital analysis, provides the data required to screen for germplasm that can maintain more aggressive silking during drought stress (Anderson *et al.*, 2010). This technique has been successfully used to characterize vastly different silk exertion kinetics of drought-tolerant versus drought-susceptible hybrids in flowering stress environments. For example, it was found that quantification of exerted silk number is highly correlated with the final number of kernels per ear as well as yield (g per plant) in a drought-stress environment (Fig. 17.1). As an optional phenotypic measure, Oury *et al.* (2016b) reported an association between leaf and silk growth, and they suggest that leaf growth rate could be used as a surrogate trait to select for enhanced silking dynamics under drought stress. Use of these approaches, or others, could help identify genotypes that more effectively exert silks under flowering drought stress.

### 17.2.3   Insight 3: manipulate pollination timing to improve kernel development synchrony

The pattern of silk exertion among ovary cohorts is well-documented (Oury *et al.*, 2016a), and generally proceeds from base to tip of the ear. Synchrony of silk exertion and pollen shed from tassels is nearly always adequate for optimum kernel set in the basal one third of the ear. When moderate to severe plant water deficits occur at the time of first silk exertion, however, exertion of the more apical silk cohorts will be delayed such that some will never extend past the husks and others that do exert will not be pollinated due to arrested development (Oury *et al.*, 2016a). It has been proposed that under drought stress, carbon assimilate availability may limit the development of acropetal ovary cohorts (Westgate and Boyer, 1986; Zinselmeier, 1991). Alternatively, reduced water potential of the ovaries themselves may have a direct negative impact on growth and metabolism in these organs (Schussler and Westgate, 1991; Zinselmeier, 1991).

With typical commercial plant populations in well-watered environments, manually controlled synchronous pollination can increase kernel numbers per ear by 20% or more (Carcova *et al.*, 2000). Thus, kernel set can be increased significantly with the same amount of C and N assimilates and,

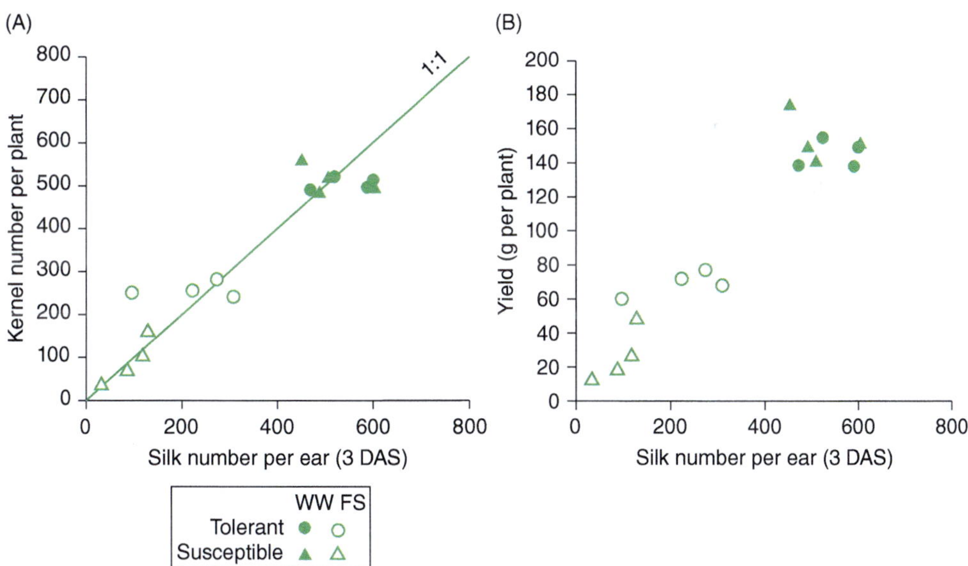

Fig. 17.1. (A) Silks exerted at 3 days after silking (DAS) versus final kernel number per plant. Four drought tolerant and four drought susceptible hybrids were evaluated. Performance was compared in both a well-watered (WW) and flowering stress (FS) environment. For FS, water was withheld 4 weeks prior to silking, plots were fully rehydrated at 7 days after silking, and then were fully irrigated throughout grain fill. (B) Silks exerted at 3 days after silking (DAS) versus final yield (g per plant). Silks were cut with a silk cutting device and counted via digital analysis. Day 0 DAS = the day silks were first observed in 50% of plants in the plot. Silks were cut and counted at 3 DAS.

presumably, with water availability remaining constant. Schussler *et al.* (2002) demonstrated that synchronous pollination could also increase kernel number per ear up to 23% under water-limiting conditions; this result was associated with a yield increase of 16% even at a high plant population (110,000 plants ha$^{-1}$ or 45,000 plants acre$^{-1}$) (Fig. 17.2). In this experiment, a transient drought stress was imposed before flowering, reducing the rate of silk exertion at the time of flowering, and pollination was delayed for 3–4 days by covering the ears with shoot bags, followed by synchronously pollinating all exposed silks by hand with excess pollen. At 7 days after manual synchronous pollination, all plants were rehydrated so that source activity was largely recovered to pre-stress levels. This enhancement of sink size (as kernels per ear) during stress became valuable when the source activity was recovered via irrigation, an effect that mimics natural environments where transient flowering stress is relieved by rainfall early in grain fill. If rainfall does not occur after flowering to revive source activity, any increase in sink size with synchronous pollination would likely not be beneficial due to kernel mass reduction from limited assimilate supply.

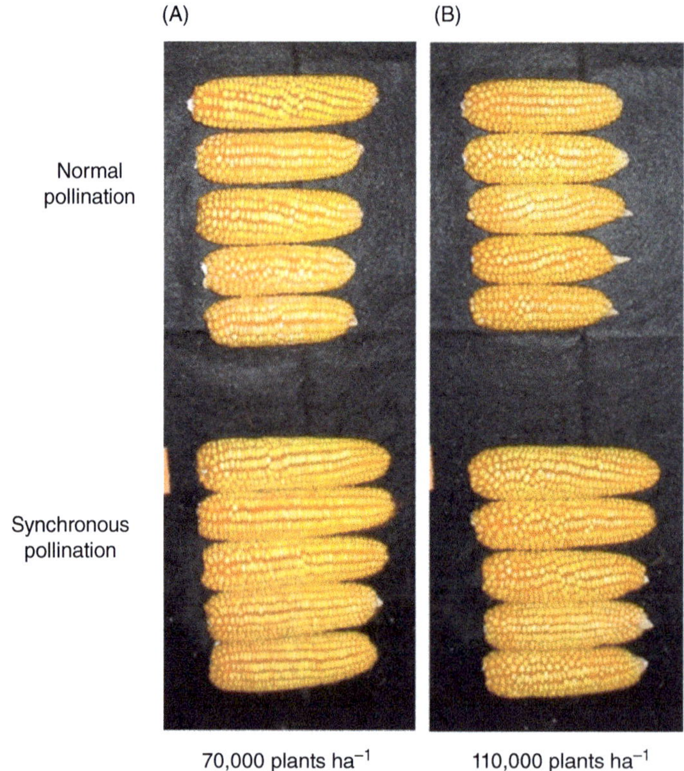


**Fig. 17.2.** Impact of delayed synchronous pollination on kernel set of maize under drought-stress conditions at flowering. Water was withheld 4 weeks prior to onset of silk exertion. In the normal pollination treatment, ears were open pollinated within the solid stand of the hybrid. In the synchronous pollination treatment, ears were covered with a shoot bag prior to silk exertion. At 3 to 4 days after first silk exertion of each ear, bags were removed and fresh viable pollen was used to hand pollinate all exerted silks. Silks were left exposed for any possible natural pollination after initial bag removal. Plants were grown in two densities: (A) 70,000 plants ha$^{-1}$ (28,000 plants acre$^{-1}$) and (B) 110,000 plants ha$^{-1}$ (45,000 plants acre$^{-1}$). The severity of stress tended to be greater at the higher plant population and a proportional increase in kernel numbers per ear via synchronous pollination tended to be greater at this population.

These controlled pollination studies suggest that a regulatory mechanism is in place that inhibits the development of acropetal ovaries/kernels once successful pollination has occurred in more basal cohorts. This response is likely related to an evolutionary survival process that ensures a limited number of "fit" individuals survive at the expense of later developing cohorts. Unfortunately, this process is counterproductive for optimum grain yield in modern-day intensely managed high-production fields. Thus, a critical question for this relationship is: What process induces kernel abortion in ear tips under drought stress? Is a deleterious growth factor released from early pollinated basal ovary cohorts that negatively impacts late-emerging silk cohorts in the top half of the ear (Carcova et al., 2000; Oury et al., 2016a)? It has been hypothesized that the inhibition of more acropetal cohorts is due to reduced water potential, assimilate starvation, and/or hormone imbalance. The net effect is fewer silks exerting from husks under drought-stress conditions, and/or the irreversible loss of kernels to accept assimilates for yield formation resulting in tip-kernel abortion. If this factor(s) could be identified and manipulated, grain yield under transient drought-stress environments could be enhanced, resulting in improved grain yield for maize growers in areas with variable rainfall patterns. Given our ability to experimentally manipulate the acropetal phenomenon in the ear, combined with the availability of a plethora of omics tools, identification of this mechanism could have real benefits that not only increase maize grain yields, but also increase the seed yield in other crops where this phenomenon occurs.

### 17.2.4 Insight 4: alter phytohormone pathways to improve silk/kernel development

Given the role of phytohormones in controlling multiple aspects of plant growth and development, their modulation provides an opportunity to improve silk exertion and/or kernel set. Unlike trait selection via plant breeding, modifying plant hormone synthesis, transport and/or signaling assumes that a particular rate-limiting step has been identified and can be altered. What follows are a few examples of published research using this approach.

Ethylene has been linked to various aspects of reproductive growth in maize (Cheng and Lur, 1996; Young et al., 1997; Feng et al., 2011). Transgenic modification of elite maize hybrids with a Zm-ACS6 (a gene that encodes a protein involved in ethylene biosynthesis) RNAi construct resulted in transgenic events with reduced (~50%) ethylene emission in leaf tissues (Habben et al., 2014). When events were evaluated in a field flowering-stress environment, which significantly reduced silk exertion, they yielded up to 0.58 Mg ha$^{-1}$ (9.3 bushels acre$^{-1}$) more than the wild-type comparator. A small reduction in ASI combined with an increase in kernel number per ear suggested that the reduction in ethylene biosynthesis in the transgenic events improved silk exertion in the drought-stress environment. A positive yield benefit was also observed under the different abiotic stress condition of low nitrogen (N); when water was not limiting, reduced N levels were associated with a reduction in silk exertion. The importance of the ethylene pathway in maize abiotic stress was confirmed when a transgene involved in ethylene signaling (Zm-ARGOS8) was constitutively overexpressed, resulting in a modulation of ethylene signaling (Shi et al., 2015). When the transgenic events were field tested, their grain yield was increased under both drought stress and well-watered field conditions, relative to the non-transgenic control. To further verify the relationship between manipulation of the ethylene pathway and grain yield increases, Shi et al. (2016) employed a CRISPR-Cas-enabled advanced breeding technology to generate novel variants of Zm-ARGOS8. A native maize promoter (GOS2), which confers a moderate level of constitutive expression, was inserted into the 5'-untranslated region of the native Zm-ARGOS8 gene or was used to replace the native promoter of Zm-ARGOS8. A field study showed that relative to the comparator, both of the Zm-ARGOS8 variants increased grain yield by approximately five bushels per acre under 0.31 Hgha⁻¹ flowering stress conditions with no yield loss under well-watered conditions.

Abscisic acid (ABA) is another well-studied phytohormone linked to drought stress. Increased accumulation of abscisic acid induced by plant water deficits has been associated with alterations in post-pollination kernel development and kernel abortion in maize (Ober *et al.*, 1991; Ober and Setter, 1992; Setter *et al.*, 2001). Water deficits increase the accumulation of ABA in developing kernels, and this was most dramatic in apical kernels where greater arrested development/increased abortion typically occurs under these conditions (Ober *et al.*, 1991). There is evidence that the source of ABA accumulation in kernels is primarily leaves (Ober and Setter, 1990). This would provide a logical regulatory connection between the plant's response to water deficit conditions (leaf rolling, reduced leaf expansion, and reduced photosynthetic rates) and subsequent sink size reductions in the ear. This "abortion" signal may be partially responsible for post-pollination loss of kernels, which occurs predominantly in the apical region and may be expressed primarily through an inhibition of endosperm cell division (Ober *et al.*, 1991). While this relationship has not been conclusively demonstrated, it remains a clear target for transgenic modification or genome editing. Cell division and protein synthesis in pollinated kernels may be sensitive to ABA accumulation, and so reductions in synthesis or signaling of ABA in kernel tissues could improve kernel set under abiotic stress.

In contrast to ABA, enhanced cytokinin levels are believed to play a positive role in maintaining sink activity in maize kernels following pollination (reviewed in Jones and Setter, 2000). Peak levels of cytokinin typically occur at 8–12 days after pollination (Cheikh and Jones, 1994), coinciding with the peak in mitotic index during the lag phase of kernel development (see Chapter 10, this volume). Cytokinins appear to be critical for thermal stability and maintaining cell division during the lag phase.

Unlike ABA, *de novo* cytokinin synthesis does occur in kernels (Schreiber and Jones, 1995). Heat stress treatments can reduce accumulation of key cytokinins, resulting in greater kernel abortion (Cheikh and Jones, 1994). Stem infusion of the synthetic cytokinin benzyl adenine into maize plants subjected to heat stress at pollination substantially reduces the amount of kernel abortion by improving sink activity (Cheikh and Jones, 1994); this is most likely the consequence of maintaining endosperm cell division and amyloplast biogenesis. Manipulation of cytokinin levels in apical kernels, and/or improving the ratio of cytokinin to ABA are possible targets for manipulation that would lead to better sink establishment for maize plants growing in water-limited environments.

Nuccio *et al.* (2015) reported the impact of overexpression of rice trehalose-6-phosphate phosphatase on maize yield after a flowering drought stress. Ectopic expression of this transgene increased the concentration of sucrose in ear spikelets and resulted in more rapid silking, increased kernel number per ear, and increased grain yield (an average of 0.50 Mg ha⁻¹ (8.0 bushels acre⁻¹) in five drought-stress environments. The researchers' explanation of the proposed mechanism for increased yield is not directly linked to ovary or silk water status, but rather is due to enhanced assimilate flux into the ovaries. In summary, numerous physiological, transgenic, and pharmacological studies demonstrate the potential and utility for manipulating plant hormone pathways to improve silk exertion and kernel set in maize.

## 17.3 Question 2: Can Changes in Maize Water Utilization Enhance Kernel Set under Drought Stress?

Oury *et al.* (2016a) demonstrated that ovary and silk growth are restricted by water-induced expansive growth, rather than a carbon limitation during drought stress. Reduced water availability caused a loss of $\psi_w$ in silks and ovaries significant enough to arrest development (Schussler and West-gate, 1991; Bassetti and Westgate, 1993b), or trigger rapid post-pollination abortion within 2 days after pollination (Schussler and Westgate, 1994). Schussler and Westgate (1991) demonstrated that a decreased ovary $\psi_w$ reduced the capacity of ovaries to take up and utilize $^{14}C$-sucrose *in vitro*, thus providing evidence for the direct impact of reduced $\psi_w$ on kernel set. This outcome suggests an advantage to a plant type that can increase water availability to the developing ear. But what mechanisms are available to enhance water availability to growing maize reproductive structures, particularly under drought stress?

### 17.3.1 Insight 1: reduce evapotranspiration during vegetative growth to allow for increased soil water availability during flowering

Does genetic variation exist in maize to conserve water prior to flowering and in effect "save" water for the critical flowering period? Genetic variation in maximum transpiration rates has been observed across different maize hybrids, especially at high vapor pressure deficit (VPD) levels (Gholipoor *et al.*, 2013; Messina *et al.*, 2015). Thus, indirect selection for reduced maximum transpiration during vegetative development, leading to a more favorable ovary and silk water status at the onset of flowering, should result in more aggressive silking under drought stress. With small reductions in maximum transpiration rate under high VPD (typically occurring during only part of the day), a maize hybrid can slightly reduce pre-anthesis water use. For example, in hybrids expressing a reduced transpiration trait, Messina *et al.* (2015) simulated pre-flowering water savings of 17–25 mm, and predicted yield improvements of $60\ g\ m^{-2}$ (10 bushels acre$^{-1}$) in environments with flowering and grain fill stress. It is likely that this yield response is a direct result of improved water availability for cell growth and expansion in reproductive organs at the time of flowering. Maintenance

of ovary growth and volume expansion is associated with more rapid and synchronous silk exertion (Oury *et al.*, 2016a) and thus greater kernel set under moderate drought-stress levels, which can occur in major maize growing regions. This represents a drought avoidance mechanism rather than drought tolerance per se, but the outcome to a maize grower is just as attractive. In fact, this approach has proven to be successful in a commercial breeding program in which a set of drought tolerant AQUAmax® hybrids were released. Through restricted transpiration, these hybrids conserve water in the soil profile, allowing greater water availability to support silk exertion during the flowering period (Cooper *et al.*, 2014; Gaffney *et al.*, 2015). The key to success of these hybrids is a water conservation phenotype that does not create yield drag (loss) under well-watered conditions (Gaffney *et al.*, 2015). Importantly, it remains to be determined how far this water conservation mechanism can be exploited before the reduction in transpiration induces yield drag under optimal growth conditions.

## 17.4 Question 3: Can Male Sterility Systems Increase Kernel Set under Drought Stress?

Besides fixed carbon and water, nitrogen assimilates are also critical to support normal maize reproductive development under abiotic stress conditions. Loussaert *et al.* (2017) completed a N balance study of maize and demonstrated that under limiting N conditions, the amount of nitrogen actually declined in ears of fertile plants during the R1 (silk reproductive stage) to R2 (kernel blister reproductive stage) growth stages. In contrast, N accumulated in tassels during these growth stages, suggesting prioritized N partitioning to the tassel, compared to the ear; again, this is likely a result of residual apical dominance. These findings argue for an approach to shunt N away from the tassel, to make it available to the developing ear.

### 17.4.1 Insight 1: use genetic male sterile germplasm to improve kernel set

Researchers hypothesized that a genetic male sterile (GMS) gene (*Ms44* allele) could be used to eliminate pollen production in the tassel, thus shunting N away from the tassel and making it available for ear growth and development (Loussaert *et al.*, 2017). Previously, CMS (cytoplasmic male sterile) hybrids had been shown to save 10 to 30 kg N ha$^{-1}$ (Weingartner *et al.*, 2002), a significant value especially in drought-stress or N-deficient environments, where N availability can be one of the primary limitations to grain yield. One could propose that GMS could have an even more significant impact on N savings, since GMS sterility is induced shortly after pollen tetrad release (Albertson and Phillips, 1981) versus CMS mutants where sterility is not determined until 10 days prior to anthesis (Weider *et al.*, 2009). When evaluated in field experiments, GMS plants increased exerted silk number, final kernel numbers per ear and grain yield under limited N and drought-stress environments, as well as in some favorable environments (Loussaert *et al.*, 2017). Accordingly, a seed production system was proposed that would generate a 50:50 blend of GMS:wild-type fertile that when planted in commercial fields would provide the opportunity to stabilize kernel set and grain yield in abiotic stress environments (Loussaert *et al.*, 2017).

### 17.4.2 Insight 2: use male sterility combined with xenia to enhance kernel set

Xenia is defined as the direct effect of an unrelated pollinator on developing kernels (Kiesselbach, 1960). Combinations of GMS or CMS with xenia provide an opportunity for even more robust yield stability in maize grown in abiotic stress environments. In this situation the yield component most likely to be affected is kernel mass. In commercial single-cross maize fields, the ovaries can be sib-pollinated and subject to inbreeding depression (pollinated kernels are the $F_2$ generation). In contrast, pollination by a non-related single-cross pollinator leads to double-cross hybrid kernels, which are assumed to have higher heterozygosity, compared to sib-pollinated $F_2$ kernels. This heterozygosity results in transient increases in the activity of carbohydrate metabolism enzymes in the growing kernels shortly after pollination, leading to increased sucrose and starch content in developing kernels (Bulant *et al.*, 2000). As a result, kernel mass increases of ~ 7% for hybrids (Bulant *et al.*, 2000) and 11–13% for inbreds (Bulant and Gallais, 1998) occurred, presumably due to improved sink strength.

A case study for the commercial potential of this combination of male sterility plus xenia has been developed. The Plus-Hybrid system combines CMS hybrids grown in a blend with male-fertile, non-isogenic pollinator hybrids (Weingartner *et al.*, 2002). In yield trials, the Plus-Hybrids provided a 9% increase in yield across all environments and all pollinators, with increases in both kernel numbers per ear and mass per kernel. Male sterility tends to increase the number of kernels per ear while the xenia effect primarily increases mass per kernel (Bulant and Gallais, 1998; Munsch *et al.*, 2010). In summary, utilization of a male sterile hybrid, blended with a non-related pollinator, appears to be a realistic method to improve energy reallocation within the maize plant not only in favorable conditions, but also in abiotic stress environments. The physiological mechanisms responsible for the improved productivity in this system would not be easily optimized within typical breeding programs. Validation of this approach remains to be demonstrated, but introgression of GMS into elite germplasm followed by wide area testing in favorable and abiotic stress environments to confirm additive commercial value seems justified.

## 17.5 Conclusion

In the next 50 years, the yield of maize and other cereals is predicted to become a more prominent issue, as the demand for grain will surge, driven by the increased size of the human population and increased risk to cereal production from climate change (Ray *et al.*, 2013). To date, tremendous progress has been made by both breeders and agronomists to increase the yield potential of maize under favorable conditions. We petition for an equally focused research effort in understanding and enhancing maize yield stability under water-limited conditions. Towards this end, we propose that the most critical target to improve grain yield stability is increased kernel set.

Intriguingly, when calculating the impact of kernel set on maize yield, one quickly realizes that even modest improvements in the number of kernels per ear can result in a significant positive effect on grain yield. At a standard U.S. Corn Belt plant population of 80,000 plants ha$^{-1}$ (32,000 plants acre$^{-1}$), one only needs to add one additional ring of kernels per ear (assuming 16 kernels per row) to increase grain yield by 0.31 Mg ha$^{-1}$ (5 bushels acre$^{-1}$). If the research community (both private and academic) can address and decipher the mechanisms controlling kernel set, it should be possible to consistently increase the number of kernel rings per ear under drought stress. The achievement of a step change in kernel set could lead to improved productivity and ultimately lessen the yield gap that exists between contest winners and the average yield of the typical maize grower.

# References


Albertson, M.C. and Phillips, R.L. (1981) Developmental cytology of 13 genetic male sterile loci in maize. *Canadian Journal of Genetics and Cytology* 23, 195–208.

Anderson, S.R., Farrington, R.L., Goldman, D.M., Hanselman, T.A., Hausmann, N.J., *et al.* (2010) Methods for counting corn silks or other plural elongated strands and use of the count for characterizing the strands or their origins. US Patent WO 2010/022346 A3.

Araus, J.J., Serret, M.D. and Edmeades, G.O. (2012) Phenotyping maize for adaptation to drought. *Frontiers in Physiology* 3, 305.

Bangerth, F. (1989) Dominance among fruits/sinks and the search for a correlative signal. *Physiologia Plantarum* 76, 608–614.

Barker, T., Campos, H., Cooper, M., Dolan, D., Edmeades, G., *et al.* (2005) Improving drought tolerance in maize. *Plant Breeding Reviews* 25, 173–253.

Bassetti, P. and Westgate, M.E. (1993a) Emergence, elongation and senescence of maize silks. *Crop Science* 33, 271–275.

Bassetti, P. and Westgate, M.E. (1993b) Water deficit affects receptivity of maize silks. *Crop Science* 33, 279–282.

Borras, L., Astini, J.P., Westgate, M.E. and Severini, A.D. (2009) Modeling anthesis to silking in maize using a plant biomass framework. *Crop Science* 49, 937–948.

Boyer, J.S. (1982) Plant productivity and environment. *Science* 218, 443–448.

Bulant, C. and Gallais, A. (1998) Xenia effects in maize with normal endosperm. I. Importance and stability. *Crop Science* 38, 1517–1525.

Bulant, C., Gallais, A., Matthys-Rochon, E. and Prioul, J.L. (2000) Xenia effects in maize with normal endosperm. II. Kernel growth and enzyme activities during grain filling. *Crop Science* 40, 182–189.

Campos, H., Cooper, M., Habben, J.E., Edmeades, G.O. and Schussler, J.R. (2004) Improving drought tolerance in maize: a view from industry. *Field Crop Research* 90, 19–34.

Carcova, J., Uribelarrea, M., Borras, L., Otegui, M.E. and Westgate, M.E. (2000) Synchronous pollination within and between ears improves kernel set in maize. *Crop Science* 40, 1056–1061.

Cheikh, N. and Jones, R.J. (1994) Disruption of maize kernel growth and development by heat stress. *Plant Physiology* 106, 45–51.

Cheng, P.C., Greyson, R.I. and Walden, D.B. (1983) Organ initiation and the development of unisexual flowers in the tassel and ear of *Zea mays*. *American Journal of Botany* 70, 450–462.

Cheng, C.-Y. and Lur, H.S. (1996) Ethylene may be involved in abortion of the maize caryopsis. *Physiology of Plants* 98, 245–252.

Cooper, M., Gho, C., Leafgren, R., Tang, T. and Messina, C. (2014) Breeding drought-tolerant maize hybrids for the US corn-belt: discovery to product. *Journal of Experimental Botany* 65, 6191–6204.

Crabb, A.R. (1947) *The Hybrid Corn Makers: Prophets of Plenty*. Rutgers University Press, New Brunswick, New Jersey.

Duvick, D.N. (2005) Genetic progress in yield of United States maize (*Zea mays* L.). *Maydica* 50, 193–202.

Duvick, D.N. and Cassman, K.G. (1999) Post-green revolution trends in yield potential of temperate maize in the north-central United States. *Crop Science* 39, 1622–1630.

Duvick, D.N., Smith, J.C.S. and Cooper, M. (2004) Long-term selection in a commercial hybrid maize breeding program. *Plant Breeding Review* 24, 109–151.

Edmeades, G.O., Bolanos, J., Elings, A., Ribaut, J.-M., Banziger, M. and Westgate, M.E. (2000) The role and regulation of the anthesis-silking interval in maize. In: Westgate, M.E. and Boote, K.J. (eds.) *Physiology and Modeling Kernel Set in Maize: CSSA Special Publication 29*. Crop Science Society of America, Madison, Wisconsin, pp. 43–73.

Feng, H.Y., Wang, Z.M., Kong, F.N., Zhang, M.J. and Zhou, S.L. (2011) Roles of carbohydrate supply and ethylene, polyamines in maize kernel set. *Journal of Integrative Plant Biology* 53, 388–398.

Gaffney, J., Schussler, J., Loffler, C., Cai, W., Paszkiewicz, S., *et al.* (2015) Industry-scale evaluation of maize hybrids selected for increased yield in drought-stress conditions of the US corn belt. *Crop Science* 55, 1608–1618.

Gholipoor, M., Sinclair, T.R., Raza, M.A.S., Loffler, C., Cooper, M. and Messina, C. (2013) Maize hybrid variability for transpiration decrease with progressive soil drying. *Journal of Agronomy and Crop Science* 199, 23–29.

Habben, J.E., Bao, X., Bate, N.J., DeBruin, J.L., Dolan, D., *et al.* (2014) Transgenic alteration of ethylene biosynthesis increases grain yield in maize under field drought-stress conditions. *Plant Biotechnology Journal* 12, 685–693.

Ho, L.C. (1988) Metabolism and compartmentation of imported sugars in sink organs in relation to sink strength. *Annual Review of Plant Physiology and Plant Molecular Biology* 39, 355–378.

Jeschke, M. (2016) Managing Corn for Greater Yield: Crop Insights. DuPont Pioneer Agronomy Sciences. Available at: https://www.pioneer.com/home/site/us/agronomy/library/corn-greater-yield (accessed June 6, 2017).

Jones, R.J. and Setter, T.L. (2000) Hormonal regulation of early kernel development. In: Westgate, M.E. and Boote, K.J. (eds.) *Physiology and Modeling Kernel Set in Maize: CSSA Special Publication 29*. Crop Science Society of America, Madison, Wisconsin, pp. 25–42.

Kiesselbach, T.A. (1960) The significance of xenia effects on the kernel weight of corn. *Nebraska Agricultural Station Research Bulletin* 191, 1–30.

Loussaert, D., DeBruin, J., San Martin, J.P., Schussler, J., Pape, R. *et al.* (2017) Genetic male sterility (Ms44) increases maize grain yield. *Crop Science* 57, September–October.

Messina, C.D., Sinclair, T.R., Hammer, G.L., Curan, D., Thompson, J., *et al.* (2015) Limited-transpiration trait may increase maize drought tolerance in the US corn belt. *Agronomy Journal* 107, 1978–1986.

Munsch, M.A., Stamp, P., Christov, N.K., Foueillassar, X., Husken, A., Camp, K.H. and Weider, C. (2010) Grain yield increase and pollen containment by Plus-Hybrids could improve acceptance of transgenic maize. *Crop Science* 50, 909–919.

Nuccio, M.L., Wu, J., Mowers, R., Zhou, H., Meghji, M., *et al.* (2015) Expression of trehalose-6-phosphate phosphatase in maize ears improves yield in well-watered and drought conditions. *Nature Biotechnology* 33, 862–874.

Ober, E.S. and Setter, T.L. (1990) Timing of kernel development in water stressed maize: water potentials and abscisic acid concentrations. *Annals of Botany* 66, 665–672.

Ober, E.S. and Setter, T.L. (1992) Water deficit induces abscisic acid accumulation in endosperm of maize *viviparous* mutants. *Plant Physiology* 98, 353–356.

Ober, E.S., Setter, T.L., Madison, J.T., Thompson, J.F. and Shapiro, P.S. (1991) Influence of water deficit on maize endosperm development. *Plant Physiology* 97, 154–164.

Otegui, M.E., Nicolini, M.G., Ruiz, R.A. and Dodds, P. (1995) Sowing date effects on grain yield components for different maize genotypes. *Agronomy Journal* 87, 29–33.

Oury, V., Tardieu, F. and Turc, O. (2016a) Ovary apical abortion under water deficit is caused by changes in sequential development of ovaries and in silk growth rate in maize. *Plant Physiology* 171, 986–996.

Oury, V., Caldeira, C.F., Prodhomme, D., Pichon, J.-P., Gibon, Y., Tardieu, F. and Turc, O. (2016b) Is change in ovary carbon status a cause or a consequence of maize ovary abortion in water deficit during flowering? *Plant Physiology* 171, 997–1008.

Ray, D.K., Mueller, N.D., West, P.C. and Foley, J.A. (2013) Yield trends are insufficient to double global crop production by 2050. *PLOS ONE* 8, e66428. Available at: http://dx.doi.org/10.1371/journal. pone.0066428 (accessed June 6, 2017).

Schreiber, B.M.N. and Jones, R.J. (1995) Incorporation of $^{14}$C-adenine into cytokinins in developing maize kernels. In: *1995 International Plant Growth Substances Association Abstracts*. IPGSA, St. Louis, Missouri, p. 150.

Schussler, J.R. and Westgate, M.E. (1991) Maize kernel set at low water potential. II. Sensitivity to reduced assimilates at pollination. *Crop Science* 31, 1196–1203.

Schussler, J.R. and Westgate, M.E. (1994) Increasing assimilate reserves does not prevent kernel abortion at low water potential in maize. *Crop Science* 34, 1569–1576.

Schussler, J.R., Edmeades, G.O., Campos, H., Wink, B. and Ibanez, M. (2002) Use of synchronous pollination to investigate kernel set in drought-stressed maize. *Annual Meeting Abstracts 2002 (CD-ROM)*. ASA-CSSA-SSSA, Madison, Wisconsin.

Setter, T.L., Flannigan, B.A. and Melkonian, J. (2001) Loss of kernel set due to water deficit and shade in maize: carbohydrate supplies, abscisic acid, and cytokinins. *Plant Physiology* 41, 1530–1540.

Shaw, R.H. (1983) Soil moisture stress prediction for corn in a western corn belt state. *Korean Journal of Crop Science* 28, 1–11.

Shi, J., Habben, J.E., Archibald, R.L., Drummond, B.J., Chamberlin, M.A., *et al.* (2015) Overexpression of ARGOS genes modifies plant sensitivity to ethylene, leading to improved drought tolerance in both Arabidopsis and maize. *Plant Physiology* 169, 266–282.

Shi, J., Gao, H., Wang, H., Lafitte, H.R., Archibald, R.L., *et al.* (2016) ARGOS8 variants generated by CRISPR-Cas9 improve maize grain yield under field drought stress conditions. *Plant Biotechnology Journal* 15, 207–216. DOI:10.1111/pbi.12603

Sinclair, T. (2011) Maximum Crop Yields. No longer available at: https://www.youtube.com/watch? v=MUkC64JyZxk (accessed June 6, 2017).

Tanaka, W., Pautler, M., Jackson, D. and Hirano, H.Y. (2013) Grass meristems II: inflorescence architecture, flower development and meristem fate. *Plant and Cell Physiology* 54, 313–324.

Weider, C., Stamp, P., Christov, N., Husken, A., Foueillassar, X., Camp, K.H. and Munsch, M. (2009) Stability of cytoplasmic male sterility in maize under different environmental conditions. *Crop Science* 49, 77–84.

Weingartner, U., Kaeser, O., Long, M. and Stamp, P. (2002) Combining cytoplasmic male sterility and xenia increases grain yield of maize hybrids. *Crop Science* 42, 1848–1856.

Westgate, M.E. (1997) Physiology of flowering in maize: identifying avenues to improve kernel set during drought. In: Edmeades, G.O., Banziger, M., Mickelson, H.R. and Pena-Valdivia, C.B. (eds.) *Developing Drought- and Low N-Tolerant Maize*. CIMMYT, El Batan, Mexico, pp. 136–141.

Westgate, M.E. and Boyer, J.S. (1986) Reproduction at low silk and pollen water potentials in maize. *Crop Science* 26, 951–956.

Young, T.E., Gallie, D.R. and DeMason, D.A. (1997) Ethylene-mediated programmed cell death during maize endosperm development of wild-type and *shrunken2* genotypes. *Plant Physiology* 115, 737–751.

Zinselmeier, C. (1991) The role of assimilate supply, partitioning and metabolism in maize kernel development at low water potential. Ph.D. thesis, University of Minnesota, St. Paul, Minnesota.

# Index